Systems Analysis and Computer Applications in Health Information Management

Kathleen A. Waters, R.R.A., M.Ed.
Gretchen Frederick Murphy, R.R.A., M.Ed.

Seattle University
Seattle, Washington

AN ASPEN PUBLICATION®
Aspen Systems Corporation
Rockville, Maryland
Royal Tunbridge Wells
1983

Library of Congress Cataloging in Publication Data

Waters, Kathleen A.
Systems analysis and computer applications in
health information management.

Includes bibliographical references and index.
1. Medical records—Management—Data processing.
2. Medical records—Data processing. 3. Systems
analysis.   I. Murphy, Gretchen.   II. Title.
[DNLM: 1. Systems analysis. 2. Office management.
3. Computers. 4. Records.   SX 173 W329s]
RA976.W37   1983   651.5′04261′02854   82-18468
ISBN: 0-89443-838-7

Publisher: John Marozsan
Editorial Director: Michael Brown
Managing Editor: Margot Raphael
Editorial Services: Jane Coyle
Printing and Manufacturing: Debbie Collins

Library of Congress Catalog Card Number: 82-18468
ISBN: 0-89443-838-7

Printed in the United States of America

2   3   4   5

To Marion Ball
An educator and advocate
of excellence in computer technology
for health care

Table of Contents

Preface

Through the intelligent use of knowledge, experience, and in some cases, unexpected discovery, advances in medicine have been made. The human effort that has brought about such advances in the latter part of this century represents a great collaborative effort of researchers, clinicians, and scores of allied health professionals all working together to achieve optimum patient care. These groups of individuals have extended their efforts beyond their own professional horizons to better use the knowledge and tools available from nonmedical industries. Therapeutic and diagnostic equipment available in contemporary health care settings exemplifies technology that originated in such nonmedical areas as space exploration and national defense. Such seemingly unrelated technological innovations now almost equal those developed by the medical profession specifically for health-care purposes. This fact demonstrates the collaborative effect achieved by the medical profession in its efforts to use the latest and most effective methods, equipment, and medicine for the care of patients.

This book addresses one of the most significant of these collaborative advances: computers and their unique role in patient-information handling. Our premise is that the computer is a powerful tool that offers all health professionals an opportunity to contribute to and benefit from improved medical care. Computers have been used to streamline, improve, and extend patient-care services. Today, hospital and clinic administration, medical record systems, diagnostic investigation, patient-care assessment, and many other areas are supported extensively by computer applications. To some extent, all health professionals use and rely on computers. It is clear that this technology is a major force in health care delivery today and will be in the future.

As medical record administrators, we have witnessed the development of remarkable changes in health care over the past 20 years. The fact that the patient record has provided a foundation for many of the most advanced medical developments has not escaped our attention. For it is the patient record that is the primary source of information about the details and activities of an individual's medical care encounter. As such, the patient record will remain a constant resource for future care and research. Except for format and storage media, we believe the medical record will never change in its basic characteristic as a device that can link past events to present and future actions. Its characteristics as memory source, repository of personal and private information, and source document for planning and research purposes all highlight the permanency of the patient record. Less permanent are the methods and techniques, equipment and systems used to collect, develop, process, store, and distribute the data contained in an institution's medical records. This book attempts to address the milieu in which contemporary medical records are developed and used, based on a recognition that the individuals and methods, physical facilities and systems that develop and distribute the permanent data provided by patient records must be amenable to change so that medical records may be comparable in quality to the other advances in medical care. Computer technology will direct much of that change.

Medical record professionals have an opportunity to work with the computer and to be challenged by the many and as-yet-undeveloped applications it offers to their area of expertise. We address this book to them. The computer applications described herein are only the beginning of what will eventually be accomplished to improve the flow and quality of the data provided by medical records. They demonstrate what has been achieved to date primarily in the area of hospital records. The book focuses on the process of systems analysis. It emphasizes our strong conviction that the capability most needed by contemporary medical record professionals in their relationship to computers and patient records is the understanding of and ability to carry out systems analysis.

Chapter objectives are included to preview chapter content and identify goals for the readers. Questions and problems are provided at the conclusion of each chapter to direct students in synthesizing and applying chapter content and references.

KATHLEEN A. WATERS
GRETCHEN F. MURPHY

Acknowledgments

We are grateful to the following individuals for their assistance, advice, support, and inspiration during the preparation and completion of the manuscript: Mervat Abdelhak; Ardis Alfrey; Darla Betz; Leslie Blide; Meryl Bloomrosen; Becky Davenport; Pat Dolan; Arden Forey; Elemer Gabrieli, M.D.; Malcolm Glaser, M.D.; Marsha Harris; Henry Heffernan, S.J.; John Lewis; Arthur McNeil, S.J.; Lorraine Matsumoto; Helmuth Orthner; Sue Potter; Mary Potts; Gaye Prewitt; Cheryl Smith; Marsha Steele; Sue Stubbs; Marlene Van Noy; and Mike Weidemann.

Chapter 1

How the Computer Has Redirected People Who Work with Medical Records

Objectives

1. Describe the rationale for the use of the term *patient data*.
2. Differentiate between *primary* and *secondary* patient data.
3. Define and describe *computerization of patient data*.
4. Define and describe *patient data system*.
5. Define and describe *patient data subsystem*.
6. Define and give examples of the term *system*.
7. Define and describe the systems team.
8. Define and describe the terms *users, user-manager, operations personnel*, and *vendor*.
9. Describe the process of change in computerization of medical record applications.
10. Describe the role of medical record professionals in computerization of patient data.

Patient records are the primary information source for all clinical and analytical computer applications in medical care. Accurate diagnosis, optimum therapy, adequate physical facilities, availability of selected specialty personnel and equipment are all examples of information-dependent decisions of a medical or administrative nature. The information dependency of all aspects of medical and health care is of particular interest to the medical record profession. It is the development and use of patient data that give purpose to the profession. The challenge for the profession is to utilize contemporary methods and systems to achieve optimum development and use of the patient's medical record. For it is the source document from which medical and hospital information systems retrieve information that relates to demographic, diagnostic, therapeutic, and all other categories of services and statistics.

As a source document the medical record comprises a wide variety of data. In fact the term *medical record* is frequently used to describe both partial and complete medical record documentation. Often a single report such as a history or operative report is referred to as a medical record. The same term is also used to describe the complete record of an individual patient. But the medical record professional refers to the complete record

1

as the medical record and to single reports by their designated title. Since this book concerns the computerization of medical record operations and the processes involved in providing useful records, the terms chosen for use are the ones that most appropriately represent accepted and acceptable usage. Medical record data is often referred to as patient data. There are two reasons for this. First, the term *medical record* by tradition connotes the individual medical record of a patient treated in an acute care hospital. But an exploration of the many and varied applications of computerized systems and subsystems touches on data entries, data elements, processes, and even health-care-related organizations that are associated with both traditional and nontraditional medical records. The term *patient data* can be clearly identified in these applications. Second, the term is used because it more precisely describes the unprocessed nature of the formats, forms, and documents that constitute the medical record. Individual items of information such as laboratory data or radiology findings or electrocardiogram interpretation are all forms of patient data that need to be considered individually as data elements in computer applications.

Since this book is directed to medical record professionals in particular, the descriptions, definitions, explanations, and opinions it contains reflect medical record processing as it relates to computerization. Primary patient data is targeted.

Primary records are the source documents prepared in the care of the patient, whereas *secondary records* are documents prepared for use in analyzing or summarizing primary data. *Primary record* is synonymous with *medical record*. Examples of secondary records include those retained or retrieved by Medicare, professional standards review organizations (PSROs), insurance companies, and local health departments. Although secondary data is included in the discussions here, primary data is still the most developed area. The discussions therefore mention it more often than secondary data. Another reason for focusing on primary data is that the traditional domain of medical record administration has been the patient record in acute care. The medical record profession has expanded greatly, in the 1970s and 1980s, to include administration of patient records in ambulatory care, health maintenance organizations (HMOs), third party payers, PSROs and other diverse institutions that retain and retrieve patient data. During this expansion, the major emphasis of computerization was on record-related functions in the hospital setting.

Developmental computerization activities have had many of the characteristics of similar pioneering efforts. They have been fragmented as to standards or objectives and not united in their involvement of appropriate people to understand and work with the various stages of analysis and implementation. Most interesting, the activities themselves—computerization of the data in their many forms—have taken on different titles for similar sounding projects.

DESCRIBING COMPUTERIZATION OF PATIENT DATA

Computerization of patient data and medical records has many different names in current practice and literature. But *medical computing* is the term most commonly used to encompass computerization of all aspects of medicine, including clinical medicine, clinical algorithms, and even medical record locator systems. This generic term is widely accepted. It is used by scholars and writers who are recognized as leaders in the computerization of medically related applications. Since we medical record practitioners are not physicians, information scientists, or systems analysts we prefer to use synonyms close to our own discipline and experience. The term *patient data computerization* is the choice for this book because it implies automation of all facts related to an individual patient's care. It includes facts about the complete array of clinical and administrative functions associated with the care of a patient. Clinical medicine, statistical analysis, retention and retrieval, and audit functions exemplify the functions it covers.

The remainder of this chapter covers topics essential to the computerization of patient data. Patient data are described as well as the various roles of those involved with computerization of medical records. Special skills needed by medical record professionals in order to achieve optimum patient data computerization are listed.

PATIENT DATA AS AN INFORMATION SOURCE

The medical record as it is currently known in its primary or secondary stage is the focus of this book. The record merits this attention because it is the sole document that provides the following information about a patient's care and treatment:

- Dates and times of events regarding diagnosis and treatment of an individual patient;
- Descriptive narrative data entries regarding the diagnosis, therapy, and progress during care to an individual patient;
- Identification of individual care providers who treated the patient.

Individual patient records represent all patient care departments in the hospital because the documentation that describes patient care reflects the services these departments provide. The patient record is a collective central repository from which all departments derive data to verify services rendered, utilization of facilities, plans for staffing and budgets, evaluation of departmental objectives, and so on. It is the collective retrieval of medical record data that has furnished impetus for the development of information systems. For the collective data provides a picture that fairly and objectively represents the status of health care in any particular health care facility. It can be accumulated, divided, and analyzed into units that represent any given segment of health care. This benefit of cumulative data is most efficiently achieved through the use of computers.

Automated information systems, especially patient record systems, have the capacity to facilitate greatly clinical research. This potential derives from the relative ease with which computer systems can examine records from many points of view, concatenating and combining the observed and measured variables in whatever way desired.[1]

An excellent example of a computer's ability to handle data is the Automated Hospital Information System (AHIS). It is an on-line, real-time, computer-based system that receives, stores, processes, transmits, and displays information from and for all major patient care, administrative, and financial departments in a hospital.[2] Another example is the Clinical Information System (CIS). It is an on-line, real-time, computer-based system that contains in its files an appropriate medical knowledge data base and the rule base or structure for creating higher level medical information from its resident clinical data base or from clinical data entered at the time of inquiry.[3] Whether one views AHIS with its on-line capacity designed to facilitate day-to-day operations or CIS with its on-line, real-time, medical knowledge data base, it is apparent that both ends of the computer system norm contain and represent appropriate and essential components of the clinical and management responsibilities of the medical record administrator (MRA) and the medical record department (MRD).

DEFINING PATIENT DATA SYSTEMS

One premise of this book is that the medical record and, hence, medical record professionals are keystones in the development of a functionally adequate computerized patient data system or subsystem. A computerized patient data system includes the collection, storage, and processing of primary patient data. During treatment of the patient, the patient record provides a communication link between the providers of care because it is the source of information regarding what has taken place during diagnostic and therapeutic activities. After treatment, the record serves as the source of information for all those who need to know what took place during care. It is a source that can reveal all the various minute details regarding care, treatment, financial reimbursement, and any other particular detail regarding the patient or the providers' interaction with the patient. The patient data system, then, is made up of all components that are coordinated to provide a useful medical record. It includes medical records, functions and procedures, processes and methods, analyses and access, all coordinated and aimed at a central objective: complete, accurate, timely, and accessible patient information.

A patient data subsystem includes specific units of work or tasks carried out by personnel in their efforts to achieve the objectives of the MRD. It includes tasks related to the collection, storage, and processing of primary or secondary patient data.

Commonly utilized work units for an effective MRD include the master patient index, numbering and filing methods, retention and retrieval policies and procedures, word processing, and quality assurance programs. The activities, tasks, procedures, policies, and objectives involved cross an organization's departmental boundaries. For example, the development and maintenance of the master patient index directly affects not only the medical records department, but also admitting and the business office. It can also affect the nursing service, laboratory, and x-ray departments. This networking is true of almost all of the patient data subsystems. Personnel who work in the various departments recognize the interdependence of their work tasks and the proper flow of data. Note that for the purposes of this text the terms *patient data system* and *medical record system* are sometimes interchanged. Patient data subsystems will be further defined in Chapter 3.

THE ROLE OF A SYSTEMS APPROACH

A second premise of this book is that patient data computerization requires a systems approach to design and development. A system can be defined as related elements that are coordinated to form a unified result, or, more specifically, people, activities, equipment, materials, plans, and controls working together to achieve a unified objective or whole. A system can also be defined as an array of components that interact to achieve some objective through a network of integrated procedures designed to carry out a major activity. The components of the system may themselves be systems, depending on the complexity of the parent system.[4]

An example of a system is a hospital. High-quality patient care is the goal of the hospital. It is the reason that all departments, physical facilities, and personnel are organized into functional units. The system includes organizational structure, financial structure, community standing and purpose, physical environment and equipment, patients, appropriate personnel and staff, policies, and procedures coordinated to carry out specific objectives. All of the components of the system are in accordance in their efforts to achieve the system's purpose—quality patient care. Together, the departments, from the patient-care centered such as nursing, to supportive operations such as housekeeping, are the subsystems that make up the total hospital system.

Figure 1–1 demonstrates the interdependent characteristics of a system, in this case, a hospital. The interdependence and the interdisciplinary nature of the jobs of those who work in the facility to carry out the prime organizational objective, patient care, lead one to another important aspect of systems analysis: the systems team or systems committee. Espousing the systems approach to design and implementation of the new computer system means believing that a successful systems approach requires the active participation of a responsible, knowledgeable group of individuals (see the next section).

Figure 1–1 The Hospital As a System

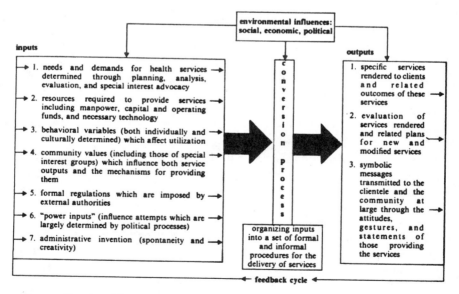

Source: Reprinted from "Current Status of Education for Health Administration," by Charles J. Austin, in *A Future Agenda: Education for Health Administration,* vol. III, by permission of Health Administration Press, Ann Arbor, Mich., © 1981.

An example of another, smaller, system is an MRD. Access to accurate and complete patient data by authorized users is the purpose of a medical record system. The system is made up of the providers of care who enter data into the patient record, users who retrieve data from patient records, patients, personnel, and staff of MRD, as well as procedures, processes, methods, and all other related objectives that achieve the purpose or goal of the department. Since patient data computerization cuts across all departmental boundaries or has the capacity to do so, the design of a system to computerize patient data requires those who intend to use such a system to recognize the system concepts described herein. Interdisciplinary personnel who are involved in patient care and support staff are the people who will be affected by the information processing function. They are the individuals who were affected by the information processing function when the data was processed, retrieved, and analyzed under old, manual methods. With the advent of computers and the accompanying mythology so often associated with any highly touted new technology, there is serious need for those who wish to improve services by use of the new technology to inform and educate all potential users of the new system. In addition, this educational process should include even those who may not be users but who, because of their job responsibilities, will be affected by the new system.

The systems approach to computer design and development takes into account the interrelationships among the activities, objectives, individuals, and organizational units in a facility. It recognizes that the total organization is only as strong as its weakest part. The systems approach to design and development of a new computerized operation includes from the outset discussion and work with all departments and all individuals who have any potential relationship with the new system. It also includes all individuals who have any potential relationship with the new procedures and equipment. A plan to carry out the tasks needed to achieve the purpose or major goal of the system is necessary in a systems approach to design and development of a computerized system just as it is in the design and development of any system. The planning process should involve all individuals in the organization who will be affected by the new technology in any way.

The systems analysis process presented in detail in Chapters 3–6 is the method proposed for use by those who are planning, developing, and implementing partial or complete information systems. Whether one is interested in computerizing a simple, single "stand-alone" function or a comprehensive, institutionwide information system, the process of systems analysis is applicable. The state of the art of computerization is such that both ends of the norm are available in current practice. With this factual background, we attempt throughout this text to cover as much information as possible about all types and sizes of computerized patient data systems and subsystems in our descriptions and explanations.

In the remainder of this chapter the role of the systems team in the computerization of patient data is described, as are the roles of others directly involved in computerization of patient information. The thorough description of the role of the medical record professional concludes with a list of exactly what the medical record professional should be prepared to do in the design and development of a computerized application that in any way relates to the medical record or the MRD.

SYSTEMS TEAM DEFINED AND DESCRIBED

The purpose for appointing a group of individuals to serve as the systems team or committee is to assure that the design and development of a computer project will have the benefit of the ideas and experiences of knowledgeable people responsible for the end uses of the system. Determining which individuals or disciplines will be selected to work on the systems team means making sure to include representatives of all levels and specialties of organizational and medical staff personnel who will be affected by the computerized system. Also included are those who have expertise or experience with computer systems. Although the team size must not become unwieldy it should be large enough to accommodate appropriate representatives. The various departments and service areas that will be the end users should be represented by individuals appointed to serve short term or long term as necessary. The institution determines the duration of appointment based on the size, scope, and extent of the represented department's involvement in the project.

Many organizations find it useful to select a full-time, ongoing committee to oversee all the computer project activities and to serve as a permanent communication channel for the coordination and evaluation of the interactions that occur during the project. Individual representatives can then be appointed to serve on an ad hoc basis. These individuals are selected to serve before, during, and immediately after the implementation in their respective areas.

Reflecting back on the computer projects that abounded in the hospital industry during the late 1960s and early 1970s, one can easily see that the evolution of the team concept has been a boon to computerized systems. The sixties and seventies, without the benefit of systems design teams or systems teams to overlook project development, found many hospitals spending inordinate amounts of money on hardware and software development. These expenditures and the projects were doomed to fail in many cases for the reason that the proper people among the user group were not fairly represented initially or at appropriate decision-making times. When properly organized to represent major users, the systems team can be a tremendous vehicle to inform and motivate all potential users. It can create interest and generate support for all aspects of systems design and development.

The team concept affords directed group cohesiveness right from the start. Some organizations use the team as a working body with power to make decisions. Others use the team in an advisory role to the information department. Many organizations use one main team and form ad hoc committees or subcommittees to function as specialists in collaboration with the team. The purpose of the team is to provide direction to the project by those people who have either expertise with the application under consideration or expertise in computer design or development or a combination of these qualifications. This purpose is a reflection of organizational attitude and expectation regarding the team's power and influence on the computerization of medical information or patient data.

An organization that delegates decision making to the team is making a clear and strong statement in support of the interdisciplinary and interdepartmental nature of the computer project under consideration. If the team is designated to act only in an advisory capacity it will not have the power that is embodied in a team whose authority includes decision making. A team whose authority is advisory only can be strengthened in its endeavors, however, if it can channel its deliberations and recommendations to one administrative individual or unit with line authority for departments affected by the new system. Some organizations demonstrate a disregard for the interdisciplinary and interdepartmental nature of information systems and their interactive character. They do this by designating the information department to carry out the role of a systems team. Another example of disregard is requiring the systems team to report to the information department, which in turn, makes final decisions regarding the new system.

Candidates for the Systems Team

Individuals who are considered vital to the successful implementation of a computerized system belong on the systems team. Appropriate candidates are department heads or selected members of departmental professional staff who represent the technical and professional knowledge and skill that are necessary to develop new and innovative technology successfully. During the process of systems analysis and of design and development of a new computer system, these individuals can double as members of the systems team while continuing their regular duties. It may also be possible for them to devote portions of time on a full-time basis to the systems teams' work. Appropriate budgeting and organizational delegation are necessary to achieve continuity of departmental direction during such periods. The concept of calling on individual department heads or selected members of departmental professional staff is relatively recent and once again reflects the systems approach.

An example of a professional member of a departmental staff nominated for the systems team is a head nurse on a unit of a hospital that is planning to design and develop a computerized pharmacy system. The nurse represents both the unit and the nursing department regarding the aspects of computerization that will affect that department. Another example is a medical record administrator whose department is to develop a computerized discharge abstracting system. Lack of previous experience or education in the use of computers need not disqualify such individuals from membership on the systems team, for there is still compelling reason for them to be members. Primary among the reasons for appointing such individuals to the team is that they represent management level knowledge of the processes that are under consideration for changeover to computerization. Further, any deficiencies they may have in computer knowledge can be greatly reduced by directed readings or attendance at short courses, discussions with vendors, visits to other facilities where computers are in place, and so on.

Even with a background in computers or systems design individual health professionals, administrators, or physicians will have to communicate the objectives of their own areas of responsibility anyway and should be doing so from the outset of the project. No one individual computer programmer, systems designer, systems analyst, or vendor can make decisions in place of the appropriate health professional end user. Nor do these individuals want to do so. In addition to the obvious benefits of the professional staff working in a proactive role with the systems designers and systems analysts, programmers, and other computer professionals, there is the cost benefit. Duplication of effort in the form of conversations, meetings, and other communication activities is reduced when the users are involved directly; reduction in errors of design and streamlining of flowcharting are two other obvious benefits that equate with dollars to the organization that is paying for the computer project.

Whether the facility has an information services department or not, an individual whose background includes systems analysis or programming should also be considered for the team. If the facility does not have such an individual in-house, these services can be provided either on a consultative basis or as part of a service provided by a vendor (see Chapters 6 and 7). The programming expert can be a valuable asset from the outset of the project. This person brings to the other team members early and essential collaboration that can save steps, streamline program and output document design, and effectively establish communication channels that optimize the understanding of departmental jargon, flow of data, and interdepartmental data overlaps. Specific methods that can be used to clarify and specify information requirements and facilitate project progress will be detailed in later chapters. The systems team should be appointed and commence its deliberations as soon as the problem is well defined and has been clearly described to administration.

USERS, USER-MANAGERS, OPERATIONS PERSONNEL, AND VENDORS

Terms used to identify those who work with and on the computer project need to be defined. In conjunction with the *systems team,* which has already been discussed, are several other terms, some of which identify members of the team.

Users are the individuals whose job responsibility requires them to use or employ the products of the computer system. The users' responsibility to maintain their organizational unit and achieve the objectives of that unit requires that the computer system be responsive to the needs of information flow and appropriate documentation just as the manual system was.

User-managers are the department managers or administrators of departments in which at least one prime objective is the use of data or information. Medical record administrators obviously fall into this category.

Operations personnel is a term that refers to those employees whose job responsibility is the day-to-day, task-oriented work level that achieves the objectives of individual job descriptions. Operations personnel perform nonmanagement tasks.

Vendor is the term that denotes individuals or companies who sell computer-related products. It encompasses those who sell software and hardware and those who sell partial or complete systems. The term is used interchangeably to denote companies or individuals. Some companies whose prime product is computer systems and equipment also offer training programs to teach concepts and practices of computers. Others offer consultative services with or without purchase of their products. There are also companies whose major purpose is to provide consultative services in the area of systems design, systems analysis, and selection of computer systems, including hardware and software.

PEOPLE, COMMUNICATION, AND CHANGE

Three major considerations for medical record professionals who anticipate the use of computers are people, communication, and change.

> The key ingredient in every computerized system is people. People design, develop, operate and maintain the system and they utilize the output generated by the system. If a computerized information system is to be implemented successfully, then everyone who is affected by the system must be made aware, first, of their individual responsibilities to the system; second, of what the system provides to that person and third, of the impact of change on the personnel involved. . . .
>
> The major factor involved in making the impact of computer technology on operational personnel a positive occurrence is communication. Employees need to understand why the changes are being instituted. They need to understand the benefits that the system can bring to them. They need to be thoroughly trained in new skills required to operate the system and to be assured that though their jobs may be changed they will not necessarily be eliminated.[5]

The systems team can work with the individual employees and others who will be affected by the new system to establish goals so that employees at the operational level, in particular, do not have distorted expectations of computerization. People usually accept change if they recognize a need for it and if their advice and help is genuinely sought during the development of the new system. Perhaps the significance of two major developments of the 1970s—the great progress made through the use of computers and the refinement of distinct computer systems for hospitals—is the recognition that we are all part of the constant, unrelenting process of change.

> Change has become the dominant concern of top management, and growth plans are geared to projected changes in wealth, technology, demand patterns, birth rate, habit, taste, population distribution, power supply, raw material production, and other such considerations. As these factors change, so the firm's activities must change to meet them. Change can be of only two

kinds—imitative or creative. You can change the way other people have changed already, or you can change in a new way. You can follow, or you can lead. You can wait until you find out how other firms have coped with or exploited the projected changes, and then copy them, or you can think up original ideas that they have not hit on. And if you do that, you are being creative in the fullest sense. Change is not a sideline in the business of leadership, it is integral to the whole idea: to describe a man who left things exactly as he found them as a "great leader" would be a contradiction in terms. A leader may change the map of Europe, or the breakfast habits of a nation, or the capital structure of an engineering corporation; but changing things is central to leadership, and changing them before anyone else is creativeness.[6]

Change is inevitable. The medical record, the primary source of information for all those involved in the care of the patient, reflects the change process, for as the modes and technologies of care change, so does the record. The record also serves those who must administer programs and facilities, those who are responsible for fiscal equilibrium and those who evaluate the services and care provided by facilities. There is a direct link between the record itself; the medical record professional whose responsibility encompasses development, processing, retention, and retrieval of the record; and all these users of the data. Because of this linkage the medical record profession and its practitioners need to understand the process of change and the need to manage and utilize change much as modern professionals have learned to manage time.

Instances continually occur that reflect the medical record professionals' direct relationship with change. MRAs are commonly consulted about problems that cross disciplinary lines because the utilization of the data contained in the medical record affects the patient, the hospital or health care facility, the health care system, and medical science. MRAs are in a position to have an increasing impact on the management practices of the organizations where they work. They must endeavor to facilitate all the uses and users of primary patient data. Secondary data is also increasingly under their control and direction. To prepare them for their role as change agents, schools that educate MRAs are faced with the challenge of curriculum redesign. Students must be aware that the changes being brought about by computers will continue, as they prepare for future job opportunities.

Familiarity with the techniques and problems of computerization and computer use helps prevent the pain and waste of reinventing the wheel when it comes time to institute a computer system. We medical record professionals must all acknowledge our need to accept a role in the complex, technologically dynamic areas of computer medicine, where change and our ability to adjust to its many directions is necessary for professional survival and growth. Adjusting to change creatively is challenging and in the end rewarding. We can prepare for that challenge by participating in imitative change. This is one of the major purposes of this book: to provide a look at what has already been done and a review of change that has already taken place.

MRAs AND COMPUTERIZATION OF PATIENT DATA

The MRA is responsible for the personnel and operations that constitute the medical records unit or department in a health care facility. The objectives and outcomes of that unit include all processes necessary to retain or retrieve accurate and up-to-date patient records.

Contemporary designations for medical record professionals and work units reflect the broad scope of patient records. In many organizations where patient records are used, titles for those who work with the records include health information manager (HIM) and health record manager (HRM), and organizational units may be known as health record departments (HRDs). These designations will be used interchangeably throughout the text with those titles previously identified.

As computer technology is increasingly utilized to process, store, and provide data for clinical medicine, medical record professionals will find themselves working more and more with methods and processes to consolidate and reformat clinical medicine reports. The dynamics of computer technology and medicine itself provide an enormous opportunity for MRAs to assist in the development of new directions for data linkage and use. New treatment modalities, new methods to link and combine laboratory tests, drugs, ultrasound scans, and other processes and techniques yet to come are examples of the information components that clinicians, physicians, computer experts, and medical record professionals will need to address for efficient and maximally beneficial computer processing.

Perhaps the greatest incentive for medical record professionals to shed their old manual, middle-management, nondirective approach to the data they have worked with in the past is computer technology. Computers provide the medical record profession a unique opportunity to take a proactive role in health care delivery. Computers are the greatest tool the medical record has ever had. MRAs now have an opportunity and a responsibility to utilize this great tool in the development of patient data systems that are responsive to the vast amount of information now available to the individual patient record, to the clinicians who enter data and retrieve data in their daily treatment of human illness, and to the researchers and planners who must commit money and resources to better programs and hence rely on the quality and availability of data.

MRAs have already pioneered in the development of computer use to achieve the objectives of the management of their departments and the objectives of providing a more easily accessed and accurate patient record. They must now continue that pioneering to achieve some coordination and communication with the many other individuals who are developing stand-alone or single-purpose computer systems with no interface between the systems. The MRA needs solid communication skills, an understanding of clinical medicine and its needs and uses regarding data processing, an understanding of the change process and how to work effectively as a change agent, and a working vocabulary and knowledge of systems analysis in order to bring all the people together to achieve coordinated computer usage in a given setting or facility. Because the hospital information system depends on the MRD for successful existence, the health care organization depends on the MRA to define patient information components, information structure, and information flow so that coordinated and cost effective use of computer technology results.

> Information processing activities are increasing in importance . . . giving rise to a need for qualified, trained professionals who can work closely with the health care team toward defining their information needs, interpreting the capabilities of information systems, setting forth the alternatives available and guiding the appropriate diffusion of information science technology into the health care system.[7]

Many MRAs have worked with others or alone to develop interim records. The interim record is a computer printout of a particular, traditional, paper record form. The original paper form, such as a patient history or laboratory report, when processed by the computer is reformatted to allow automated processing, and the printout is also a reformatted version of the original. It is the computer-stored record available to users in the form of a printout. *Interim record,* then, is a developmental term used to describe a document that represents the meshing of the old, manual record and the new, computer record. Interim records are necessary for continuation of documented data use when a manual medical record is being transformed into a computer record. Usually only one or two parts of the record are computerized at a time. The interim record, then, serves as a replacement for the parts of the paper record that are stored in the computer. In the future, when all parts of the patient record are computerized, the interim record will lose its title but retain its use as a back-up, or hard-copy record. In that sense the future of the interim record includes replacement of the old familiar, dog-eared paper record. The development and use of interim records by competent MRAs is an example of the need for new skills.

Medical record educators and medical record practitioners are faced with some serious decisions. Practitioners must ask themselves if they already have the essential skills and attributes to work in the computerized world. They must ask if they are prepared to expand their current responsibilities into the computer field. Educators must ask whether the knowledge and skills needed to work with computers can be transferred into course content that will fit into already overcrowded undergraduate degree programs. The following quote from one medical record educator's perspective on this serious topic answers these questions very well:

> Having identified the areas in which there would be a need for a health information system specialist, one must proceed to define the capabilities and attributes of the professional who would fill these needs. The knowledge and abilities necessary to work effectively in this field may be obtained by integrating concepts relating to people, the health care system, information systems, [and] technology.
>
> *People*—knowledge of personality, learning and motivation theories; ability to communicate, to listen and hear others; ability to describe individual and group behavior and the variables that affect behavior; ability to identify information users; ability to interact effectively with multi-level users; ability to formulate and articulate information needs and describe information seeking behavior; ability to articulate the conditions that are operative in generating an information need; ability to motivate and introduce change.
>
> *The Health Care System*—knowledge of the existing philosophy patterns and modes of health care delivery; knowledge of the shifting and changing concept of health care; knowledge of the existing organizational structures of the health care system; knowledge of the political, social and economic framework that influences the health care system; knowledge of the financing of health care; knowledge of the regulating/planning agencies' requirements; knowledge of the health care team concept and its interrelationships; knowledge of the disease process, diagnostic

and therapeutic procedures and the health sciences; knowledge of the multi-sources and generators of health data in the health care system; knowledge of the usage of health data in the health care system.

Information Systems—ability to view, describe and define any information need situation as a system (boundaries, components, elements, and interactions); ability to apply this "system viewpoint" of informations system to the health care system; ability to formulate and identify alternatives to satisfy and meet the users' information needs; ability to recognize, transfer, diffuse and apply in context the appropriate models for situations commonly encountered.

Technology—knowledge of basic hardware/software components of computer and telecommunication systems; knowledge of I/O [input/output] storage medium and devices, sources for updating knowledge of technology; ability to develop several logical structures for a specified problem; ability to transfer logical plan into physical design specifications; knowledge of programming language; ability to develop the major alternatives (assuming current technology) for the problem at hand; ability to evaluate (in terms of effectiveness and efficiency) proposed new techniques or applications of current technology; ability to select among alternatives, including the identification of necessary information for making that decision; knowledge of the capabilities and limitations of computer and telecommunication systems relative to the logistical and functional requirements of an organization.[8]

MAJOR PREMISES FOR COMPUTERIZATION

The 1980s provide expanded alternatives to the rapid technological developments that directed computer applications of health information functions of the 1970s. Professional health information managers are faced with computer decisions based on four major premises.

- Technology is now and will continue to be a necessary, inevitable, dynamic tool for information managers. Equipment, information processing, and data reporting will be computer based. The design and development of effective computer applications in traditional patient information settings will require extensive management planning based on a working knowledge of computer technology.

- Patients will be increasingly active in their own health care. Already an integral part of programs in mental health, social and mental rehabilitation, and treatment of the physical and developmentally disabled, the practice of patient participation will be extended to the traditional medical environment of physicians' office and hospital. Patient-carried personal health records and increased patient education on alternatives in therapeutics will be further developed and their use expanded.

- Transfer of patient information from one setting to another will be improved to provide timely, accurate data upon patients' transfer from one facility to another. Ongoing information exchange from levels of care ranging from acute care in hospitals to home-based ambulatory care settings will be a reality. Health data banks will be developed through existing medical care delivery resources such as the federal government, HMOs, and through voluntary participation in specialized patient information systems. Cancer surveillance programs were a forerunner of the latter example.

- Intensive critiques of patient care data in order to improve the quality of that data will be implemented along with proposals to improve the methods used to verify data accuracy in as many settings as it is collected. Financial, not medical or social forces will continue to drive this movement. It will affect all who work directly with data at both the retention and the retrieval ends of the function. They all will take part in studies and developmental processes aimed at providing accurate, timely data elements for both primary and secondary patient data related documentation. The ability to manipulate myriad factors rapidly and to produce high-level statistical graphs makes the computer the perfect candidate to carry out the basic work needed to verify relations among various elements of the data and to forecast economic and administrative planning schedules.

MRAs must learn more about communications, telecommunications, computer science, medicine, and change itself. Medical record professionals currently work with physicians, computer vendors and systems analysts, business managers, financial managers, and researchers and in the future will do so more and more. They will also increase their activities as liaison between systems teams and information specialists and between systems teams and clinicians. They will play an increasing role as communicators of the data stored in the user rich data files they develop and control. "Hospital information systems, on all levels of sophistication, are dependent on the medical record department for the core of their successful existence"[9]—this quote rightly describes the vital importance of the medical record department. As the center or core of hospital information systems (HIS) the MRD provides key information components upon which the total system is dependent. This system affects everyone. Starting with the patient, its effects never really end but affect all providers and users of data. It will affect providers and users as long as data are needed for research, planning, and evaluation of care and services provided.

KNOWLEDGE OF COMPUTERIZATION FOR MEDICAL RECORD ADMINISTRATORS

Medical record professionals have demonstrated an ability to meet the challenge presented by computer technology. The continuing challenge for the medical record profession in computer technology is to acquire specialized knowledge. Special knowledge is needed to develop further the computer's capacity to enhance the accuracy, linkage, analysis, and access of patient data and medical records. The purpose of this book is to present technical and management information that medical record professionals can use to carry out computerization of patient data. It includes the identification and description of skills needed by practitioners to participate actively in the process of systems analysis.

The following list represents skills needed by MRAs when working on computer applications. It represents the basis for the remainder of the book and the processes described in detail throughout the following chapters. Medical record professionals must develop special knowledge in order to

- Identify elements of data processing, including tracking information flow through a computer system beginning with a source document and continuing with data entry, data processing, storage, access, printout, and information distribution.
- Develop and maintain an understanding of data coding, information storage, and file design in medical computing including logical structure in application programs and physical structure in data storage.
- Communicate in data and teleprocessing terms that will facilitate working with computer systems professionals in cooperative planning and in operating particular computer applications.
- Analyze computer programming to read and follow the logic flow that the programmer uses in writing a program; this entails the ability to prepare procedures in a precomputer format and to assist programmers to identify the programming activity.

- Monitor developing computerized patient records, focusing on problem-oriented medical record computerization and its effect on patient record processes and evaluations and in particular on longitudinal records and how computers augment and enhance the development of such records.
- Work with clinical algorithms and decision support system models and their interface with patient records as this embryonic component of medical computing emerges so as to anticipate its impact on the patient record and incorporate it into the management of patient record and patient information systems.
- Develop an awareness of current technology that includes functional operations within medical record departments and other health information departments in a variety of organizations. This includes hospital computer systems, health data networks, ambulatory care, and HMOs.

Questions and Problems for Discussion

1. Why can the patient record be considered the primary source for all clinical and analytical computer applications in medical care?

2. What is generally understood by the term *medical computing?* Why is it important for health record practitioners to understand this concept?

3. What is meant by the term *patient data computerization?* Develop a list of examples of patient data computerization from acute and ambulatory care settings.

4. Prepare a brief report to a hospital administrator commenting on the role of patient data as a significant information source in the ongoing development of computer applications.

5. What is the role of patient data systems? How do they fit into overall developments in AHIS and CIS?

6. Why is a systems approach important in the effective development of computerized information systems in medical care?

7. Prepare a report recommending the formulation of a systems team to develop an automated master patient index for a 200-bed hospital. Include a rationale for the recommended members of the team.

8. Explain the difference between users and user-managers.

9. What is the change agent role for health record professionals? What knowledge and skills will be required?

10. Your hospital is embarking on an automated hospital information system to be implemented over the next three years. Prepare a report for the hospital administrator requesting support to pursue some continuing education in medical computing. Include a description of any increased knowledge and skills expected from this process.

NOTES

1. Donald A. B. Lindberg, *The Growth of Medical Information Systems in the United States* (Lexington, Mass.: D.C. Heath, Lexington Books, 1979).

2. Richard M. Dubois, "Clinical Information Systems: Current Trends and Outlook for the '80s," *Computers in Hospitals,* January–February 1981.

3. Ibid.

4. Kathleen A. Waters and Gretchen F. Murphy, *Medical Records in Health Information* (Rockville, Md.: Aspen Systems Corp., 1979).

5. Leslie Blide, Unpublished paper presented to the Society of Computer Medicine, San Diego, Calif., September 1980.

6. Antony Jay, *Management and Machiavelli* (New York: Holt, Rinehart & Winston, 1968).

7. Marion J. Ball and Stanley Jacobs, "Hospital Information Systems As We Enter the Decade of the '80s," *Proceedings, IEEE 4th Annual Symposium on Computers in Medicine,* 1980.

8. Mervat Abdelhak, "Health Information Specialist," *Journal of AMRA,* April 1980.

9. Ball and Jacobs, "Hospital Information Systems."

How Computer Structure Directs Data Processing, Retention, and Retrieval

Objectives

1. To understand and explain the need for a knowledge of basic information processing terms as a foundation to effective performance in computer applications developments.
2. To identify and explain the functions of the major hardware devices that make up a computer system, including data entry devices, input devices, central processing unit, storage, output devices, and communications equipment.
3. To contrast batch processing and interactive processing and illustrate their use in health information processing.
4. To describe and explain how stored programs (software) are used in computer systems operations.
5. To explain the general characteristics of high-level programming languages and their applicability to the medical environment.
6. To trace information flow from a source document through a computer system to an output form.
7. To relate descriptions of computer systems and operations to appropriate health information systems applications.

The purpose of this chapter is to look at computers and see how they operate. If the professional knowledge and competencies identified in Chapter 1 are to be translated into effective action, gaining a foundation of information processing terms and a knowledge of computer architecture must be the first step. Practitioners require a clear picture of the capabilities of computers. Hence this chapter provides a brief review of these items and relates technical terminology to health information processing applications. A close look is taken at computer structure, with emphasis on describing the equipment or hardware devices that make up a computer and computer systems generally. The role of stored programs or software will be explained, including a review of major high-level languages and their applicability to health care. Finally, we will examine the steps in information processing from a source document, such as the patient record, through a computer system, to an appropriate output form. Let us begin with a look at the computer itself.

DEFINING THE COMPUTER

What exactly is a computer and how does it work? "A computer is a device that has the capacity to accept data, manipulate or process it in a prescribed way, and when instructed or ordered, display or store the results of that process," one definition says.[1] Another calls it "a device that is by electronic, electro-mechanical, or other means capable of recording and processing data according to mathematical and logical rules and of reproducing that data or mathematical or logical consequences thereof."[2] Still another definition says that "A computer is a piece of electronic equipment that (1) performs large numbers of mathematical calculations at very high speeds, (2) operates under the command of a set of changeable instructions called programs, and (3) stores both programs and data in electronic and electromagnetic devices called memories."[3]

These definitions provide descriptions that help to create a picture of computers as concrete devices. In the next step, we will look at a picture of computer functions.

Two Ways to Process Data: Batch Process and Interactive Mode

Computers process data in two basic ways. We refer to these as batch processing and interactive or real-time processing. In batch processing, data are prepared, organized, and collected in such a way that they can be fed into the computer in related groups. See Figure 2–1. The monthly processing of hospital discharge abstracts may be done in a batch. Medical record practitioners work with batch processing when they subscribe to a discharge abstracting service and send in groups of completed discharge abstracts in certain numerical sets. For instance, an abstracting service may require the monthly abstracts to be submitted in sets of 100. These groups are held by the data processing center until one month's discharges can be accumulated. Then they are entered into the computer and processed in a single batch.

Interactive processing is also called the "conversational mode." This is a method of using the computer by means of an "on-line" terminal. The individual "converses" with the computer to send and receive information. *On-line* means the terminal is actively connected to the computer. This allows the computer to prompt the user for data and respond to questions. See Figure 2–2. On-line master patient indexes may function in this fashion. The user enters an inquiry about patient identification by typing a patient name into the computer through a typewriterlike keyboard connected to a TV-like screen called a cathode ray tube (CRT). Together the keyboard and the screen constitute the external parts of a CRT ter-

Figure 2–1 Flow of Data in a Batch Computer System

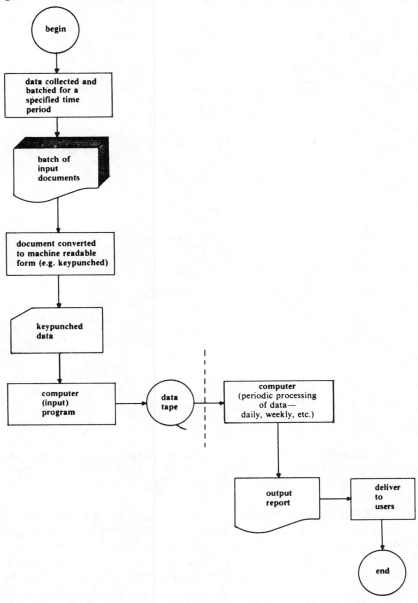

Source: Reprinted from *Information Systems for Hospital Administration* by Charles J. Austin by permission of Health Administration Press, Ann Arbor, Mich., © 1979.

Figure 2–2 Flow of Information through an Interactive System

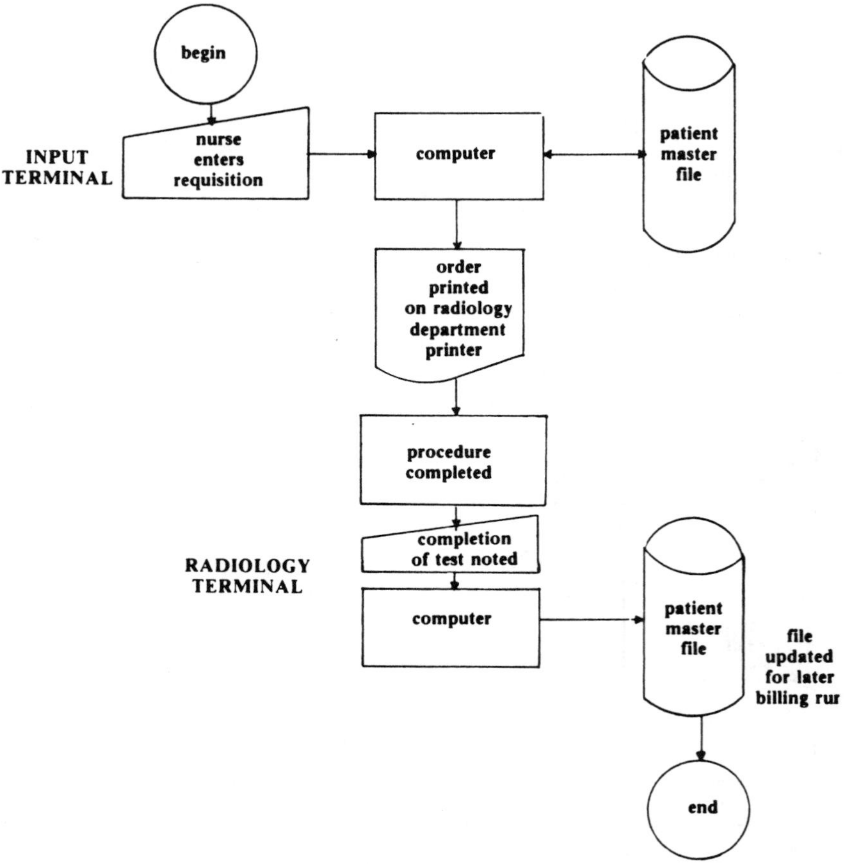

Source: Reprinted from Information Systems for Hospital Administration by Charles J. Austin by permission of Health Administration Press, Ann Arbor, Mich., © 1979.

minal. The computer responds to a user inquiry by searching a stored index and returns a response by listing the patient name and hospital number. In medical computing, computers perform in both batch mode and interactive mode. We will describe examples related to patient records and health information as we continue to describe the computer and how it works.

Before we move into a description of computer structure, we need to again ask the question, "How does the computer work?" We have described how it processes data but not how specific processing is accomplished. Specific processing, whether in batch or interactive mode, is accomplished under the command of stored programs. Programs are coded sets of instructions that are entered and stored in the computer to wait until the planned data is entered for manipulation. In computer systems the equipment is called hardware and the programs are called software.

Computer Equipment Structure

Let us turn now to computer structure. A knowledge of computer system structure, also known as architecture, provides the foundation for understanding computer potential in information processing. It is helpful to understand exactly how different kinds of computer equipment or hardware are connected together to provide a physical base on which the instructions or software are placed to perform data processing. We can begin by viewing the schematic in Figure 2–3. We will define the components in the figure and describe them in detail in the following pages.

Figure 2–3 Basic Computer Hardware

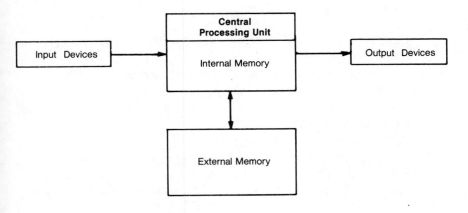

Generally, three primary functions occur. Data is entered into the computer through input devices. It is processed by the central processing unit and the results are returned through output devices. First we will define these terms. Then we will go on to discuss particular types of equipment and see how they fit into a functioning computer system.

Input devices are defined as "a device or collective set of devices used for bringing data into another device; for transferring data from an external storage device to an internal storage device; to transfer data from any device to the computer or to any of its operating registers. Card readers and tape readers are examples of input devices."[4] The *central processing unit* (CPU) is ". . . the part of the computing system that contains the circuits that control the interpretation and execution of instructions, including the necessary arithmetic, logic, and control circuits to execute the instructions."[5] *Output devices* are defined as ". . . a device or set of devices used to take data out of another device; equipment used to remove data from the computer."[6]

Internal memory is the storage facilities forming an integral physical part of the computer and directly controlled by the computer. It includes all memory or storage that is automatically accessible to the computer without human intervention. See also *primary storage* in the glossary. *External memory* refers to a device that is not an integral part of a computer. Off-line magnetic disks, magnetic tape units, or punch card devices are used to store data usable by the computer and are examples of external memory.

The general functions of a computer system—input, processing, and output—are divided into the following specific components for handling information:

- Off-line data entry and collection devices used to prepare data for input.
- Basic input devices that are used to enter data into the computer.
- The CPU, the part of the computer that performs the processing activities, consists of the control unit, the primary storage unit, and the arithmetic logic unit.
- Secondary or auxiliary storage devices.
- Basic output devices used to return the results of the computer manipulations to the users. These devices produce such things as "hard copy"—that is, paper printouts—and computer terminal screen displays.
- Data communications devices, which are communications equipment such as telephones, radio links, satellites, and others used to transfer

information to and from the computer. These devices enable computer systems to work with data entry and retrieval from remote locations.

As we discuss these components, we will add to Figure 2–3 to illustrate the concepts involved.

Data Entry and Collection

Off-Line

Because the computer can process large amounts of data very quickly, data entry and collection devices specifically designed to handle fast processing are used to collect or process the data. For instance, keypunching is a process whereby information is punched into patterns on standard punch cards so that the information on the cards can be entered into the computer system through a card reader. The card reader is a device that quickly "reads" the patterns of the punched holes. For many years the primary method for preparing information for computer processing was the keypunch. Today, while many organizations still use the keypunch method, more emphasis is placed on alternative ways to prepare data.

Several other methods of entering data into a computer also exist. Referred to as data entry techniques, these methods include

- Keyboard-to-storage
- Bar code recognition
- Optical character recognition or optical mark sense-recognition
- Magnetic ink character recognition
- Voice recognition

Medical record practitioners should be familiar with these methods. Of particular interest are the keyboard-to-storage devices, bar codes, and the optical character recognition.

Keyboard-to-storage consists of several CRTs connected to a minicomputer or programmed controller that collects the data input, verifies it, provides various other functions, and writes it on tape or disk for processing.[7] The CRTs in this description are often referred to as keystations.

The typical keyboard-to-storage system generally is one of two types: key-to-tape or key-to-disk. These devices are very appropriate for data programs that collect health information as transaction records from several sources. For instance, a small municipal hospital association may collect abstracts of discharges from local hospitals. A key-to-disk system (see Figure 2–4) could support CRT terminals in each hospital. Abstracts

Figure 2–4 Keyboard-to-Storage

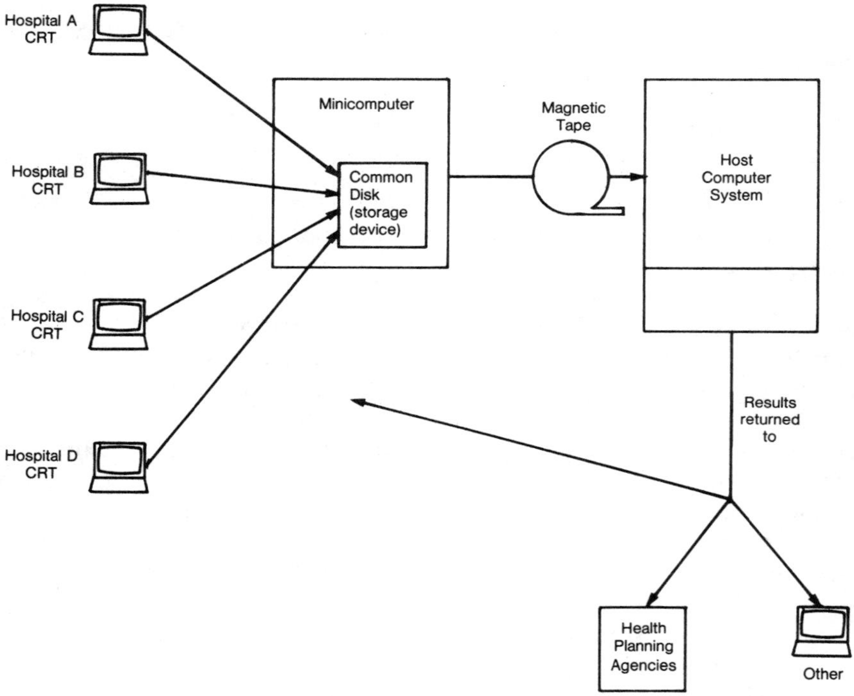

would be typed in via the CRT terminals to a common disk. The hospital association then would analyze the aggregate health information for community health planning.

The optical mark sense-recognition device is used by organizations such as national discharge abstracting companies for data collection. Client hospitals are provided forms for hospital discharges, outpatient services, or emergency visits. These forms are completed using special pencils so that they can be used for direct data entry via optical scanning devices. Pencils must be used when completing these forms so that they can, in turn, become the direct data entry source via optical scanning equipment. Figure 2–5 illustrates the optical scanning process. Optical character recognition devices use special type fonts. Two of the most common are OCR-A and OCR-B. These fonts allow human-readable information to be

Figure 2–5 Optical Scanning Process

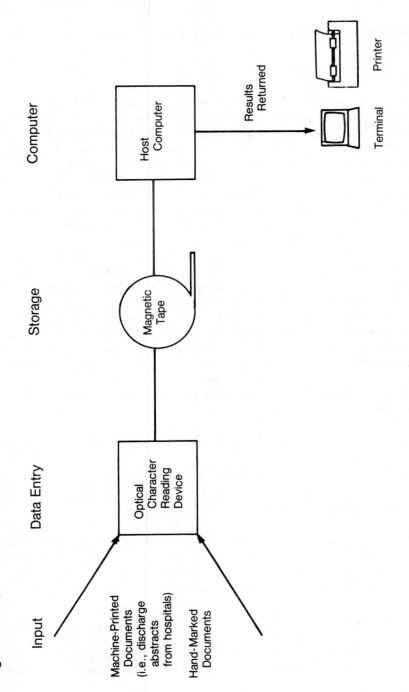
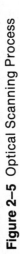

machine read, for example, by using standard typewriters with OCR type balls.

On-Line

Data entry devices may also be operated on-line. This means they are connected directly to the computer so that the data can be entered into the system. An interactive hospital admission, discharge, transfer (ADT) system could use a computer terminal to enter each new admission on-line and have the computer act on the entry by updating the census roster and admission lists accordingly. The same application (hospital ADT) could be designed using a computer terminal with storage capability so that all the admissions are entered and stored temporarily at the admission site. Then at a standard or scheduled time of the day, the terminal could be switched to on-line status and all the admissions for that period would then be transferred to the computer for processing. In the first case the system would be considered on-line, interactive, and real time. *Real time* refers to a setting in which events are accepted and acted upon by the computer as they happen. Real time also indicates that the data is processed in such a rapid manner that events affected by the processed data can be altered to reflect the data as soon as it is processed. For example, the hospital ADT system in real time could process the entries and provide departmental personnel with information that directs patient bed assignment, meal preparation, laboratory tests, and so on.

The reader can see there are a number of configurations possible with data entry equipment:

- Off-line preparation of data for computer processing with subsequent transfer to input devices
- On-line entry of new cases for immediate processing
- Local data collection through data entry devices for regular, often daily, direct transfer to the computer

On-line data entry will be further discussed from another perspective when we consider the role of computer terminals and interactive applications. To conclude the description of data entry devices, Table 2–1 provides a comparative description of such devices.

Basic Input Devices

Basic input devices are used to read data into the computer system. Input devices are primarily responsible for making information (data and

Table 2–1 Data Entry Devices

System	Equipment	Applications
Keypunch	Keyboard to punched card transcription and verifying device. Buffered versions permit performing both functions on the same machine.	Ideal for turnaround applications where the source document and the data entry medium are one and the same. Batched medical record abstracts for retrospective data collection is an example of this application.
Key-to-tape	Stand alone, keyboard-to-magnetic tape transcription and verifying devices. Tape is computer-compatible.	Designed for and adaptable to the same range of applications as keypunched (except for turnaround applications, etc.). PSRO and other external reporting agencies may use this.
Key-to-cassette	Stand-alone, keyboard-to-magnetic tape transcription and verifying devices. Tape must be converted to computer-compatible tape off-line or read by special-purpose computer peripherals.	Adaptable to most document transcription applications. Specialized studies and medical record reports are examples of appropriate use.
Key-to-disk	Also called shared-processor systems. Under control of a dedicated minicomputer, multiple keyboards enter data onto a disk or other secondary storage device. Information is subsequently transferred to computer-compatible tape. Numerous peripherals can be offered. Printers, card equipment, and arrangements for communication with remote central computers and with other key-to-disk systems are examples. Data formatting and accuracy validation are absorbed by software, leaving a highly simplified interface for the operator.	Suitable for virtually any data entry application in Medical Record Department Information Systems. A patient could be registered into a hospital system through terminals at several locations.

Table 2–1 continued

System	Equipment	Applications
Key-to-diskette	One- or two-operator keyboard entry and verifying devices. Information is recorded on a small removable disk storing about the same amount of data as a box of punched cards. Diskette data can be transmitted by wire, converted to computer-compatible tape in an off-line operation, or processed directly through suitable computer peripherals. See floppy disks.	Like the keypunch, well suited for isolated low-volume applications requiring one or two work stations. With comparable accuracy assurance and operator assistance features, key-to-diskette machines are currently less cost effective than shared-processor systems in terms of employee efficiency. See Chapter 9 on micro-computers.
OCR (Optical character recognition)	Machines that create computer-compatible media by scanning human-readable printed characters, bars, handwriting, or other symbols on the source document itself.	Securely established in very high-volume applications involving machine-imprinted documents. Handprint recognition is emerging. Type-then-scan methods to replace keypunching have frequently proven cost ineffective.
MICR (Magnetic ink character recognition)	Similar in principle to OCR; characters are detected by magnetic rather than optical means and scannable data must appear in a single line in a single industry-standard font.	By preprocessing the documents, organizations that receive and deposit large volumes of checks can substantially expedite availability of funds. Not appropriate for medical record applications.
On-line data entry	Operators at CRT terminals enter data directly into the same central computer that will eventually process the data. This category excludes inquiry/response or interactive applications that involve interaction between the operator and the files in the CPU.	Theoretically adapted to a wide range of applications. Data entry for medical record department tumor registry programs is one medical record example.

Intelligent terminals

Consist of a processor for which the user can write and store application programs: secondary storage such as cassette or diskette, keyboard, video display or printer, and communication interface.

Equipment flexibility permits use theoretically, for practically any application. Actual limits are set by peripheral device capabilities; software competence (vendor's and user's); and the organizational impact of decentralization. Terminal directed data and edit checking features in an in-house hospital discharge abstract system could illustrate these devices.

Remote batch terminals

Typical terminal consists of a card reader and/or magnetic tape drive for remote data entry, a high-speed printer for remote output, a controller for system and communications functions, and software for system control, job control, and a limited amount of applications programming.

Used in distributed data networks consisting of a central computer and satellite data input output stations. In the same installations, terminals have replaced satellite computers, along with programmers and operators of remote locations, with considerable cost savings. Systems with full RJE capability can substantially reduce processing turnaround time by eliminating job queues, and the transporting of card decks, tapes, and listings. Time share applications may employ these in health settings.

instructions) available to the system. Included are optical scanners, card readers for punched cards, and paper tape readers and tape drives, which read magnetic tape or write magnetic tape as output. This equipment operates by direction of the programmed instructions executed by the central processing unit. Once entered, data are often stored on magnetic tape or disk for later processing. In the previous example of a batch of discharge abstracts, for example, the key-punched cards that contain the abstract data could be entered into the computer by card readers. This example extends to another health information application such as processing patient record abstracts for transfer to a cancer surveillance program. Figure 2–6 illustrates typical input devices.

Figure 2–6 Some Typical Computer Input Devices

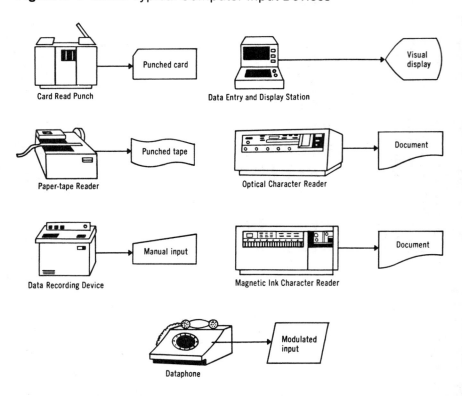

Source: Reprinted from p. 190 of *Elements of Systems Analysis for Business Data Processing* by Marvin Gore and John Stubbe by permission of Wm. C. Brown Company Publishers, Dubuque, Iowa, © 1975, 1979.

Central Processing Unit

The CPU is the core of the computer system. It is responsible for reorganizing data, performing calculations, and summarizing and reporting results. Let us look at the CPU more closely by examining its parts. Figure 2–7 illustrates the relationship between off-line data entry, data input, and the CPU.

The Control Unit of the CPU

The CPU is generally made up of a control unit and an arithmetic/logic unit. It is the control unit of the computer system and may consist of various registers depending on its design. A register is a device in the computer that temporarily stores a specific amount of information. Registers are placed where they have immediate access to the internal memory and any input/output bus structures. Buses are high-speed communication channels. They are called buses because they carry pieces of information much the way buses carry passengers from place to place. The CPU can control both input and output. It directs information into or out of memory.

Figure 2–7 Off-Line Data Entry, Data Input, and the CPU

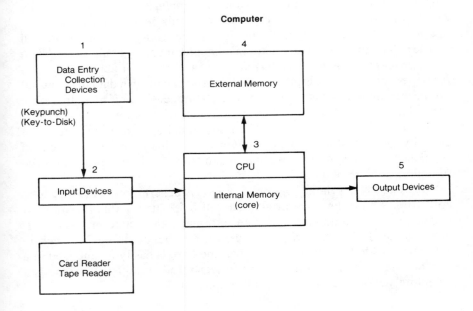

It also controls the arithmetic/logic circuits so that processing will occur. The hardware control is maintained by means of a designed machine instruction set. These machine instructions are generated by high-level language processors from a computer program written by a (human) programmer.

Before any program code is executed, the instructions are stored in primary storage, which is the main memory. When the program begins to run, one instruction at a time is brought from memory to the control unit, where the instruction will be executed. The control section does the following:

- Notifies data entry and/or input devices what data to enter into primary storage and when to enter it
- Notifies the primary storage unit where to place the data
- Tells the arithmetic/logic unit what operations are to be performed, where the data will be found, and where to send the results
- Instructs the system to access the appropriate storage devices, the correct files, and the corresponding data within those files
- Specifies to which output devices or media the processing results should be sent

Primary Storage/Main Memory

Until the late 1970s primary storage and main memory were usually referred to as *core storage*. Let us see how this central storage actually operates. After describing this kind of primary storage operation, we will consider how the newest advances in computer memory compare. *All data to be processed must pass through primary storage.* It accepts the data from the input unit, holds processed data, and can furnish data to an output unit. Since all data pass through primary storage, the unit must therefore have the capacity to retain a usable amount of data and the necessary instructions for processing it. When additional storage is required, the capacity of core storage is augmented by auxiliary storage, also called secondary storage. All information to and from auxiliary storage must be routed through primary storage.

What exactly is primary or core storage? A description of the magnetic core from which the term originated will help answer this question. A magnetic core is a doughnut-shaped ferromagnetic-coated material, vertically aligned. The tiny ring measures a few hundredths of an inch in diameter. Aside from its compact size, the core is easily magnetized in a few millionths of a second and unless changed, retains its magnetism indefinitely. Figure 2–8 illustrates this concept.

Figure 2–8 Magnetic Core

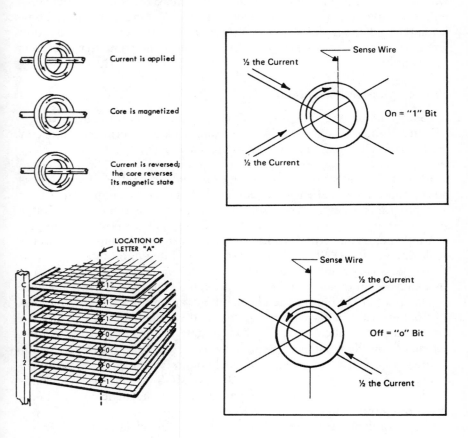

Cores are placed on a wire in a series like a string of beads. An electric current is sent through, magnetizing the individual cores. The direction of the current determines the polarity of the magnetic state of the core. By reversing the current, the magnetic state of the core can be reversed. In a binary system of information, only one of two conditions need be tested.

Core storage lends itself perfectly to binary coding in that each position will have a reading of 1 or 0 depending upon the state of the magnetism—either positive or negative.[8] (We will look again at binary coding and how it relates to data and information further along in this chapter.) The foregoing description of core storage describes how the computer interprets the binary signals.

Primary Storage Innovations

Today's technology has resulted in expanded uses of primary storage. "Virtual Storage Techniques is the dynamic linking of primary storage of the processor to auxiliary storage so that each user appears to have very large primary storage, usually measured in megabytes."[9] Virtual storage provides faster, more powerful use of primary storage. It enables applications programmers to be less concerned with the storage requirements for their programs. Interested readers may want to review techniques of buffer storage, another method for extending the capability of primary storage.

Another current method of adding to primary storage is employing Large-Scale Integration (LSI) or Very Large Scale Integration (VLSI), which "provides an accumulation of many (100, 1000 or more) switching circuits on a single chip of semiconductor. Through a series of processes, which include photolithography, chemical etching, and diffusion, a microminiature pattern of circuits with transistors and conductors is created along the surface of a silicon chip."[10] The volume of data stored per chip is dependent on the number of circuits on each chip. With the development of integrated circuits, replacement technology based on transistors has emerged. This technology is referred to as MOS (for metal-oxide semiconductor). When VLSI is used for main storage, magnetic core technology is rendered obsolete.

Bubbles and charged-coupled devices (CCD) are the latest form of main memory technology. While it is beyond the scope of this text to explain the technical advantages of this technology, we recommend that interested readers investigate these concepts using data processing texts. It is useful to note that bubble technology has slower data movement than the MOS technology but up to 10 times the storage density at a lower cost. In addition, bubbles are "nonvolatile." This means that if the power fails, the data in main memory is not lost. It is clear that these technological advances are improving the computer's ability to deal with information.[11]

The following names of storage media are currently used in computer centers:

- Bubble
- Charge-coupled Devices (CCD)

- Metal-oxide semiconductor (MOS)
- Magnetic Core
- Thin films
- Holographic
- Plated wire
- Magnetic drum
- Magnetic disks

(See the glossary for specific definitions and consult data processing texts for more detailed information.)

Let us continue our discussion of primary storage.

The primary storage unit of the CPU is the workhorse of a computer system. All processing takes place in it. Primary storage, or main memory, is the most expensive storage device in terms of cost per storage location, but it provides the fastest access time and thus may be the least expensive per machine calculation. Primary storage comes in different sizes. The size, which is measured in terms of the number of characters that it can retain, determines how much information can be held within a system at one time. The larger the capacity, the more powerful and expensive the computer. For instance, a computer containing 256,000 bytes is commonly referred to as 256K byte system (K stands for 1,000 approximately. 256K = 256 × 1,000, or 256,000). In some computers the storage capacity is measured in millions of characters, whereas other systems have smaller storage facilities. The capacity and design of the storage unit has an effect upon the method of processing data through the computer.

Normally the instructions in the program and the data necessary for immediate processing are located in primary storage. When the storage is limited, the length of the programmed instructions will also be limited. For example, it is not unusual for one program to require 200K memory locations. Many computers are unable to hold a program of this size in main memory. This means that the program has to be segmented into multiple job steps. The primary storage innovations listed in the previous section have been developed to overcome such limitations and extend the capability of the main storage function to the users. It is important to understand, however, that many computers are limited by the size of their main memory.

Each primary storage location has a unique address. It identifies the location of data for storing and accessing operations. Although all data are represented as binary digits, or *bits,* the smallest addressable location will differ from one computer to another. (Storage location designations are defined in the glossary.)

Access Time and Cycle Time

The speed with which data can be retrieved from memory is another important factor to be considered. Speed is defined as the time it takes for a character of information to travel from its location in memory to the control section of the computer system. The name for this is *access time*. The operating speed of the entire CPU depends on the speed of the access. During processing, both instructions and data move from memory to control and back again. The faster these moves can be accomplished, the faster the processing operation of the system.

Cycle time refers to the cycle of the operation. It can be broken down into two facets, the instruction cycle and the execution cycle. An instruction is obtained from the primary storage location and transferred to the arithmetic/logic unit, where it is decoded. It is decoded by its operation code, which specifies what is to be done, and its operand, which specifies the address of the data to be processed. The computer then moves into the execution cycle, in which the decoded instruction performs the operation on the specified data. Then it moves back to the instruction cycle and begins again. The amount of data that can be accessed in one cycle depends on the computer design. One model may have slower access time but may be able to access a greater amount of data each time. Consequently the amount of data accessible during each cycle and the time required to perform each cycle combine to determine the effective access time for a given system.

The ability of the computer to perform operations while maintaining optimum response to user personnel is dependent upon capabilities such as this. As the number of users increases, the demands to move and process each user's data increases. The functional performance of a computer system for admission clerks who are admitting patients through a computer terminal connected to the computer is expressed in how fast the admission information can be entered and processed by the computer. As users, health information managers are constantly measuring the computer's performance through its ability to support daily departmental operations. The foregoing discussion helps illustrate some of the factors involved in a particular computer system's capability.

Arithmetic/Logic Functions

The third unit of the CPU performs the arithmetic and logic functions, which can be viewed separately. The arithmetic section consists of circuits in the computer system that perform the arithmetic operations (add, subtract, multiply, and divide). Data can be taken from memory, then added, subtracted, multiplied, or divided, and the resulting answer returned to memory. These operations are done in the arithmetic section.

The logic section consists of those circuits in the computer system that can be used to make what are considered to be simple decisions. These decisions are based on the ability to sense if two characters are the same; if certain numbers are greater than, equal to, or less than others; or if certain conditions exist in other parts of the system. An example of this last decision might be that of determining if the input unit contained additional information to be processed. All these are decisions that would be made in the circuitry of the logic section of the CPU.

Today's technology includes many permutations of the CPU just described. Interested readers may want to pursue additional developments, for example, innovations in storage, in other texts. Some of the alternative capabilities of minicomputers and microcomputers will be featured at the end of this chapter.

Auxiliary or Secondary Storage Devices

Primary storage is very expensive and is economical for use only with the data and programs the computer is currently processing. Auxiliary or secondary storage is used to increase the storage capacity of the computer. It can include magnetic drum, magnetic tape, magnetic disk, mass storage devices, and floppy disks. Secondary storage may be used to store data and programs permanently or it may be used by the CPU to transfer data temporarily when carrying out operations. Remember that all data must be transferred from secondary storage into main memory before it can be processed. Table 2–2 features details of two common types of auxiliary storage.

The Floppy Disk

A floppy disk is a flat 5¼- or an 8-inch circular magnetic diskette that rotates within a jacket. It is analogous to a 45 RPM record and jacket. Floppy disks can have a storage capacity of up to 1 million bytes, depending on the size and density. They are used increasingly in microcomputers and have excellent potential for health information applications. For instance, chart location systems are available for microcomputers that use floppy disks for storage. Computerized word processing systems also use floppy disks. The current PAS-Plus program offered by the Commission on Professional Hospital Activities (CPHA) uses data entry onto floppy disks

Table 2–2 Comparison of Magnetic Tape Storage and Magnetic Disk Storage

Storage Medium	Major Characteristics	Advantages	Disadvantages	$
Magnetic tape	Widely used in batch processing. Stores record sequentially. More data is stored on tape than on punch cards. Data is recorded on 7-, 8-, 9-, or 10-channel tape in form of magnetized spots. Density is the number of characters, or bits, recorded per inch, (556, 800, 1,600, or 6,250). Processing with tape depends on the speed of the tape drive and the density of the data on the tape. Data can occur in blocks: `IBG	PT REC 1	PT REC 2	PT REC 3	IBG`  or unblocked:  `IBG	PT REC 1	IBG	PT REC 2	IBG`	1. More information can be stored on tape than on cards. 2. Processing is faster on tape than on cards. 3. Tape reels are easy to handle. 4. Tapes provide excellent, low-cost back-up storage. 5. Tapes are erasable and can be used over again for many applications. 6. Tapes are inexpensive; i.e., one cent can buy enough tape capacity to store tens of thousands of characters.	1. Records to be processed must be accessed sequentially. 2. When tape records are changed, the entire tape must be recreated. 3. Tape storage is not applicable for jobs requiring direct access to specific records. 4. Since reels are indistinguishable from one another, tapes require external and internal labeling. 5. The magnetized spots are unreadable by a human and must be translated to human-readable printouts. 6. Temperature, dust, and humidity content of the storage environment must be carefully controlled.	Tape reels range between $10 and $100. Tapes cost approx. $25 each.

in which the blocks or individual records are separated by spaces called interblock gaps (IBGs).

The tape reader starts and stops according to these IBGs. As tape processing is dependent upon the speed of reading/writing the tape, the greater the blocking factor (number of records per block), the faster the processing can occur.

Tapes are identified by file numbers and, frequently, cross-indexed with the appropriate programs.

| Magnetic disk | Considered the medium of choice. Widely used in interactive and batch processing.

Are commonly referred to as direct access storage devices (DASDs) because they allow random access to records. If 300 patient records are stored on disk, any one may be read without reading the entire file.

Analogous to a record player, many disks are made in a pack with recording surfaces on top and bottom. There is a corresponding set of "arms" with read/write heads that move in unison in and out between the surfaces. | 1. Direct access to individual records makes processing faster.
2. Can also be processed sequentially, if desired.
3. Each transaction or event can be processed as it occurs, which keeps the files up to date.
4. Storage capability is greater than magnetic tape's.
5. Records may be on-line to the computer when needed. | 1. Maintaining records is more expensive.
2. For many applications sequential tapes will work as well.
3. In updating files on magnetic disks, the record is read, updated, and written back to the same location, thus destroying the original contents. |

Table 2–2 continued

Storage Medium	Major Characteristics	Advantages	Disadvantages	$
	Disks come in several forms, among them stationary and removable disk packs. Use of this medium includes accessing those needed for moving the read/write heads in position; rotational delay, which allows colater to proper alignment on the disk; and data transfer, which is the time taken to transfer the data with primary storage for execution. We can view this as follows: 5000 byte (character) records on IBM 3330 disk storage facility Per Record average access 30.0 msec average rotation delay 8.4 msec data transfer time 6.2 msec total 44.6 msec 1000 records— 44.6 seconds			

at the hospital medical record department. The hospital medical record department staff enter discharge abstracts directly through a computer terminal onto the floppy disk. The hospital then mails the disk back to CPHA for routine processing. We will look more at floppy disks in other chapters.

Output Devices

After information processing has occurred, the results are made available to users via output devices. The output unit is designed to get information out of the computer in a form that is readable and usable by human beings. Computer output can be defined as a statement or representation (whether in written, pictorial, graphical, or other form) that is a statement or representation of fact either (1) produced by a computer or (2) accurately translated from a statement or representation so produced. Many input devices also serve as output devices. Examples are tape drives, computer card punch machines that punch information from the memory of the computer onto blank cards, and hard-copy printers, which print the output in words or numbers. Hard-copy printouts might be a printed page, a lab test result, or pharmacy label. For health information practitioners, the most commonly used output devices are hard-copy printers and computer terminal screens. One illustration is a chart request system in which doctors and other providers request patient records through computer terminals located throughout the facility. The request is printed out in the form of a duplicate charge-out slip on a printer in a medical record file room. File room staff may attach one copy of the charge-out slip to the patient record for routing and place one in the out-guide on the shelf.

Computer Output Microfilm

Computer output microfilm (COM) is a medium produced when data output from the central processor is read into a microfilm recorder that is connected to a film developer. The final product may be either microfiche or roll film. Both can be viewed directly through a special reader. Computer output microfilm is commonly used as a paper replacement. It can also be used as a replacement for magnetic tape. That is, data presently stored in machine-readable form on a magnetic tape such as the previous year's disease index may be transferred to COM to make the access more readily available and reduce cost. Health information managers routinely use COM as a back-up to a computerized master patient index (MPI). Some of the advantages of COM devices are as follows:

- COM devices print at computer tape speed.

- Since COM forms can be printed simultaneously with the data, the need for separate printing and storage of costly forms is avoided.
- Some COM recorders possess graphic ability.
- Retrieval coding is placed on the records as they are created, therefore easy, fast, pushbutton retrieval of data on microfilm may be made by the microfilm reader.
- With microfilm material, costs are reduced. One roll of 16-millimeter microfilm in a cartridge costs $4.50, whereas one carton of computer printout paper costs $35.
- The weight of reports and data is reduced.
- No decollating, bursting, or binding is required.
- COM copies are economical and of a higher quality than multiple carbon sets.
- Microfilm copies are clean, whereas impact printer carbon copies are dirty.
- COM devices are virtually noiseless in operation compared to impact printers.

Some of the disadvantages of COM are:

- Particular records are difficult to change, since the microfilm must be recreated.
- Data cannot be written in or penciled in on the microfilm.
- Microfilm viewers must be located near users. Sharing of viewers is sometimes an awkward process.

Telecommunications and Data Communications Devices

Telecommunication, the merging of telephone and other communications with computer technology, is the foundation of data communications devices operations. Introducing computers into the communications field generated significant changes in computer systems developments. These changes extended the power of computing into the medical field extensively. What is now possible? In the words of Donald Lindberg,

> The new communications network technology makes it possible for any user, medical or otherwise, to use his computer terminal to connect either to his local computer within the hospital, or to connect to a regional service bureau computer, or to send to or receive data from a computer terminal or network anywhere in

this country (and others with whom bilateral arrangements have been made). The communication charges with some arrangements are essentially independent of the number of miles between the user and the host computer, and are based solely on volume of data transmitted. Likewise the user is unaware of the transmission conventions and modalities employed. The efficiency of the communication is such that the costs are frequently less for special or rarely used functions than if they had been maintained locally.[12]

To discuss data communications operations, it is necessary to define four terms:

1. *Communications equipment* is equipment used to transmit signals from one location to another. It includes telegraph lines, telephone lines, radio links, coaxial cable, microwave, satellite, laser beam, and waveguides. It also includes the equipment necessary for sending and receiving the transmissions and changing them back to a form understandable by human beings.
2. *Peripherals* is a term denoting all computer equipment attached to the CPU in a given computer system, including input and output devices, tape disks, CRT terminals, and so on.
3. *Remote job entry* is the term used to describe the capability of computer systems to accept data from remote locations (geographically distant from the CPU) and store them for later batch processing. Remote job entry uses such things as on-line CRTs and key-to-disk units to collect and organize the data and the telephone or other communications equipment to transmit the data to the CPU.
4. *Front-end processors* are minicomputers or programmable machines used to accept the data from communications lines in one or more locations, further organize it, and store it in the computer or route it over additional communications equipment to the host computer.

These ideas come together as follows. Data communications devices are the equipment used for managing the communications, such as telephone lines, to enable the computer equipment to function as a system that features interactive processing and remote job entry. Data can be entered in Seattle and transmitted to a computer in Washington, D.C., for example. Communications devices transform and transmit data between similar devices, as defined in the beginning of this section, and the CPU. Five kinds of devices are involved in data communications.

1. *Direct communications channels* are the circuits or lines used to provide transmittal of electrical impulse, either voltage or current. Telephone lines are one form of communications channel.
2. The *modem* (modulator-demodulator) is a device used to translate impulses from electrical to audio (telephone) signals and back to electrical for reception by the computer system. One modem must be present to translate signals from the sending location and one present to translate signals at the receiving location.
3. *Multiplexors* and *concentrators* combine or merge several separate data signals into a single signal for transmission; multiplexors may feed into a modem.
4. *Programmable communication processors* are usually minicomputers or microcomputers that perform high-speed, high-volume message switching and front-end processing over major communications trunk lines or for data communications with a computerized information system. These act as coordinators of data entry operations from many locations. Figure 2-4 illustrates their role.
5. *Data communications terminals* and video display terminals (VDTs), also called cathode-ray terminals (CRTs), are typewriter or touch screen and televisionlike screen devices that are located away from the central computer. In hospitals terminals are commonly used for hospital admissions, outpatient registrations, clinical appointments, and MPIs. Terminals can be hard copy (printer or plotter), video, or audio response types; computer systems in medicine employ all types of terminals.

This concludes our brief discussion of the computer equipment structure. It has focused on the hardware or physical components. Figure 2–9 illustrates the structural model presented thus far in this chapter.

COMPUTER SOFTWARE AND PROGRAMMING LANGUAGES

Operating instructions that run the hardware and direct the processing of the data make up the software of computer systems. Let us explore the role of computer software and programming languages.

Computers and Programming Languages

Computers read data by receiving and interpreting patterns of electrical currents in on (1) or off (0) state. When the current is on, it is interpreted as 1. When it is off, it is interpreted as 0. Original computers used instruc-

Figure 2–9 Computer Hardware

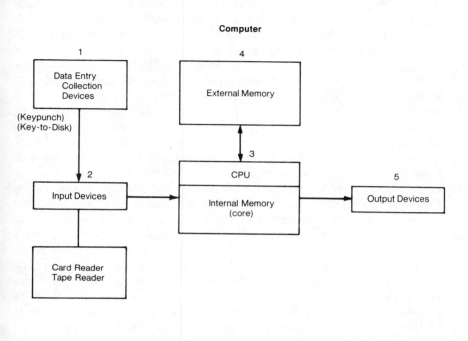

tions written in binary code. The following table shows how digits are represented by a combination of binary digits:

Binary	Decimal		Binary	Decimal
0000	0		0101	5
0001	1		0110	6
0010	2		0111	7
0011	3		1000	8
0100	4		1001	9

Originally, computers had to be programmed in binary code. This was painfully slow because everything had to be translated into patterns such as those just shown. In the early years a major drive was made by programmers to establish the assembly language, which allows the programmer to work with mnemonic expressions for binary code. For instance, instead of writing 0001, the programmer could use the mnemonic ADD in the program. Although this represented a considerable advance in programming, it still was very slow because the programmer still had to break the problem down into a sequence of machine operations. By 1958 the first higher level languages such as COBOL and FORTRAN began to appear. Today probably 95 percent of computer programming is performed in one or another of these high-level languages.

The high-level languages were developed to make the preparation of the instructions by programmers more rapid, readable, efficient, and easier to understand. Although it is not the intent here to explain languages in detail, we will consider some characteristics of the high-level languages. There are a number of high-level languages in use. Many were developed for specific applications or for one particular type of computer. A small number of languages are relatively standardized and have widespread use, however. Each of these is explained and documented in separate textbooks. There is an overall consensus within these languages concerning the language specification, and for each programmers are readily available. For health information managers this is an important fact. It is more important to use a language in which a programmer is accurate, highly trained, and efficient than it is to use a language that may be inherently more efficient but difficult to get highly trained and accurate programmers. It is reasonable to assume that health information managers will be working with programmers within their departments or in a liaison capacity in an increasing variety of applications in the next decade. Some knowledge of high-level programming languages will be an essential commodity.

One of the most widely used high-level languages is COBOL. An English-like language, it was developed primarily for financial and administrative applications. It has major strengths in report generation, storage, and retrieval and can perform simple calculations on vast amounts of data. It is not necessarily the most appropriate language for medicine.

One of the oldest and most popular high-level languages that is currently taught in universities is FORTRAN. It is a language oriented to numerical computation and performs complicated computations efficiently. PL/1 and ALGOL are newer languages that were intended to add new capabilities to those available in COBOL and FORTRAN. They both represent attempts to develop a completely general language but have been only partially successful. APL is another mathematical language developed by IBM and it is implemented on both larger and smaller scale computer systems. Another common language is BASIC, which was developed at Dartmouth University primarily as a tool for familiarizing students with programming. It is very well suited for interactive systems in which the user communicates directly with the data files and is involved in numerical computations. In research applications where health information managers are directing staff to assist in the analysis of patient statistical information, such as tumor registry systems and PSRO health data programs, the use of BASIC and the interactive communication mode may be particularly appropriate.

MUMPS was developed specifically for health care applications. It has analogies to BASIC but springs from the very first interactive language, OOSS. It is easy to make a transition from BASIC to MUMPS. MUMPS is a product of the Laboratory of Computer Science of Massachusetts General Hospital and stands for Massachusetts General Hospital Utility Multiprogramming System. It is now being used in applications in well over 4,000 medical settings in this country, including hospitals, health maintenance organizations, and group practices. Another 1,000 sites worldwide have adopted this inexpensive, versatile computer language.

Its supporters, the MUMPS Users' Group (MUG), consider its approach to the interactive handling of input data in applications and its file structure to be the factors that makes the language particularly well suited for medical use. These capabilities are not limited to medicine, and many business and administrative functions have effectively utilized this language. Although MUMPS was designed to meet specific medical requirements, these turned out to be very general requirements with widespread applicability. MUMPS differs from structured languages such as PASCAL, FORTRAN, and COBOL. These last three presume that the processes and data structures that may arise can be anticipated in advance. In medicine, this is seldom the case. MUMPS accepts definitional ambiguity at programming time by using an interpretive mode of converting program code to machine instructions at the time of actual use. It also relies on special capabilities to screen for errors and control the dialogue. The slowness of some early MUMPS implementations has been gradually overcome by improved operating systems and by the increased capabilities of modern hardware.

Two special purpose programming languages are also in common use. The report program generator (RPG) is used primarily for retrieval of statistical information from master data files and then formatting into printed management reports (tables). It is a feature often provided by manufacturers as a portion or a unique marketing feature of a computer system. General Purpose Simulation System (GPSS) is a discrete simulation language used for the simulation modeling of management processes and of organizations. The details of these languages are not intended to be covered in this text but awareness of how they fit into the overall structure of computer processing is important for the student. Table 2–3 is a comparative analysis of computer programming languages.

Now that computer structure and some programming languages have been described, we are ready for a description of how data moves through a computer system. Remember that computers require data and pro-

Table 2–3 High-Level Languages

Each instruction in a high-level language corresponds to several machine instructions. These languages were developed as easier tools for programmers. It is often estimated that an application can be developed and modified three to ten times as quickly in a high-level language as in assembler. The more common of the many languages are the following:

ALGOL, the first major language to be designed by an international committee of people from different organizations. . . . It is used for numerical and logical problems, particularly in Europe, and is available on many computers.

APL, a general language with complex notation and powerful operators, . . . The unusual symbols it uses make it seem obscure and difficult to those unfamiliar with it. Subsets are available on IBM machines.

BASIC, a simple and particularly learnable language with some character-manipulation capabilities. . . . It is available on many computers, usually in a considerably extended form such as BASIC-PLUS on the PDP-11 manufactured by Digital Equipment Corporation.

COBOL, a language for solving business problems. It is quite similar to natural English and is available on some minicomputers and most large computers. It was used by 12 health care institutions out of the 13 surveyed by Phillipakis and was the most popular. . . .

FORTRAN, a language for scientific and technical work, particularly involving numerical computation. It is available on most computers, and was used by 4 institutions of the 13 surveyed by Phillipakis. . . .

GEMISCH, a language developed for medical record applications. . . . It is implemented on only one computer, the PDP-11.

MUMPS, a language designed specifically for handling medical records. The simple but powerful language is strong in data-management and text-manipulation features. . . . It is available on about 20 types of computer, especially minicomputers.

PASCAL, a language that is of particular interest to computer scientists and is used increasingly for applications. . . . It is available on about 60 types of computer.

PL/I, an advanced, powerful programming language for scientific, commercial, and other applications. . . . It is available largely but not exclusively on large IBM computers and was used by 2 of Phillipakis's 13.

RPG, a language developed specifically to facilitate the specification of report formats to be printed from stored files. It is available on several types of computer.

Source: Reprinted from *Computers for the Physician's Office* by Joan Zimmerman and Alan Rector with permission of Research Studies Press, © 1978.

grammed instructions to manipulate or process the data. In the early computers, programs and the data to be processed were tied together. That is, the program, or set of instructions about what to do with the data, was specific to the file of data within the computer. As a consequence, the use of any particular application was extremely limited. Other programs could be applied to a particular data file only by rewriting the program or the data file. As computer technology evolved, it became possible to improve computer-reported data through exception reporting. Programmers and system designers carried out thorough analyses and planning of the various uses of data, thus allowing them to develop data files that could either accommodate several sets of instructions or different programs that used the same data files. This situation provided better use of data and more efficient storage of the information but was still limited in its ability to manipulate the information.

Computers and Data Base Management

Today data base management (DBM) systems are used. A data base is defined as "a collection of interrelated data stored together with controlled redundancy to serve one or more applications in an optimal fashion; the data are stored so that they are independent of programs which use the data; a common and controlled approach is used in adding new data and modifying and retrieving existing data within the data base." Sippl defines data base management as "A systematic approach to storing, updating, and retrieval of information stored as data items, usually in the form of a record in a file, where many users, or even many remote installations, will use common data banks."[13]

How does this work? A data file is physically stored in the computer. It is structured in such a manner that a number of programs can be written now or in the future to access the data file and to utilize the information for a variety of users who may present it in a variety of formats. Data base management systems can accommodate a number of users simultaneously, but they are still in an evolutionary stage and have not been challenged to their limit in the medical arena. Currently DBM systems are quite effective for certain kinds of data such as identification and financial information and even certain sets of specific medical data such as some clinical diagnostic decision-making models. They are not yet able to solve the puzzle for the total gamut of information needed and provided by the original paper record, however. This is primarily because of the overwhelming overload of dynamic data that is found in medicine. Medical computing must develop lower cost dynamic systems to deal with these difficulties in order to establish computerized patient records that are efficient enough to satisfy the clinical, analytical, and administrative needs of patient information providers and users.

Logical and Physical Data

The way or form in which the data are actually stored in the computer does not necessarily correspond to the form in which they are presented or used in the actual application program. The structure the application program uses is commonly referred to as the logical structure. The data structure, which is actually stored on tape, disks, and other media, is called the physical structure. The reader can see from the brief description of data base management concepts how the physical structure of data stored in the data base may be matched with the logical structures referenced by the various application programs. Logical and physical are only terms for describing the different aspects of the data.

Application Software and Patient Records

Before the entry of the paper record data into a computer is described, two facts should be considered. The first is that computers have brought a whole new image to the traditional paper record. The second is that medical record professionals must now be familiar with computer terminology and technology to allow them to interact intelligently with a wide variety of computer personnel such as vendors, systems analysts and programmers.

We deal today with a new and different structure of patient records. Previously we were concerned with paper documents filed in manila folders or binders, stored on shelves and in drawers that had to be organized in such a way as to be available for the users in a particular clinical care center. Today we are talking about hardware that emulates the procedures and equipment previously needed for physically storing and using medical information, and about software that includes the set of instructions that operates the entire process itself. Software is generally divided into a set of instructions that directs actual computer operating functions and is called the "operating system" and a set of instructions or programs that does a particular operation for a unique application, called "applications software." More and more applications software is being developed by vendor companies. This means that programs known as software packages can be purchased to run on an organization's existing computer. In hospitals and other health facilities, however, many applications programs are still being developed by data processing department personnel and therefore tend to be tailored to the individual hospitals' unique information needs. This often makes such software nontransportable to other settings, even for the same applications.

Some examples of MR applications for which software exists are the following:

- MPI
- Discharge analysis

- Major indices
- Census
- Physicians' incomplete files
- Record control
- Clinic appointment schedules
- Automatic tape-to-tape third party billing
- Specialized statistical research and analysis
- Special registries maintenance and follow-up
- Administrative reporting systems for the health organizations and facilities

Among the reasons that it is imperative for medical record professionals to be conversant with computer science is the growing reliance on computers experienced by all of us who are involved in the retention and retrieval of clinical and administrative use of patient data. A great deal of time must be spent in the pursuit of current and applied information regarding the cost, administrative impact, and user acceptance of computers on the retention/retrieval function. Without such knowledge, the medical record practitioner (MRP) will not be able to make sound decisions regarding software development and hardware selection. This will be a decided handicap in dealing with vendors who sell entire computer systems or component portions, and service companies that offer subscriptions for individual operations. The MRP may be asked to help select time share systems that provide operational capability to users. These systems provide access to sophisticated software that might otherwise be too expensive for an individual user and the associated facilities' in-house computers. Choosing time sharing, of course, will put the MRA in the role of working with other management members in decisions regarding the feasibility and use of a computer for the total facility and, in the case of the MRA, in particular, should be tempered with a knowledge of the applicability of alternatives. Microprocessors, for example, offer still other options in computerizing the functional operations of the MRD.

HOW THE COMPUTER PROCESSES INFORMATION

Now that we have examined the computer structure and identified some information concepts related to information processing and information users, we will look at how the information is processed from a source document through a computer system. This process involves changing the form of data. Look at the six steps below. At each step, the form of the data is changed. It continues to change until it has been translated into a form that the computer can process.

Steps in Processing Information through the Computer

The steps below illustrate one pathway. Alternative paths may be used when alternative equipment is used. For instance, the reader may remember that data input can occur with an optical scanner. This equipment allows the users to skip step 3 in the example.

1. Patient information is expressed in language and medical terminology on the *source document* patient record.
2. Patient information is abstracted or transcribed from the source document into the data fields on input forms in preparation for entering the data into the computer system.
3. Information transferred into the data fields is then keyed onto a data input medium such as punch cards or magnetic tape for entry into the computer system.
4. Data is entered into the computer system from the data input medium and by the programmed instructions already stored in the computer and stored in internal storage for later access and processing. The programs have been expressed in a high-level programming language.
5. Programs expressed in the high-level programming language are translated by compilers into machine-level programming.
6. The program is expressed in machine-level programming in binary form, which is then processed by the computer.

Remember that the computer can be operating in both batch and interactive modes. Approximately the same information-changing steps will apply in both cases.

Beginning with the Source Document

Let us look at these steps in greater detail. Step 1 occurs at the source document level. This is the source where the information is originally captured. Patient information is expressed in English and medical terminology on the record. Information is also expressed in this form on other source documents. A preregistration form completed by patients requesting services in an ambulatory care clinic is another example of a source document. Continuing with the example drawn from the discharge abstracting process, we can see exactly how the information is changed in step 2. The fields illustrated in Exhibit 2–1 are found on discharge abstract forms.

Exhibit 2-1 Abstract of Medical Record (MED-ART Case Abstract)

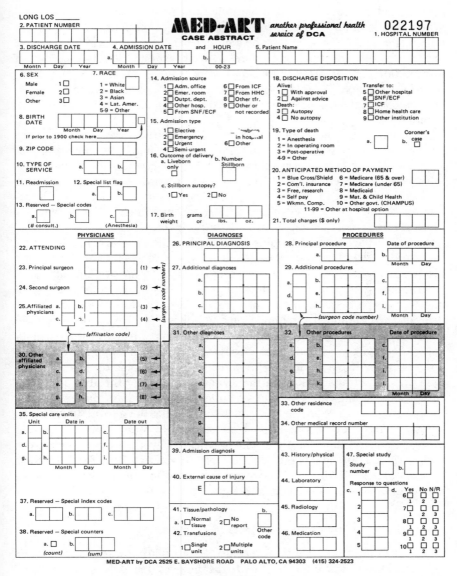

Source: The MED-ART Case Abstract is a product of IMI Health Systems, 2525 E. Bayshore Road, Palo Alto, California; IMI Health Systems is a Division of Information Management International, Inc., San Jose, California. © IMI Health Systems.

From these forms information will be keypunched for data entry. Notice that the information is condensed at this step. Diagnoses are translated into diagnostic codes and demographic and physical entries coded and grouped. It can be seen that coding may be a formal number assigned to an informational item or it may be a numerical grouping designated by the user to express information in a condensed fashion. Effective coding of information can facilitate more rapid processing of information in a computerized process. An interesting view of data change is seen in Figure 2–10, which illustrates how one character is sequentially translated into computer files.

Step 3 occurs when the information is prepared from the data input preparation form for direct entry into the computer. In the example, keypunching or similar procedure will take place when all the abstracts have been prepared from all the discharge records for a specified time period. In addition to keypunching, data entry clerks may enter the information by means of key-to-tape or key-to-floppy-disk. (See Table 2–1 again for additional examples of data entry devices.)

As MRDs become more sophisticated in the use of computer technology, they may streamline the steps listed. A growing application in on-line discharge abstracting is evidence of this fact. In an on-line discharge abstracting system, an MRD data entry clerk sitting at a computer terminal may request a display of the information already stored on the patient through the computerized admission and billing files. This information, in turn, may be displayed on the screen as a discharge abstract form partially completed. The clerk may then directly enter any remaining information on the screen displayed form. When the form is completed, the clerk may then press the transmit key on the terminal to transfer the abstract to the computer system. The computer system will store the information for later processing. In this example the reader can see that the information-changing steps have been streamlined considerably. Many data processing professionals predict that the "state of the art" will continue to progress until organization managers will be able to interact directly with their information in all kinds of ways via computer terminals. This progress may eventually include voice-activated applications.

Entering Information into the Computer System

Step 4 is concerned with entering the information into the computer system. This is accomplished with the help of the data input devices already described in this chapter. Card and tape readers may be used to enter the data into the system where it will be held for future processing as the data may come from locally stored files via data communications

Figure 2–10 Changing Data from Characters to Computer Files

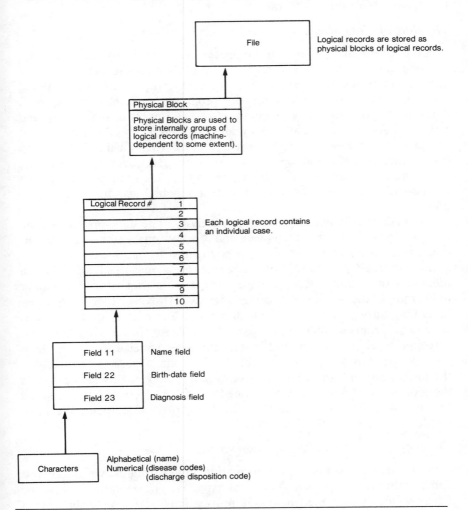

terminals. Remember that the control unit commands and directs all data entry into the CPU. Under program control the operating systems software directs the data to be entered and stored in a particular location. These programs also check and verify the information. Each field is identified, perhaps, by a particular identification code or by position in the entry data stream. Any codes stored in preestablished user files indicate whether or not the information is in appropriate form for entry into the system. Once verified, the program accepts the block of information and it is often sent to a location in auxiliary storage.

The Role of Main Memory

If the information is going to be processed immediately, the computer will maintain it in main memory. The control then goes to the appropriate application program necessary for processing that information. When processing is complete, the processing programs begin the cycle again. A computer-stored directory keeps track of valid users in the system and knows where their application programs are stored. This directory is part of the system for protecting the privacy of information that is built into the organization of the computer system. Placing the data file or directing it into the main memory and selecting the appropriate application programs are part of step 4 in information processing, which has now been completed.

The fifth step is actually a parallel activity. The development of the processing program has taken place and it is now changed into an executable form. In this step the program is changed from a higher level programming language into machine language. How does this happen? Each computer operating system provides or acts as a translator or compiler interpreter of the high-level language such as MUMPS or BASIC. It translates that through compilers, which are the interpreters or themselves special programs contained in the machine. These special programs analyze high-level instructions and emit machine code or else convert the high level program into machine-readable binary code. Figure 2–11 illustrates the programming translation process, which is step 5 and which parallels the process identified in step 4. Notice that with each step we have moved further away from the original medical record or document. We have condensed the original written data.

Completing the Information Processing

We have described how the information moves from the written record and on into the processing. Then what happens? The operating system we have previously described maintains information on each job or application being done. Because the computer works very fast, it may do one or two steps from one job and then move on to another. Because it operates so fast the program instructions continue so that users are unaware they are sharing computer processing activities.

As particular data are finished, the operating system may route it to a spooling device. The spooler then allows the data to be independently read out by an output device without holding up the job. Figure 2–12 shows how this is done.

Figure 2–11 Programming in a High-Level Language (e.g., COBOL)

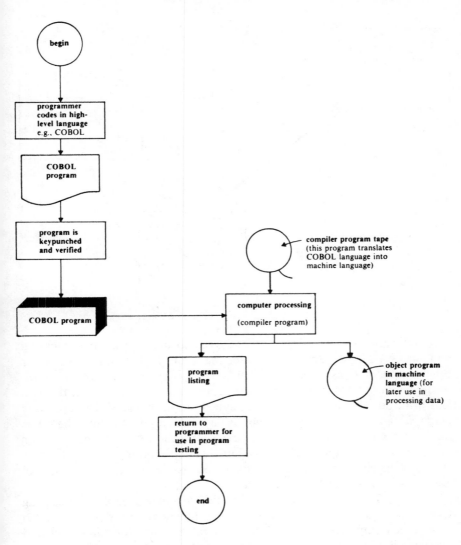

Source: Reprinted from *Information Systems for Hospital Administration* by Charles J. Austin by permission of Health Administration Press, Ann Arbor, Mich., © 1979.

Figure 2–12 Employing Spooling Devices

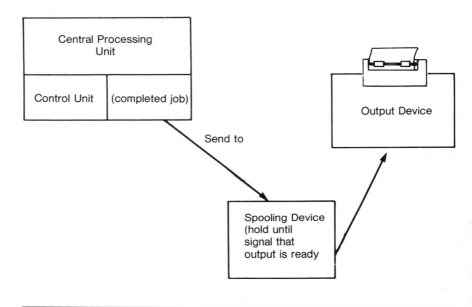

Spooling or similar devices collect the results of the program processing and act as a slowing down bucket for the job to be held until printers or other output devices are available. When that occurs the control program, through another program, instructs the spooling device to transfer the data and print out the information on assigned output devices. The output is cards, a paper printout, or a video display on a terminal. When the job is ready to be read out to the output device, the control unit directs it to its place in line for output. Priority of jobs is built into the master programs stored in the control unit. Let us look at how data communications control devices fit into the picture at this point.

When CRTs are used, then the operating system coordinates processing with data communications. The data communications devices are an important component of computer systems because they interpret and manage all data coming in through terminals. Sometimes such data communications devices are actually microcomputers with their own set of programs for coordinating and managing the information coming in and going out. The operating system acts as an overall manager of all input, processing, and output for the computer system. In many cases this means coordinating with microcomputers that handle preliminary processing or preparing of data. When microprocessors carry out preliminary processing tasks they are referred to as front-end processors. Figure 2–13 indicates how an overall computer system looks. It is also a map showing how patient or medical information is entered and processed in such a computer system. Notice the variety of ways that information can be handled with this technology.

Remember that the operating system controls programs and regulates the data entry devices and the input devices. It also manages and records data entered and swaps the application programs and the data files back and forth between the secondary storage and primary storage in the central processing unit as the jobs need to be accomplished. It directs the completed jobs to the output devices, as well as coordinating information back through data communications into CRTs. For many health information managers, CRTs and printers are the doorway to the overall computer system. Now let us move on to view computer systems as they appear in health organizations.

COMPUTER SYSTEMS IN HEALTH FACILITIES: LARGE, MINI-, AND MICROCOMPUTERS

Computer systems in hospitals and other health facilities are constructed in a variety of ways. The organizations may use computers of different size and capabilities. Along with large computer systems, such organizations may use minicomputers and microcomputers for particular applications. Minicomputers, the most rapidly developing segment of the computer market, are designed to fit the needs of small- to moderate-sized operations.

A minicomputer has the same components as a full-sized system. It can be programmed to perform many of the tasks of larger computers. Minicomputers generally have main storage capacities ranging between 16,000 and 1 million characters. They are physically small machines, often weighing less than 50 pounds. The term "mini" usually refers to the low cost of the minicomputer system. The purchase price can range from $20,000 to $100,000. Cost reductions of recent years increased the potential applications so that today minicomputers can be considered general purpose machines readily suitable for medical applications.

Microcomputers offer another opportunity. A microcomputer is made up of a microprocessor, memory, and input/output devices. The microprocessor on a microcomputer is the equivalent of a CPU on a silicon chip. The chip is about the size of a nail head. The microcomputer is a small, limited-capability, low-cost computer, relatively slower than a minicom-

Figure 2–13 A Data Processing Center

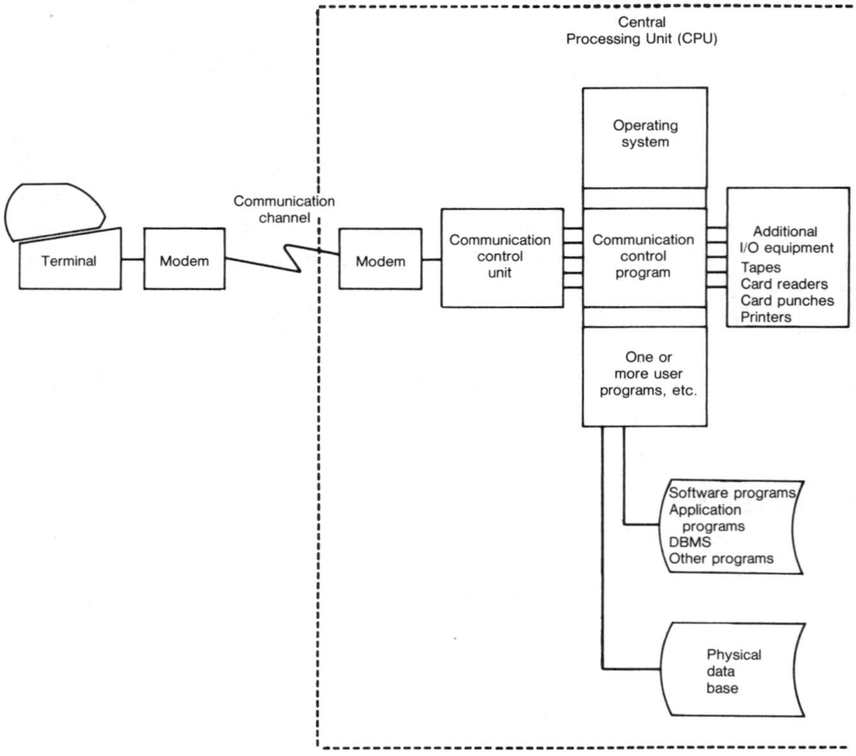

Source: Reprinted from p. 8 of *Information Systems: Theory and Practice* by John G. Burch, Jr., Felix R. Strater, and Gary Grudnitski by permission of John Wiley & Sons, New York, © 1979.

puter. It has extensive potential in physician office practice systems, MRD operations, and other related medical tasks.

Determining Appropriate Computer Configuration.

There is no one "best" computer configuration to use in any setting. Each organization must identify its own unique information needs and select the correct or most appropriate computer package for itself. We will discuss the process involved in arriving at identification of needs throughout the rest of the text. Let us conclude the present view of computer structure and data information flow by examining three figures that illustrate alternative computer system configurations for hospitals. These include employing different-sized computers in alternative systems. (See Figures 2–14, 2–15, and 2–16.)

Figure 2–14 Communicating Star Network

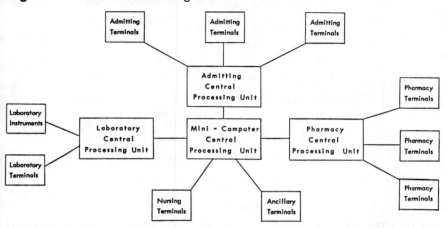

Source: Reprinted from *Hospital Information Systems* by Homer H. Schmitz by permission of Aspen Systems Corporation, Rockville, Md., © 1979.

Figure 2–15 Distributed Star

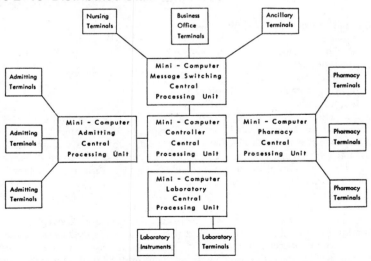

Source: Reprinted from *Hospital Information Systems* by Homer H. Schmitz by permission of Aspen Systems Corporation, Rockville, Md., © 1979.

Figure 2–16 Distributed Network

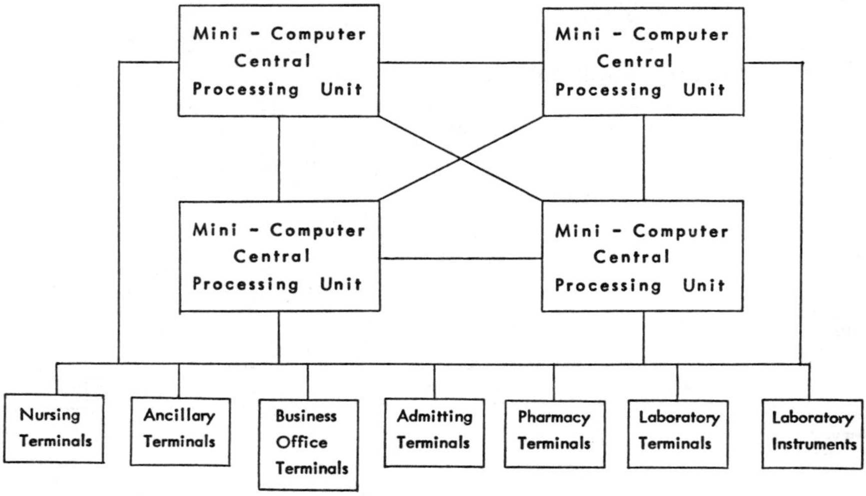

Source: Reprinted from *Hospital Information Systems* by Homer H. Schmitz by permission of Aspen Systems Corporation, Rockville, Md., © 1979.

Questions and Problems for Discussion

1. Define and describe the following terms:

computer	secondary storage	output devices
data entry devices	magnetic tape	application software
input devices	magnetic disk	printer
central processing unit	hardware	
primary storage	software	
access time	CRT	

2. Name the component parts and describe the primary functions of the CPU.
3. Which data entry devices could be used in a hospital ADT application?
4. Differentiate between batch processing and interactive mode. Explain how each could be used in performing discharge analysis operations in a medical record department.
5. Outline the steps in the processing cycle of information flow through a computer system. Begin with data entry and conclude with data output.
6. Describe two ways a card reader could be used in computerizing a medical record operation.
7. You are an RRA in a medical record department. You have been asked to comment on a plan developed by the data processing department to transfer your disease and operation index from magnetic disk to magnetic tape storage. Based on the advantages and disadvantages identified in this chapter, comment on the practicality of the plan and identify some concerns you would raise about the effect of such a change on departmental activities.

8. You are interested in exploring disease and operation indexes storage on a floppy disk for use in a departmental Apple II system. What are the advantages of such a move? How would the storage capability of the floppy disk compare with other likely storage media?
9. Describe an MRD application that would use a printer.
10. Could hard-copy printout or COM output be used for the same application? Explain your answer.
11. Which high-level language has been developed primarily for health care applications?
12. What are the benefits of data base management systems?
13. Describe applications programs that would be used for medical record department functions as featured in this chapter.
14. Should a hospital develop its own software? Why or why not?

NOTES

1. Joan Zimmerman and Alan Rector, *Computers for the Physician's Office* (Forest Grove, Ore.: Research Studies Press, 1978), p. 244.

2. Randall Jensen and Charles Torries, *Software Engineering* (Englewood Cliffs, N.J.: Prentice-Hall, 1980), p. 549.

3. Charles J. Austin, *Information Systems for Hospital Administration* (Ann Arbor, Mich.: Health Administration Press, 1979), p. 109.

4. Charles J. Sippl and Roger J. Sippl, *Computer Dictionary,* (Indianapolis, Ind.: Sams, 1980), p. 241.

5. Ibid., p. 120.

6. Ibid., p. 369.

7. John G. Burch, Jr., Felix R. Strater, and Gary Grudnitski, *Information Systems: Theory and Practice,* (New York: John Wiley & Sons, 1979), p. 516.

8. Carl Feingold, *Introduction to Data Processing* (Dubuque, Iowa: Wm. C. Brown, 1977), p. 166.

9. Burch, Strater, and Grudnitski, *Information Systems,* p. 502.

10. Ibid., p. 504.

11. William M. Taggart, Jr., *Information Systems, An Introduction to Computers in Organizations* (Boston: Allyn & Bacon, 1980), p. 100.

12. Donald A.B. Lindberg, *The Growth of Medical Information Systems in the United States* (Lexington, Mass.: D.C. Heath, Lexington Books, 1979).

13. Sippl and Sippl, *Computer Dictionary.*

How Computer Technology Has Altered the Functional Operations in Medical Record Departments

Objectives

1. Define functional operations in health record systems.
2. Summarize the major functional operations of health record services in hospitals and other health facilities.
3. Describe the impact of computer technology on medical record department (MRD) functions.
4. Identify the range of computer applications found in contemporary MRDs and associated patient information systems.
5. Explain how automation improves and streamlines manual medical record processing.
6. Relate current methods in automation to current and future activities in patient information systems.
7. Explain how alternatives in automation can be selected to improve the information processing in health information systems.

The purpose of this chapter is to provide a comprehensive picture of the impact of computer technology on MRDs and patient information systems. Twelve major functional operations commonly performed in MRDs and other health information systems are described and discussed.

First, let us define a *functional operation* in the following manner. *Function* is defined as a special purpose or characteristic action. *Operation* is defined as a process of executing a defined action. A *functional operation,* therefore, is a special purpose or characteristic process that executes the defined actions. In medical records, functional operations are common processes used to achieve specific departmental objectives. They characterize the processes commonly performed for patient record development, processing, retention, and retrieval.

OVERVIEW OF MRD FUNCTIONAL OPERATIONS

In collecting and managing patient data and records, personnel in the MRD in hospitals and other health facilities may work with the following functional operations (FOs).

Medical Record Department Functional Operations

- Maintain master patient index.
- Prepare and distribute census and other statistics.
- Screen admissions for utilization review.
- Monitor medical records for completeness.
- Analyze and abstract discharge records.
- Code and index diagnostic and procedural data.
- Prepare and distribute medical reports.
- Manage and control record accessibility.
- Screen medical records for risk management.
- Screen medical records for quality assurance.
- Develop and perform research and case mix analyses.
- Develop and manage word processing of medical reports.

The *master patient index* (MPI) is a permanent listing of all patients treated in a facility. It provides identification and location information regarding medical or health records. In manual operations, this index is usually located in file cabinets or electric rotary files that house 3- × 5-inch index cards. Admission and discharge information is updated on a daily basis. The most common items recorded on an MPI card are the patient's full name, full address, telephone number, date of birth, record identification number, admission and discharge dates, and the name of the attending physician or other responsible health care provider.

The census is the daily count of patients in the facility. *Preparing and distributing the census and related statistics* means collecting, validating, and summarizing the number of hospital admissions, the actual number of patients in hospital beds, and the numbers of discharges and transfers and presenting the data in a daily census report that is distributed to all hospital departments. The report serves as a notification and communication of patient movement into, within, and from the health facility. In many hospitals the census report is prepared manually by the medical record clerk who maintains the MPI. In other facilities the census may be prepared by a medical record technician who functions as a statistical technician for the department.

Screening admissions for utilization review is a procedure set up to certify hospital admissions. Utilization review (UR) was initiated at the federal level, primarily for Medicare patients. As federal regulations change, new groups of patients fall under UR guidelines and the percentages of certain categories of patients change. The review process checks to see that individual admissions meet a predetermined set of medical criteria and is used to determine justification for hospital admission. Although this operation is often performed in the admitting department, follow-up activities can be the responsibility of a medical record utilization review clerk. For this reason, practitioners will be interested in the utilization review information requirements and will have to keep abreast of changes in the federal rules and regulations.

Monitoring medical records for completeness means reviewing patient records to ascertain that all forms are dated, with data entries complete and accurate, signatures present, and diagnostic conclusions recorded. It is an operation that uses criteria established by the Joint Commission on Hospital Accreditation (JCAH) and the medical staff bylaws as well as standards of the particular institution. The criteria used specify patient record documentation requirements. Because the patient record is the legal record of health services rendered to the patient, it should contain sufficient information to document specific action taken on the patient's behalf. Because it is a unique record of the patient's medical problems, it should contain sufficient information to document the diagnostic investigation, treatment, and results so that ongoing and future medical care can be provided.

Analyzing and abstracting discharge records is an operation in which the records of discharged patients are abstracted to define and describe hospital services. This is carried out through completion of a uniform discharge data set on each case. Medical record technicians use this data set to record the demographic data, hospital services, and diagnostic and procedural information for each patient. Collectively, these records are used to keep account of the frequency of use of the various hospital services for a given time period, and to collect and summarize basic hospital statistics on discharged patients.

Coding diagnostic and procedural data means assigning a number or alphanumerical code to each of the various elements of a medical record. It allows for an organized approach in the later retrieval of such diagnostic and procedural data as diagnosis, operation, procedure, reason for visit, pathologic specimens, injuries, obstetric conditions and procedures, symptoms, mental conditions, congenital anomalies, physical signs, and ill-defined conditions. Medical record practitioners will be familiar with many coding systems including ICD-9CM, CPT 4, and SNOMED.

Indexing uses the code numbers assigned to compile all records classified under one code number. It enables data users to retrieve all like cases by indexing individual patient names under the disease or procedural code number that reflects the individual's health problem.

Preparing and distributing medical reports is a primary task of the MRD. Medical reports processing is a work unit established by MRDs to respond to requests for patient data and records. Included are reports pertaining to reimbursement, transfer of patients from one facility to another, patient requests for individual data, and legal requests. This operation is carried out by one person or by a departmental unit depending on the size of the institution. Personnel who perform this functional operation must understand the legal and ethical aspects of protecting the patients' privacy and apply that knowledge carefully when releasing information.

Record control, or *managing and controlling record accessibility,* is an operation used to identify and control the location of patient records. Given a departmental objective to retrieve patient records within five minutes of a valid request, medical record department personnel responsible to carry out the request must rely on a specific method of retention and retrieval. Precise tracking of charts within and outside the medical record department is necessary to achieve constant record accessibility.

Screening medical records for risk management compares the data in patient records against criteria that have been established through an organized risk management program. Such "programs are developed to control liabilities for human errors and equipment failures. Their goal is to control preventable risks and keep to a minimum the incidents for which the institutions can be held liable."[1] When screening is employed, charts of potential risk management cases are identified, summarized, and reported to a risk manager or committee for further processing. Statistics and trends identified may be summarized routinely.

Screening medical records for quality assurance is a similar procedure whereby the charts of discharged patients are reviewed against criteria established in a quality assurance program. Like risk management, this procedure enables quality assurance committees to receive specific cases and appropriate statistics relative to a topic under study. It may also enable the committee to uncover problems meriting future study.

Developing and performing research refers to activities involved in data retrieval for investigative studies of clinical data. The research is usually an activity of the medical staff. Medical record practitioners often provide assistance in code selection for individual studies as well as chart retrieval. In some cases data abstracting may also be performed. Typical research areas are clinical research, community epidemiological research, and facility services research. A more recent concept, *case mix analysis,* pertains to analysis of the use of hospital resources. Discharged patients' cases are the source of study information. Case mix analysis seeks to determine the relationship of particular diagnoses or procedures to the specific uses of the hospital resources since some illnesses require more expensive use of hospital resources than others.

Developing and managing word processing of medical reports for medical record transcription is another central function of the medical record department. Transcription of dictated reports for patient records is done by transcription units ranging in size from one part-time transcriptionist in a small hospital medical record department to an independent, multi-person unit that provides a large hospital with a wide variety and volume of transcription services. Effective transcription services provide the medical record department an opportunity to establish good rapport with the medical staff.

COMPUTER APPLICATIONS IN MRD OPERATIONS

What major medical record functional operations will be affected by computer technology in the next decade? Let us consider the twelve functional operations just identified. For each, one current computer application will be described and summarized. These summaries should provide a resource for practitioners to use in projecting the potential for their own departments. Together with effective systems analysis techniques, explained in the following chapters, these descriptions should afford medical record practitioners the ideas, examples, and methods they need to work with computer systems planning and implementation in their own setting.

Computerized Master Patient Index (MPI)

The MPI is the foundation for medical records processing, retention, and retrieval. It stores the patient's name and a unique numerical identifier so that patient records can be retrieved. Computerized "on-line" MPIs provide an opportunity to extend the reference capabilities of the manual MPI beyond departmental boundaries. All patient records, including x-rays, lab test results, electroencephalogram (EEG) tracings, and others, can be linked to the common identification data contained in the MPI. This allows more efficient retrieval of patient information for many departments in a hospital. One good way to put the MPI on-line is to link the creation of the master patient index to computerized on-line outpatient registration or inpatient reservation. In this process, a CRT terminal is used to enter and store information on a hospital computer. It does this for subsequent retrieval through that terminal and others located throughout the facility. When a patient enters the registration area, a clerk enters the person's name on a CRT terminal. If the person has had a previous admission or clinic visit at the hospital, the name appears on the screen along with the hospital or registration number previously assigned. In many systems both alphabetical and phonetic searching of the index will be performed by the computer. If the patients have not been there previously, the clerk may call for a registration screen display. Clerks may type in something like "PT/R" to indicate patient registration is about to begin. The screen then displays a format or conducts a dialogue to indicate what data elements need to be completed. The clerk enters the information provided by the patient directly into the terminal. Pushing a transmit key then sends the data directly to the computer. If the system is part of an integrated automated hospital information system, the appropriate data can be immediately communicated to other departments in the hospital. This would include dietary, housekeeping, laboratory, x-ray, business office, and any other department that may need to know when new admissions or new patients are entering the facility. Some hospitals and clinics participate in a preadmission or reservation program, in which scheduled surgery or admissions are planned ahead. Patients provide preliminary information that can be collected prior to the patient's actual physical entry into the system. When these patients enter the admitting department their names are keyed into the CRT and the previously collected information is displayed. The clerk need only verify that the patient is actually entering the hospital and again key the transmit button to disseminate the information appropriately. The admission is thereby recorded at the admitting department through a patient registration or admission process.

Design of the computer programs for an on-line MPI varies. The index may be treated as an on-line subsystem within a hospital computer system with integration to other departments as previously described. This would allow access to the MPI by the registration clerk and personnel of other departments when they need identification data on a patient. It can also be designed as a computerized operation in the MRD with no integration to other departments. Often microfiche is used as back-up for on-line systems when the computer is "down" (out of service) for scheduled maintenance or because of unscheduled mechanical or electrical problems. The microfiche can also be used to supplement the operation and store additional information on the patient that is not contained in the MPI. For example, where computer disk storage is limited, more extensive patient information such as the physician name, additional identifiers, admission dates, diagnostic information, or information flags for special studies may be incorporated in an MPI format included on the microfiche. This will allow the hospital personnel to continue to use the on-line MPI for immediate identification but will rely on the microfiche product for more extensive information related to the MPI. One of the questions that practitioners must answer is "What is the optimal amount of data to be included in a computerized master patient index?"

In another design an on-line MPI can be processed by a low-cost microcomputer. Such machines have the capacity to support more than 50,000 patient names and identifiers with two terminals for a given application. Such a system is called a *stand-alone system* because it is not connected to other computer systems in the hospital.

Figure 3–1 is a flowchart illustrating a computerized MPI operation that prints out a medical record face sheet as part of the operation. Figure 3–2 is a typical screen display for outpatient unit number assignment within a computerized MPI system. Now consider Figure 3–3, which shows how a computerized MPI operation fits into a computerized reservation–admission, discharge, transfer (R-ADT) system in a hospital information system. Flowcharts like the ones illustrated ensure that no procedural step is left out in computerizing the MPI.

Computerized Census and Statistics Preparation and Distribution (ADT System)

Each census entry is a building block in a structure that begins with the reservation or admission function itself. When a new patient is admitted, the name is added to the list that makes up the census report. If the system is a real-time on-line feature of an automated hospital information system, then each time a patient's name is added the file may be automatically updated. Someone may be admitted at 1:00 P.M. and the information transmitted to the computer. At 1:02 P.M. the radiologist, who wishes to check the current census, keys "CEN" to retrieve the census information for display on a CRT screen. This display will include the person who has just been admitted. Instead of being an extension of an on-line admitting process, computerizing the census can be done by batch. In a batch process the data on the individual patient is keypunched or typed into a terminal at the time of admission, but the data are not processed beyond the local storage medium at this time. Each night reports come from the care units that provide an accounting of the number of patients on each unit. All counts are collected and tallied by the computer in a batch mode. Recall that the census is run daily and lists of admissions, discharges, transfers, and deaths are included in the report. When the census is computerized, it benefits the entire hospital. Census printout reports can be duplicated

Figure 3–1 Flowchart for a Computerized Master Patient Index

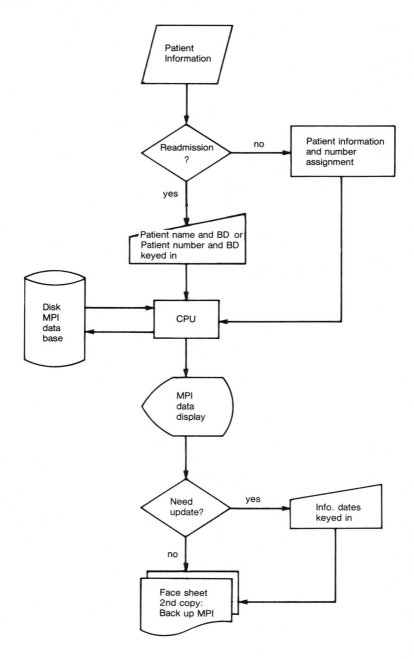

Figure 3-2 Screen Display for a Computerized MPI System

MASTER PATIENT INDEX - INQUIRY

NAME	NUMBER	DOB	SS#	DISCH	
Myers, Barbara	32 45 66	03/14/57	265-38-7406	12/14/81	IP
Myers, Burton	44 34 67	07/14/62	328-00-5406	04/32/80	OP
Myers, Burton J	44 33 68	10/09/42	537-38-2165	10/09/80	IP
Myers, Catherine	56 22 16	12/04/44	546-23-2168	06/17/78	OP
Myers, Dean	00 34 54	06/10/50	544-97-0023	08/23/79	IP
Myers, Dorinda	00 45 60	11/30/72	645-23-3133	06/26/77	IP
Myers, Egan	16 44 69	01/24/66	435-34-1965	11/23/80	OP

Key: RET. for the next screen display

MASTER PATIENT INDEX - PATIENT IDENTIFICATION

Name: Unit Number:

Address: Date of Birth:

City, State: Zip:

Sex: Social Security Number

Mother's Maiden Name: Dr's. Name:

Admission Dates: Discharge Dates:

Enter information and key:
1 if information is validated
2 if MPI update screen is required

Figure 3–3 Computerized MPI as an Element of a Registration–
Admission, Discharge, Transfer (R-ADT) System

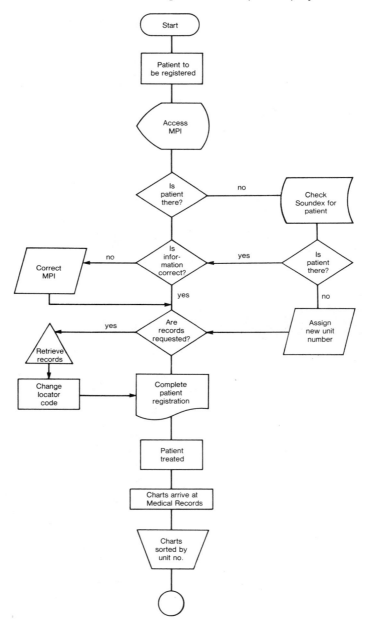

easily and sent through interhospital mail or census information can be displayed on a CRT video screen for immediate availability to those who need the information.

Census reports available on CRT displays that can be accessed on a "need to know" basis are a convenience to users. Figure 3–4 is the distribution list of a computerized census report in the U.S. Public Health Service hospital in Seattle. Many of the areas listed in Figure 3–4 could operate just as efficiently through a CRT screen display. To determine how many, medical record practitioners doing systems analysis preliminary to computerization may wish to develop comprehensive descriptive profiles identifying users (see Chapter 7).

Admission Screening for Utilization Review

A utilization review reporting system can be an integral part of data gathering in hospitals and can assist the organization and the PSRO or PRO in meeting program objectives. As we can expect to work with a medical care review program in some form, it is valuable to examine the computer potential available to carry out the operation. Along with the census, the admission screening and utilization review operation can be developed from the foundation established in a combined MPI and census function known as registration–admission, discharge, transfer (R-ADT). (A detailed explanation of R-ADT is presented in Chapter 7.) Of course, as with other systems, this operation can also be developed independent of all others. Computer systems designed to accomplish this operation should be capable of providing the data requirements necessary for the PSRO or Professional Review Organization (PRO) programs. Not only should sufficient data be collected to promote maximum use by hospitals and patient care reviewers, but patient record monitoring should be supported. Flexibility should be such that changes in criteria or norms can be easily incorporated into the system operation. Figure 3–5 flowcharts the basic admission review cycle contained in this operation. The following are system performance objectives for this application:

- Provide to the review coordinator, on a daily basis, information about each patient presently due for review, so that the review may be initiated and performed according to schedule.
- Provide a daily master file of all recorded information on each patient presently in the hospital that is easily accessible to the review coordinator and reduces the access time needed to locate review process information to less than five minutes 95 percent of the time.

Figure 3–4 Census Distribution Chart

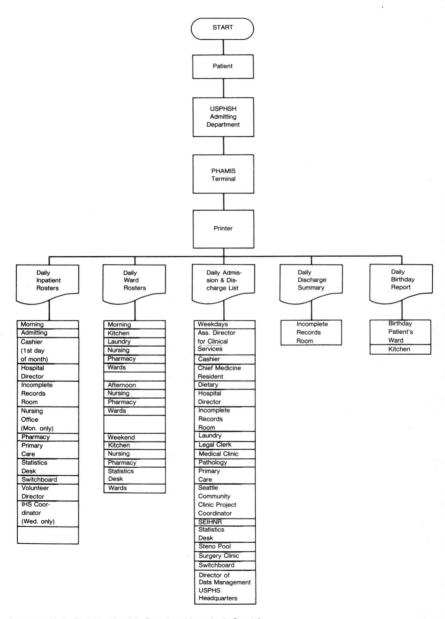

Source: U.S. Public Health Service Hospital, Seattle.

Figure 3–5 Admission Screening

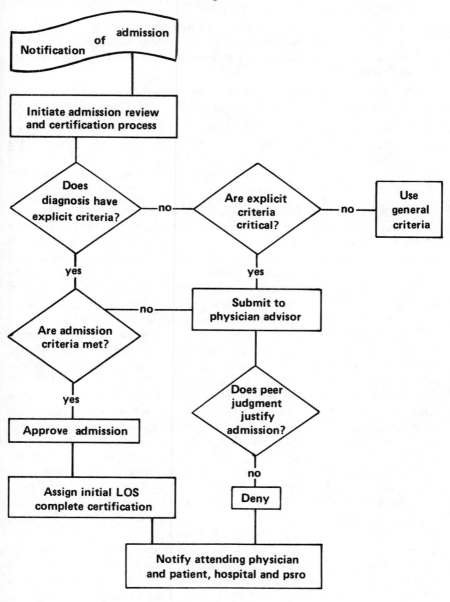

Source: Kathleen Waters and Gretchen Murphy, *Medical Records in Health Information,* © 1979 by Aspen Systems.

- Provide daily information to individual physicians about the review status of each patient under their care.
- Compile records about each review and automatically maintain statistical summaries on the review activity. This is done in each hospital for comparison among all hospitals in the area and the summaries are available within 24 hours of request.
- Produce the necessary reports in magnetic tape format compatible with federal processing requirements.
- Develop a data base containing accumulated medical information on each patient to allow identification of patient records for medical care evaluation studies and profile analyses, revision of length of stay norms and admissions screening criteria development by hospital or region.[2]

The following example is adapted from a description of an in-house computer system. At admission the admitting clerk enters patient data into the computer system on a CRT: patient's name, address, social security number, admitting physician number, admitting service number, type of federally insured benefit, health insurance number, other nonmedical data elements, and admitting diagnosis are entered. This establishes a patient data base for the medical record department. It may be used to create an MPI subsystem and provide the basis for a utilization review system.

The day after admission a computer-generated list of certification forms is printed that includes the patient's name, the admitting physician's name and number, and the patient's hospital registration number, room location, admitting diagnosis, and type of insurance (Medicare, Medicaid, or other). This list is provided to the patient review coordinator (PRC) to use as the charts are reviewed on the floors. The PRC can then enter the reason for admission, the physician-documented admitting diagnosis, and any secondary admitting diagnoses and complete the first section of the admission certification form. The admission certification form, in turn, serves as the source document for the utilization review system and as the legal documentation of certification.

In the next step, the completed certification forms can be used by the data entry clerk. This step may consist of the clerk selecting the CRT screen display labeled "Update PSRO Patient History." The clerk then enters the last name, first name initial, and other appropriate data. The update display is completed with entry of the admitting diagnosis, the diagnostic code, and the admitting criteria. The date of review, days certified, level of review, and PRC identification could be included in a review section of the screen display. Another section of the screen may be used to enter operative procedures. For example, the operating room could provide a daily log of all patients who had surgery the previous day. Medicare, Medicaid, and other federally insured patients could be identified on the printout by an "M" code. The operative information, entered by a clerk, would be the date of the procedure, the code number for the procedure and the surgeon or physician number. Notice that each data element contributes to the necessary data base.

Utilization reports could be generated as the next step. A report identifying the patients who need continued stay review could be generated daily. It may contain the patient number, name, federal insurance, patient location, service, and which review is necessary—admission certification or extended stay review. A second report could also be produced that contains a list of the federally insured patients who were discharged the day before, in order that they can be eliminated from the re-review list.

In the fifth step, when a report identifies a patient as requiring an extended stay review, appropriate reviews are performed and the information is entered into the system. For example, if an extended stay is certified for an additional eight days, the information is recorded, entered on the CRT, and the next review date is automatically stored in the computer to be used in generating new certification review forms and to print out the patient information in the scheduled review lists for the next date. This process is repeated until the patient is discharged.

If a PRC is unable to certify a stay because of lack of record documentation, a request can be placed in the record asking the physician to provide a diagnosis or reasons for hospitalization. The items that are needed in order to perform the certification function would be included on the request. After the certification information has been completed, the PRC can place a sticker in the chart to flag the next review date. Thus the physician is alerted that additional certification information will be required at that time. If additional documentation from the attending physician is not sufficient to certify the admission or continued stay, the case can be referred to a utilization review physician advisor.

Another discharge procedure often performed by a patient review clerk involves verification of the days certified with the length-of-stay assignments, accuracy of the next review dates, the recording of total extensions, total length of stay, and the total number of days certified for acute care. The discharge date is recorded at the bottom of the form and for statistical purposes discharge disposition is also entered. A billing form or electronic record may be completed indicating the hospital identification for the insurance coverage of the patient (Medicare, Medicaid, or other), the patient's name, date of admission, certified days used, total acute and nonacute days that were certified, authorized signature of the clerk or

PRC, and the date the action was completed. This record can then be sent to the patient accounts department for final disposition.

Finally, cumulative data can be provided. Monthly statistics such as total discharges of Medicare and Medicaid patients, number of concurrent reviews, number of reviews referred to physician advisors, number of continued stay reviews, total certified days of stay used, and total days of stay can be maintained. This information can then be transferred to the appropriate reporting records. The computer-supported utilization review system has proven to save time and money. In one case it was estimated that without computer support, a staff of two or three additional clerical personnel (at $16,000–$24,000 per annum) would have been necessary to maintain accurate and current statistics.[3]

A computerized utilization review application is an excellent candidate for time-sharing programs. A host computer stores the utilization screening program. Small hospitals access the computer with terminals and save the cost of computer installation at their own site. Notice how such a system could operate in Figure 3–6.

Computerized Chart Completion Systems

The chart completion system is another operation that has been computerized in many medical records departments. Such a system can include completion of an abstract form for keypunch, or it can mean directly entering the information about chart deficiencies under a physician's name that is called up on a CRT screen. When the individual patient's information is keyed in under a physician's name, this automatically stores the information so that lists of incomplete records can be quickly printed and made available upon request from the physician.

This system also produces lists of the total number of chart deficiencies for each physician, which can in turn be grouped according to services such as medicine, cardiology, and so on. Like the MPI, chart deficiency systems are also available as subsystems in an automated hospital information system, in software packages marketed by vendors who sell medical record systems, and in stand-alone microcomputer applications. It is important to note that such packages need to be evaluated before purchase to be sure they are compatible with systems planned for installation in the future.

Practitioners who are developing chart deficiency systems should perform systems analysis as described in this text to determine whether to develop the application as part of an overall chart locator and/or record control program. Appropriate attention to systems analysis will direct systems teams to identify the best computer alternative for their setting. Exhibits 3–1 and 3–2 are CRT screen displays of a physician's incomplete

Figure 3–6 Utilization Review in a Time Share System

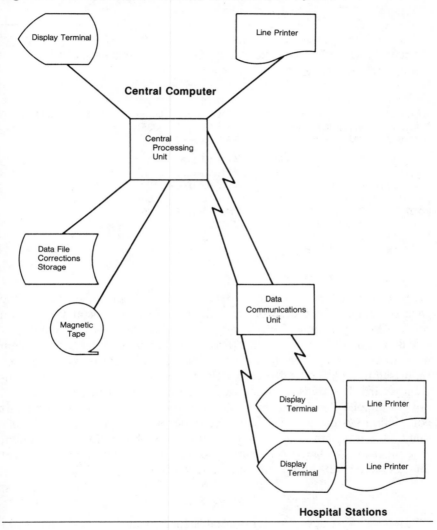

charts and a chart deficiency status report provided to physicians. These are examples of output displays that could be designed in computerized chart completion applications.

In another example, the MRD analyzes the record and completes a two-copy chart deficiency card. The deficiency cards are color-coded punch cards that are automatically keypunched on admission and routed to the

Exhibit 3–1 CRT Screen Display of Chart Deficiencies

TO : DR. Ben CASEY (Gyn) DATE OF REPORT: AUGUST 19, 1980
FROM : CHIEF, MEDICAL RECORDS DEPARTMENT, PHS HOSPITAL
SUBJECT : INCOMPLETE RECORD REPORT

Patient Number	Patient Name	Date of Discharge	Days Open	1	2	3	4	5	6	7	8	9	10	11	12	13
420043	SEASHORE, Joy	SEP-19-1979	335							X					X	X
299845	SMITH, Lesia	FEB-06-1980	195	X	X					*						
165902	EAGLE, Sharon Mae	APR-10-1980	131							X				X	X	
411947	RUSTYNAIL, Diana	APR-13-1980	128						*	X		X				
438521	LEEPIERT, Joy Ann	MAY-02-1980	109				X			X						
440295	ADAMSEN, Lorraine	MAY-16-1980	95	X						X	X					
207006	POWER, Mary Ann	MAY-31-1980	80					X		X						
439776	SEETZED, Sharon Ly	JUN-11-1980	69							*						
408544	WALK, Susie	JUN-15-1980	65	X	X					X						X
439668	HOWARD, Baker	JUN-12-1980	68							X		X		X		

medical record department to be held until the patient is discharged. One card is attached to the record and the second card is sent to the data processing department. Individual physician deficiency reports and general departmental summary statistics are produced through a batch system on a weekly basis. The data on each patient record are carried on the reports until the medical record department transmits the second deficiency card. It signals the data processing department that the record is now complete. This is an example of a lower cost computerized chart completion system that can be effective despite its low cost. Remember that computers should be used in accordance with individual department objectives and budgets. All health record departments should try to meet their objectives in simple, cost-effective ways.

Computerized Discharge Abstracting Systems

Discharge abstracting systems are designed to abstract basic data from patient records for the purpose of statistical reporting. This functional operation has been available in computerized form for a long time. The method of preparing individual abstracts upon discharge is used to collect the basic discharge data set (or UHDDS, for Uniform Hospital Discharge Data Set). Prior to standardization of discharge data, the method was used for data deemed useful by individual hospitals. Hospitals have subscribed to discharge abstracting services for many years. In these systems MRD

Exhibit 3–2 Sample Chart Deficiency Status Report

REPORT NO. MED010
DATE 12/12/81

NOTICE OF
INCOMPLETE RECORDS

NOTE***

RE: OPERATIVE REPORTS—A DELAY IN DICTATING OF AN OPERATIVE REPORT FOR 30 DAYS AFTER THE DATE OF SURGERY MAY RESULT IN SUSPENSION OF YOUR ADMITTING AND SURGICAL PRIVILEGES.

DISCHARGE SUMMARIES—A DELAY IN DICTATING OF A DISCHARGE SUMMARY FOR 30 DAYS AFTER THE DATE OF DISCHARGE MAY RESULT IN SUSPENSION OF YOUR ADMITTING AND SURGICAL PRIVILEGES.

PLEASE CALL MEDICAL RECORDS (TELEPHONE 634-5078) AND ARRANGEMENTS WILL BE MADE TO HAVE YOUR CHARTS AVAILABLE FOR COMPLETION.

TO: DOCTOR R. O. HICKMAN   DOCTOR #1190

THE FOLLOWING ARE FILED IN THE ATTENDING DOCTOR'S FILE IN THE DOCTORS' DICTATING ROOM.

PATIENT NAME	PATIENT #	ATTENDING DOCTOR	DOCTOR #	ADMITTED	OPERATION/ DISCHARGE	DAYS SINCE DISCHARGE	DEFICIENCY
ANDERSON, SALLY	0460198629	HICKMAN, R. O.	9500	2/15/81	2/28/81	12	SIGNATURE/ CC-SIGNATURE
BALLARD, STACY	0610160509	HICKMAN, ROBERT O.	1190	3/02/81	3/07/81	5	SIGNATURE/ CC-SIGNATURE
CAPLES, ALAN	0080227899	HICKMAN, ROBERT O.	1190	1/08/81	1/10/81	61	PROGRESS NOTE SIGNATURE/ CC-SIGNATURE
DOWN, KRISTA JOELL	0250210939	HICKMAN, R. O.	9500	1/25/81	2/12/81	28	SIGNATURE/ CC-SIGNATURE
SLUYTER, SUE	0520220782	HICKMAN, ROBERT O.	1190	0/00/00	2/22/81	18	OPERATION
TOWNS, DAVID	0290233924	HICKMAN, R. O.	9500	1/29/81	2/05/81	35	SIGNATURE/ CC-SIGNATURE

staff complete an abstract form for each patient upon patient discharge. The forms are sent to the abstracting company in groups as designated by the company. Abstract forms contain the data items that later are compiled to produce a standard set of statistical reports for the record department. Discharge statistical analysis, disease and operation coding and indexing, and special discharge data studies are common products of discharge abstracting service companies. The selection of these systems and participation in these programs are important concerns for medical record managers today. Managers must evaluate the nature of each system, assess the cost per discharge to their department, determine how their staff can be trained to carry out functions, and evaluate reports provided by the vendor for timeliness. They must also determine the amount of use the reports get from the medical staff and others within the facility. A partial list of discharge abstracting services and their computerized abstracting programs follow.

Effective management of data use in individual facilities requires that practitioners continually assess the reports provided by these programs to be sure that they meet the needs of the organization. In recent years discharge abstracting systems have become more flexible and the companies more willing to design custom features for individual hospitals. Magnetic disks with discharge data are mailed to the abstracting company, where they are read directly into a computer system for future processing. This represents a significant advance in data transmission by saving time at both the hospital site and the processing site. Today many hospitals elect to develop their own discharge abstracting system as a natural follow-up to an on-line hospital census and the MPI function. This is particularly so when an automated hospital information system is being developed. Notice how many feature discharge abstract modules in their programs. Today's practitioners will want to investigate fully the existing software packages and consider those that satisfy the performance definition developed in their organization's systems analysis process.

Computer-Supported Diagnostic and Procedural Coding and Indexing

Computerized coding systems are now available. Two applications are of major interest to medical records practitioners. The first is the Computerized Lexicon for Encoding and Retrieval (CLEAR) coding system, which provides for on-line encoding of the names of diseases and procedures either according to standard nomenclatures or according to the physician's own wording. It was developed by Elemer Gabrieli, M.D. The second system was developed by Code 3 Medical Systems. CLEAR is a

Table 3–1 Major Vendor Systems Developments

Department/Application	Number of Vendors	
	Now Offer	Plan to Offer
Nursing		
Order set entry	5	1
Charting	3	2
Patient care level tracking	3	2
Staffing requirements	5	2
Nursing notes*	5	0
Care plan*	1	6
Medication schedules*	6	2
Medication monitoring*	3	2
Pharmacy		
Patient medical profile*	7	1
Drug precaution/interaction	2	4
Unit dose care replenishment	5	2
Pharmacy inventory	3	4
Lab		
Specimen collection lists	6	1
Specimen collection labels	8	0
Results reporting*	7	0
Cumulative results*	6	1
Interface to lab computer	6	1
Radiology		
Scheduling	4	2
Report normals only*	3	3
Report all results*	2	5
Radiology index	4	2
Medical Records*		
Medical record index*	5	2
Medical record abstract*	5	2
Chart location delinquency tracking*	3	3
Emergency Room/Outpatient		
Registration*	8	0
Order entry*	8	0
Demand billing	7	0
Patient scheduling	6	2
Miscellaneous		
Preadmitting*	8	0
Computer-assigned patient ID*	8	0
Time-clocking	5	0
Surgery scheduling	2	3
Diet list preparation	7	1
Utilization review*	7	0
Cash receipts entry	6	0
Doctors' registry*	6	1

*Patient record impact.
**Some report Stats only.

lexicon-driven computer-based system for transforming natural language (English) terms into the following codes:

- International Classification of Disease–9th Revision–Clinical Modification (ICD-9CM)—by the World Health Organization
- Systematized Nomenclature of Medicine (SNOMED)—by the College of American Pathologists
- Current Procedural Terminology (CPT)—by the American Medical Association
- Drug coding, whereby names of generic and brand drugs are organized hierarchically according to clinical usage
- Unique code, the exact wording used by the physician

Figure 3–7 is a flowchart showing the steps in the process of encoding a single entry. In the hospital's record department, an authorized data entry clerk submits, via a computer terminal's keyboard, the data to be coded. The CLEAR system in the computer accepts the data, transforms them into appropriate codes, and returns the results to the user. The data elements and data entry format are outlined as follows:

- Level 1
  Input: Two segments, I and II
  Step I.   1. Hospital's name
             2. Initials of the data entry clerk

With step I the user logs into the system. Step II provides for entry of individual patients' data to be coded.

  Step II.   1. Patient's full name
             2. Patient's record number
             3. Date of admission
             4. Date of discharge
             5. Admission diagnoses
             6. Discharge diagnoses
             7. Procedures (surgery and other major procedures)
             8. Drugs at time of discharge
             9. Disposition

- Level 2
  Input: Three segments, I–III
  Step I.   1. Hospital's name
             2. Initials of the data entry clerk

Figure 3–7 CLEAR Coding of a Single Entry

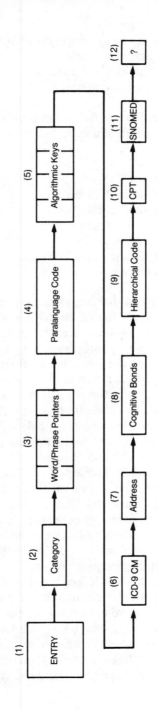

Step II.   New admissions (midnight to midnight)
1-   1. Patient's full name
1-   2. Patient's record number
1-   3. Admission diagnoses

Step II at this level provides a full list of admissions to be processed by CLEAR. The output is an alphabetic list of patients with the diagnoses fully coded.

Step III.   Discharges
1-   1. Patient's full name
1-   2. Patient's record number
1-   3. Discharge diagnoses
1-   4. Procedures (surgery and other major procedures)
1-   5. Drugs (at time of discharge)
1-   6. Disposition, etc.

Step III enables this list of discharges to be processed by CLEAR. The output is a daily alphabetic list of discharges with codes for diagnoses, procedures, and drugs; and daily, weekly, or monthly statistical reports of discharges and diagnoses by categories, length-of-stay data by diagnoses and diagnostic categories, and any other statistics requested by the user.

CLEAR also provides an automated decoding feature. When a user chooses the decoding option for a particular code, the request is made directly on the terminal. The user submits the name of the code, say, ICD-9 and the particular ICD code number to be decoded and CLEAR responds by displaying all the terms stored in the lexicon under that particular ICD code. The example in Exhibit 3–3 illustrates this function.

Another system has been developed by Code 3 Medical Systems in collaboration with the Commission on Professional Hospital Activities (CPHA). Known as "Codefinder," this system leads users through a directed process. The encoding process is semiautomatic. After the entering of key words of English text and subsequent interaction with computer requests for detail, the appropriate SNOMED/ICD-9CM/CPT codes are automatically placed in the computer record.

As each letter of the diagnostic key word is entered on the terminal, a comparison is made with the computerized medical dictionary. When enough letters of the key word have been entered to provide a manageable list of choices, the computer displays a list of choices to the medical coder. This eliminates typing in the entire lines of text.

After selecting the key diagnosis text, the coder is guided by the computer to a more accurate and specific diagnostic statement. All appropriate

Exhibit 3–3 CLEAR Decoding Screen Display

ICD-9 CODE	UNIQUE CODE	SNOMED	TERM
066.3	00059	D0070	BUNYAMWERA FEVER
	00060	D0072	BWAMBA FEVER
	00078	D0098	CHICKUNGUNYA FEVER
	02772	D0502	EPIDEMIC AUSTRALIAN POLYARTHRITIS
	02566	D0256	GUAMA FEVER
	40125		LOSS RIVER FEVER
	02664	D0360	MAYARO FEVER
	02671	D0370	MUCAMBO FEVER
	02710	D0412	O'NYONG-NYONG FEVER
	02715	D0418	OROPOUCHE FEVER
	40123		PIXUNA FEVER
	02765	D0494	RIFT VALLEY FEVER
	02771	D0502	ROSS RIVER FEVER
	02932	D0676	WESSELSBRON FEVER
	02933	D0678	WEST NILE FEVER
	02940	D0686	ZIKA FEVER

SNOMED/ICD-9CM/CPT codes produced from the guided process are stored in the patient's record. The codes are translated back to English text whenever the patient's record is printed in a report or displayed on the computer terminal.

Codefinder operates on a microcomputer housed in the medical record department. It is a feature available through the PAS+ system.

Computerized coding systems introduce a new element in data handling. Given an opportunity to employ the computer for the technical assignment of diagnoses and procedures, medical record practitioners can develop technical chart reading for more precise and accurate data retrieval to a much greater degree while assuring a consistent, accurate technical code assignment process. Today's critical need for data quality responsibility demands that coding and indexing methods be as accurate as possible. Introducing computers into this operation will provide the beginning of standardized health data.

CLEAR and Code 3 are the first systems to offer computerized coding of patient data. As such, they will be carefully studied by users who have identified coding as a functional operation that can benefit by what the computer has to offer. The computer can offer a coding system that for the first time removes the problem caused by human error. One human error inherent in coding is transposition of numbers. For instance, when filling in data elements on a computerized abstract, a clerk or technician will occasionally transpose a code number. This error is costly from an economic point of view. In manual coding a transposition is costly in terms of data quality, for it negates accuracy of the coded data and reduces the effectiveness of the disease and operation indexes. The removal of human error and, more important, the removal of human choice during data entry, as offered by CLEAR, for example, will provide users with the first standardized national morbidity and procedural data base. Such a standardized data base will make it possible to

- Organize morbidity and procedural data for local, regional, and national access. Disease outbreaks, toxic reactions, and environmentally induced syndromes or reactions are all examples of medical care problems that could be more quickly studied and remedied if there were a mechanism to (1) recognize that a case is not an isolated case and (2) draw from a pool of data on a similar group of cases available for instant study.

- Analyze the use of coding systems and associated indexes. It is not now known just what percentage of coded data is used for formal research; how much is used for financial reimbursement; how much is used for individual physicians' study of their own practice populations; and so on. There is no adequate source of information regarding how well utilized the coding indexes are. This link in the information system could have critical impact on the planning for scheduled revisions in coding systems.

- Update and revise coding systems on the basis of degree of activity among categories of users in addition to the more common practice of updating coding systems to accommodate breakthroughs in the study and diagnosis of diseases. Such activity analysis may reveal changing trends in treatment modalities from acute to ambulatory care, for example.

Computer-Supported Medical Reports Processing

Insurance reporting for reimbursement is usually carried out for each record at the time of discharge. Many hospitals now incorporate this in the coding system they use on discharge. For example, CPT coding is a system used by many medical insurance programs for determining the reimbursement amount for hospitalization. Once the diagnosis has been coded it is automatically sent to the insurance carriers along with appropriate discharge data. Today the transmittal of patient identification and the correct discharge code and diagnosis and procedure codes can be accomplished by transfer from one magnetic tape to another. This is a vast improvement on the old flow of paper records and forms. It has significance for medical record practitioners. For example, if the CLEAR coding system were employed prior to hospital discharge, the correct CPT codes could be automatically assigned and made available to the billing office very shortly after the procedure is performed or later, on discharge, and could be used to expedite the reimbursement process.

Another component of medical reports is the use of stored data sets similar to those used in word processing systems. In ambulatory care primary care physicians have utilized standard data sets—that is, paragraphs that adequately describe normal findings or particular descriptions of findings in special areas. An example would be a neurologist storing paragraphs that describe neurological signs and symptoms. These are stored in a word processor microcomputer system. When consultation reports are needed for a given patient, the doctor may draw on the pre-specified report statements or paragraphs to facilitate the dictation and reporting on a given patient. This same principle can be applied to reporting general summaries for medical information transfer as well as formats for insurance report summaries prepared by a medical reports desk in the MRD. The computer can also be used to maintain a record of information requests. Similar to a clinic chart request in a computerized appointment system, a standard patient record request display can be used to enter all requests for patient data. The request can be programmed to produce a numbered, data request log and an automatic charge-out slip for pulling records from the files. Dates of information transmittal could be entered directly in the terminal and exception reports on unanswered requests prepared at scheduled intervals. Such a system provides baseline information to review the performance of the medical reports operation itself.

Computerized Chart Location and Tracking Systems

Medical record chart location systems have been developed in many settings. The purpose of this system is to identify the exact location of the record at any given time. Chart locator systems are designed to improve access and retrieval of the medical record by inquiring through patient name or number on a CRT screen or using other computer-assisted processes. Chart locator systems can be coordinated with chart request operations for clinic appointments, patient readmission, research studies, and other record requests. The Children's Orthopedic Hospital in Seattle employs a combined appointment scheduling and chart request system. In this application chart requests are made for clinic appointments, lab appointments, and so on through a CRT screen display. The request is listed in duplicate on a printer in the file area. One copy of the request is used as a charge-out document for the individual patient's file folder and the other is attached to the patient record. The record copy serves to inform all who handle it the exact destination(s) of the record. The system provides clinic listings of appointment times and maintains clinic activity statistics for the hospital. Exhibit 3–4 is an example of the chart request form.

Chart locator systems are also excellent candidates for bar code wand or light pen systems. Pre-bar-coded folders can be purchased for use in such systems. The Ames bar code system is one example. Another is currently under development as a component of the POIS medical record chart location system. In the POIS model, several retrieval options are featured. These include:

- Chart location by medical record number
- Chart check-out by location
- Chart check-out by physician or employee number
- Chart check-in

Exhibit 3–4 Screen Display for a Computerized Chart Request
System

(Patient No.)   (Patient Name)   (Sex)   (Date of Birth)
Permanent:

1. (Pt. No.) Outp Vis on 11Feb1981 to ER

2. (Pt. No.) Outp Vis on 17Dec1980 to Oto

3. (Pt. No.) Outp Vis on 1Dec1980 to Dental

4. (Pt. No.) Outp Vis on 28Aug1980 10:34 a.m. to Pharm by 5:00

- Chart check-out display by location
- Chart check-out display by doctor or employee number

Using a menu and light pen selection process, system users select the chart locator operation desired. Individual patient data is keyed in by patient number. Because a general systems feature of POIS is to include the date and time and the internally coded identification of the user, additional exchange between user and terminal is not required. Once the patient record number is entered, verified, and transmitted, it is possible to retrieve the data individually and collectively. It can be accessed by users in any location or area. By the same token, the location of records in any area can be determined. Practitioners can determine exact chart location per individual needs. Further, the management of chart activity can be assisted by the retrievability of chart check-out by area. When the POIS program incorporates the bar-code wand method of patient record number entry rather than keyboard entry, the system will be greatly facilitated.

Two examples from the Public Health Automated Medical Information System (PHAMIS) in Seattle are illustrative of the variation available in record tracking systems. The first example of data that can be displayed by CRT is as follows:

MEDICAL RECORDS NUMBER   XXXXXX-X
PATIENT NAME   XXXXXXXXXXXXXXXXXXX
DATE OF DISCHARGE   XX/XX/XX
CHART LOCATION   XXXX
DOCTOR NUMBER   XXX

TRANSFER THIS CHART TO :_____   DOCTOR NUMBER :_____

The display thus shows, by a numerical or alphabetical code or abbreviation, just where the record is being transferred. The system can also display the individual clinic dates and names, thereby providing a patient history:

(PATIENT #)   (PATIENT NAME)   (SEX)   (DATE OF BIRTH)

PERMANENT:

1. (PT. #) OUTP VIS ON 11FEB1981 TO ER
2. (PT. #) OUTP VIS ON 17DEC1980 TO OTO
3. (PT. #) OUTP VIS ON 1DEC1980 TO DENT

4. (PT. #) OUTP VIS ON 29AUG1980 10:34 A.M. TO PHARM BY 500

The input to bring this screen display is /FETCH (pt. #). It retrieves all previous clinic visits to determine current record location, among other things.

Many chart locator packages are available. A model set of criteria to evaluate computerized word processing is featured in an exhibit in Chapter 4. Practitioners can use a similar descriptive profile documentation to determine record control or chart locator requirements for a particular organization. Packages can then be evaluated against predetermined requirements. More information on this process is provided in the remaining chapters of the text.

In 1982 individual medical record practitioners could choose a department microcomputer for a chart locator system at a cost range of $5,000 to $10,000 for the hardware and less than $2,000 for the software. When additional microcomputer applications become available, practitioners will be able to expand their application selections further. Other applications are discussed throughout this chapter and can be equated with microcomputer uses for other MRD functional operations. These modules and others like them will continue to offer substantial efficiencies in medical record operations. Farsighted practitioners will start to document present system performance statistics now in anticipation of effective computer solution selection and justification at a future date.

Using Computers to Facilitate Risk Management

Risk management is another area in which computer support can be effective. The JCAH's quality assurance standard requires problem-focused programs. This standard must be initiated with a well-developed screening procedure. In this way problem areas can be identified and properly documented to justify the concern. This in turn translates into a medical record screening program in which the discharge records are reviewed against predetermined risk criteria. In order to be maximally effective, risk screening should include integration with the incident reporting system for the entire organization.

Computer applications have already demonstrated success in this process. One example is the Hospital Risk Management program in which a standardized variance report replaces the hospital incident reports in 23 Iowa hospitals.[4] In this system the word *variance* replaces the previously recognized term *incident*. The Iowa Hospital Education and Research

Foundation offered a computer-assisted risk management program that was set up with the following objectives:

- Develop a formal reporting system to merge incident data with medical record data.
- Implement the reporting system in in-service education programs already in the hospitals.
- Quantify quality assurance and data analysis aspects of the system as a form of remedial action.
- Provide basic incident data to insurance carriers upon request.

Using a manual checklist, hospitals complete a form that includes identification of the type of variance, type of injury, extent of injury, location of variance, hospital personnel involved, and any contributing factors such as staff factors, patient factors, equipment, facility, supplies, and safety devices.

The most significant injury is recorded and the hospital employee most closely associated with the variance is listed. The system allows free text description of the variance. Free text is English language description as opposed to a numerically coded, predesigned description. It then coordinates this description with the patient identifier for the individual. This allows the screening process to be performed on unidentified recorded information because it is processed separately. It uses computer analysis of variances. The analyses are of individual hospital reports and keep track of hospital incidents. The unidentified data can be further integrated with other participating hospitals' data to identify areawide problems and related trends. The portion of the checklist that contains the free text description of the variance can be detached and the unidentified portion forwarded to the center. Locally it can still be combined with the detached identified portion to produce in-house reports for the hospital risk management program. An example of the checklist is seen in Exhibit 3–5.

It is evident that a comprehensive risk management program relates directly to the overall quality assurance standard. The example just featured is a risk management application. How would computer assistance be used in quality assurance activities generally?

Computerizing Quality Assurance Activities

Quality assurance programs encompass a number of functions. Utilization review, medical care evaluation, and profile analysis to uncover patterns of care are the major thrusts of the quality assurance standard. Quality review has evolved from comprehensive audit methods, which

Exhibit 3–5 Sample Screening Checklist Used in Computerized Risk Management System

Variance Report

0079450 8

HOSPITAL NUMBER

Mo.   Day   Yr.

DATE OF VARIANCE

DATE OF BIRTH

DATE OF ADMISSION

TIME OF VARIANCE (24 Hour)

TYPE OF PATIENT

1 = INPATIENT
2 = OUTPATIENT/ER
3 = LONG TERM CARE

1 TYPE OF VARIANCE

MEDICATIONS (INCLUDING IVs)
- 101 WRONG PATIENT
- 102 WRONG MEDICATION
- 103 WRONG TIME
- 104 WRONG DOSAGE
- 105 METHOD OF ADMINISTRATION
- 106 DUPLICATION
- 107 OMISSION
- 108 CONTRAINDICATION
- 109 OTHER

TREATMENT/TEST
- 110 PROCEDURE
- 111 PERFORMANCE
- 112 TIMELINESS
- 113 OTHER

TRAUMA
- 114 FALL
- 115 ELECTRICAL SHOCK
- 116 STRUCK BY EQUIPMENT

3 EXTENT OF INJURY
- 301 NO KNOWN INJURY
- 302 MINOR (PROBABLY WILL NOT EXTEND LENGTH OF STAY)
- 303 SEVERE (PROBABLY WILL EXTEND LENGTH OF STAY)
- 304 DEATH

4 LOCATION OF VARIANCE
- 401 PATIENT ROOM
- 402 PATIENT BATHROOM
- 403 HALL BATH/WASH ROOM
- 404 CORRIDOR
- 405 FLOOR TREATMENT ROOM
- 406 ICU/CCU/SCU

9 NURSE STATION NUMBER
(USE FOR ITEMS #401–#406 AND #420)

- 407 SURGERY AREA
- 408 POST OP RECOVERY

6 CONTRIBUTING FACTORS

STAFF FACTORS
- 601 COMMUNICATION
- 602 STAFF TRAINING/EDUCATION
- 603 POLICY/PROCEDURE UNCLEAR/INADEQUATE
- 604 POLICY/PROCEDURE NOT FOLLOWED
- 605 REQUISITION/TRANSMITTAL DELAYED
- 606 TEST RESULTS DELAYED
- 607 EQUIPMENT/SUPPLIES NOT AVAILABLE
- 608 PATIENT IDENTIFICATION
- 609 HOUSEKEEPING/MAINTENANCE
- 610 EQUIPMENT MALFUNCTION/FAILURE
- 611 CONSENT
- 612 OTHER

PATIENT FACTORS

☐ 613 CONFUSED/DISORIENTED
☐ 614 SEDATED
☐ 615 UNABLE TO UNDERSTAND INSTRUCTIONS
☐ 616 DISREGARDED INSTRUCTIONS
☐ 617 UNEXPECTED MOVEMENT
☐ 618 NOT APPLICABLE
☐ 619 OTHER _____

EQUIPMENT, FACILITY, SUPPLIES

☐ 620 NOT APPLICABLE
☐ 621 BED
☐ 622 CHAIR/COMMODE
☐ 623 CRUTCHES/WALKER
☐ 624 ELECTRICAL
☐ 625 FLOOR
☐ 626 IV EQUIPMENT/SUPPLIES
☐ 627 OXYGEN/OTHER GASES
☐ 628 RESTRAINTS
☐ 629 SIDERAILS
☐ 630 STRETCHER/CART
☐ 631 TABLE
☐ 632 TUB/SHOWER
☐ 633 WHEELCHAIR
☐ 634 OTHER _____

SAFETY DEVICES

☐ 635 UNAVAILABLE
☐ 636 NOT USED
☐ 637 INEFFECTIVE
☐ 638 DEFECTIVE
☐ 639 NOT APPLICABLE

☐ 409 LABOR/DELIVERY/RECOVERY
☐ 410 NURSERY
☐ 411 LABORATORY
☐ 412 RADIOLOGY/NUCLEAR MED
☐ 413 EMERGENCY ROOM AREA
☐ 414 OUTPATIENT AREA
☐ 415 THERAPY ALL DEPARTMENTS
☐ 416 LOBBY/CAFE/COFFEE/GIFT SHOP OR LOUNGE AREAS
☐ 417 ADMITTING AREA
☐ 418 STAIRS
☐ 419 EXTERIOR GROUNDS
☐ 420 OTHER _____

5 HOSPITAL PERSONNEL

☐ 501 RN
☐ 502 LPN
☐ 503 NURSE TECHNICIAN
☐ 504 AIDE/ORDERLY
☐ 505 WARD CLERK
☐ 506 CRNA
☐ 507 STUDENT NURSE
☐ 508 RESIDENT
☐ 509 PHYSICIAN ASSISTANT
☐ 510 PHYSICIAN
☐ 511 EMERGENCY MEDICAL TECH
☐ 512 PHYSICAL THERAPY STAFF
☐ 513 RESPIRATORY THERAPY STAFF
☐ 514 LABORATORY STAFF
☐ 515 RADIOLOGY STAFF
☐ 516 PHARMACY STAFF
☐ 517 DIETARY/FOOD SERVICES
☐ 518 HOUSEKEEPING
☐ 519 MAINTENANCE/ENGINEERING
☐ 520 VOLUNTEER
☐ 521 OTHER

5		

COMPLETE CODE FOR PERSONNEL MOST CLOSELY ASSOCIATED WITH THE VARIANCE

☐ 117 BURN
☐ 118 OTHER _____

INFECTION/NOSOCOMIAL

☐ 119 URINARY
☐ 120 WOUND
☐ 121 IV
☐ 122 RESPIRATORY
☐ 123 OTHER _____

MISCELLANEOUS

☐ 124 FIRE/SMOKE
☐ 125 CONTAMINATION EQUIPMENT/SUPPLIES
☐ 126 RADIATION EXPOSURE
☐ 127 FOOD CONTAMINATION OR POISONING
☐ 128 DIET RELATED
☐ 129 ANESTHESIA RELATED
☐ 130 BLOOD TRANSFUSION
☐ 131 UNAUTHORIZED DEPARTURE
☐ 132 OTHER

2 TYPE OF INJURY

☐ 201 ABRASION/LACERATION
☐ 202 BRUISE/HEMATOMA
☐ 203 SPRAIN/STRAIN
☐ 204 FRACTURE/DISLOCATION
☐ 205 HEAD INJURY
☐ 206 BURN
☐ 207 INFECTION
☐ 208 RETAINED FOREIGN OBJECT
☐ 209 INTERNAL INJURY
☐ 210 ASPIRATION/ANOXIA
☐ 211 CARDIAC ARREST
☐ 212 RESPIRATORY ARREST
☐ 213 AGGRAVATED PRE-EXISTING CONDITION
☐ 214 ADVERSE MEDICATION EFFECTS
☐ 215 OTHER

☐ 216 NO KNOWN INJURY

focused on patient volume, to a more careful standard that calls for identification of a focused problem that will benefit from a formal audit process.

Like the approach described in the risk management process, the medical audit operation can be initiated with a screening review. The review helps identify problems for further investigation and offers assistance with audit data retrieval itself. Medical record practitioners are familiar with audit features that are provided by discharge abstracting programs. PAS-MAP, for example, provides for abstracting (1) what were the results of admission investigations; (2) whether laboratory tests, x-rays, and other diagnostic tests were performed; (3) what categories of drugs were administered; and (4) whether other therapeutic services were provided. These data were routinely collected for every patient. Other systems, such as the Hospital Utilization Project (HUP) also provided for collection of such data, but on an optional basis. HUP programs were designed to provide data retrieval assistance to medical audit committees. The advent of in-house computers offered an alternative to HUP and PAS-MAP. We will now take a closer look at this alternative.

In one example, a computerized information system was developed to aid in the comprehensive review of care provided to patients in a 550-bed teaching hospital. The system was established for detailed, criteria-based, retrospective review of a closely defined group of patients who represented six major disease categories. For each disease entity, the medical audit committee selected criteria. Some of the criteria were routinely abstracted through the discharge abstracting program. Some were not. The extensive and time-consuming process of selecting criteria and expressing the data elements needed to specify the related medical care in computer-compatible form was a project undertaken by the medical audit committee. The system was designed to routinely collect data whenever possible. The data elements that were not available through the discharge abstracting company were abstracted separately onto an audit form developed for the disease category. These data were entered into an in-house computer and combined with abstracted data already available through the discharge abstracting company. Figure 3–8 illustrates this process. Once these data had been entered, statistical summaries and analysis could be performed through existing computer programs. Quarterly reports were produced that displayed data in exception format. The statistical results and quarterly reports were designed to focus the attention of the medical audit committee on results versus data collection and on patterns and trends versus volume of audits.[5]

Once the data elements are described and an audit abstract is developed, key data elements can be collected by means of an add-on abstract. It is an add-on to the on-line discharge abstract system. This provides a more

Figure 3–8 Computerized Medical Audit Flowchart

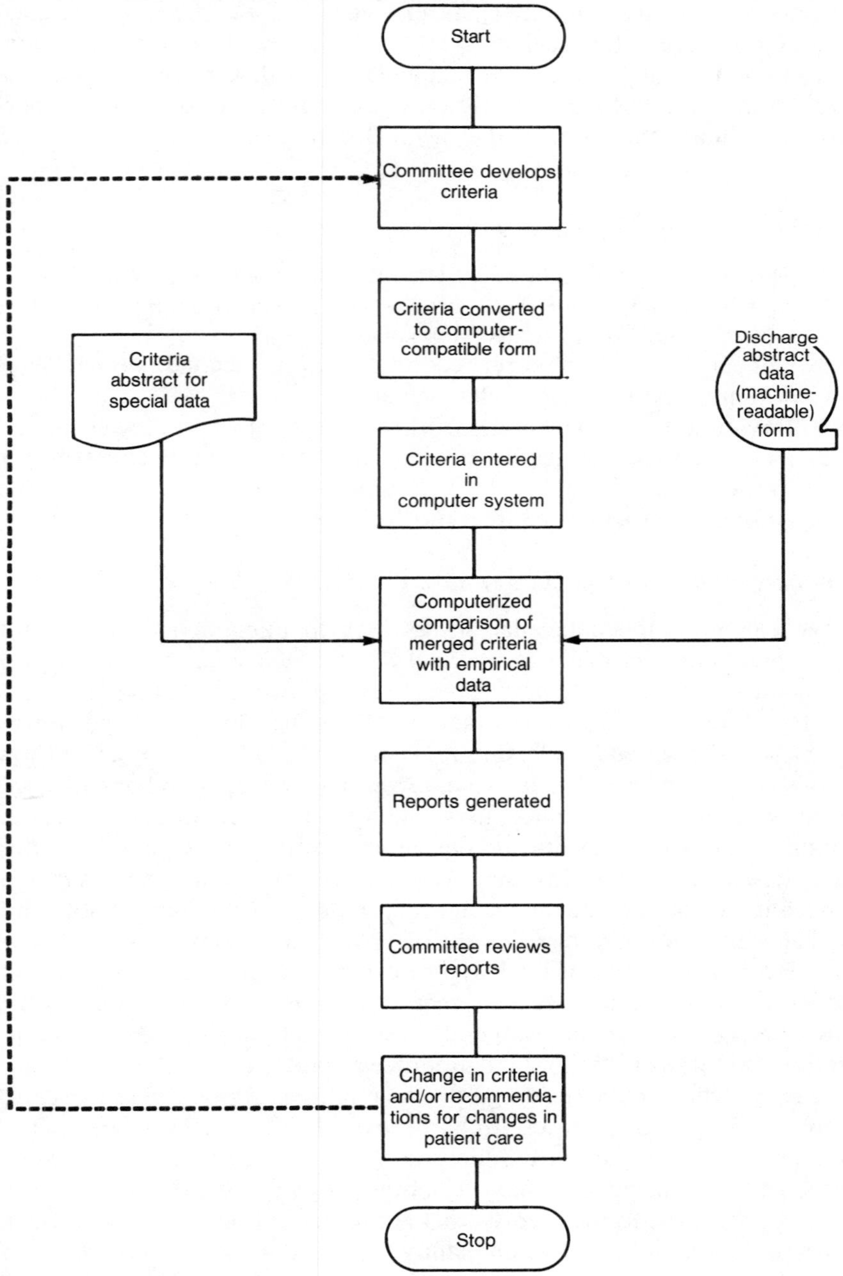

extensive data base that can be manipulated via report-generator programs and can compare and contrast significant elements. If the abstract is designed to accommodate criterion changes, through optional field use, the system design can offer a great deal of manipulation. This can free time for the audit committee to focus on solutions. Consider how computer support would facilitate the steps in the medical care evaluation process listed below:

Steps in the medical care evaluation process

1. Specify the services or care elements that should be provided.
2. Specify what percentage of patients should receive each care element with an opportunity to disallow special cases.
3. Collect data from medical records to measure actual percentage of patients who received each element.
4. Review patient charts of those who did not receive the care elements.
5. Decide on communication to medical staff to promote conformance to the elements.
6. Activate a program to effect the decision.

Computerized Case Mix Data Analysis and Research

We can view the data analysis/research function of medical record departments as a growing resource. Long familiar with record retrieval for medical staff research, practitioners currently work with a wide variety of procedures related to this function. Providing physicians and others with assistance in code selection for individual studies is one dimension of this operation. Another is providing primary records and abstracts of primary records to epidemiologists. In some cases practitioners assist in tabulating statistical results and designing data display. Computers offer extensive assistance in this area. Today MRDs that store diagnostic and procedure codes on magnetic tape can employ a number of software packages for information retrieval that extends traditional research activities. How is this accomplished? If the hospital data processing department stores the data, a simple retrieval request may be used. In these cases the MRD personnel participates in code number identification, request preparation, and transmittal to the data processing department. Possible outputs range from a patient listing by patient number for record retrieval, to automatic requests direct to the chart retrieval unit. For instance, a duplicate printout generated in the file unit could provide perforated charge-out slips to be placed in its files. The other copy could be used as a work list to be attached to the records that are to be sent out of the area. Such a list could identify the data and study topic and name the researcher. If the medical record department works with data display in other functions and the research volume supports it, a low-cost microprocessor such as a Tektronix, TRS-80, or APPLE II might be considered. Such machines enable extended research services in data manipulation and display through statistical software packages that are used with retrieved patient data. This activity could be accomplished in the medical record department. The latter two also have software packages that would enable medical record departments to promote and support more effective data use.

Effective data use is a critical issue of the 1980s and it is not restricted to individual or epidemiological research.

> The need to identify and account for case mix differences has become particularly acute in light of strong upward trends in hospital costs and recent directions in hospital reimbursement. Programs in prospective budgeting and hospital rate setting are increasingly in need of differential case mix information in order to more accurately relate reimbursement to production costs and performance.[6]

This statement refers to one of the most significant data use requirements of this decade, one that will have major impact on medical record departmental operations. In order for this identification and accounting to take place, medical record discharge data must be analyzed and assessed in light of related hospital resource expenditures on a case-by-case basis. This includes, but is not limited to, linking hospital discharge abstracts (UHDDS) with patient financial data.

One major innovation developed to meet this need are diagnosis-related groups (DRGs). This concept in health care reimbursement is designed to capture hospital costs in groups that represent like resource consumption of hospital services. Initially developed at Yale University, the system uses discharge abstract data, hospital patient bills, and hospital financial reports to determine which of 23 major diagnostic categories (MDCs) best represent the individual patient. These major categories are further broken down in 467 distinct diagnostic-related groups. Patients in one group are considered to be different from patients in another group according to their consumption of types and quantities of hospital resources. Although the DRG classification scheme has not been extensively validated as appropriate for case mix data analysis, its use illustrates the fact that methods of identifying and analyzing hospital and other health facility resource consumption are in our future.

"Medical record departments play a far more crucial role in the DRG reimbursement system than they do under traditional payment systems."[7]

We can expect that the need to link patient discharge abstracts and financial data will continue to grow. The implications for computer impact in medical record departments are threefold.

- Computerized discharge abstracting systems will be needed to link patient care data to patient billing data.
- Computer algorithms are used to convert ICDA-9CM codes into the appropriate DRG. This requires access to computer-stored codes.
- Alternative methods for case mix data analysis and resource consumption systems must depend on patient record analysis procedures. Effective, accurate chart analysis is the foundation of the development of such a system—a recognized goal for health care facilities' internal management purposes.

Such implications affect the decision regarding in-house computer potential or shared computer service programs. In 1980 two companies announced joint development of a management reporting system that processes billing, medical record abstracts, and general ledger data for periodic analysis of actual hospital data compared to DRG rates. Health care facilities use shared computer services to supply the necessary data to a computer center for generation of profit and loss statements by cost center, by physician, and by service; overall hospital profit and loss adjusted for case mix and volume; and actual charge-to-cost relationships in each department.

In another example a shared computer service provides DRG analysis aimed at utilization review. If a patient who exhibits the characteristics of a particular DRG has a longer length of stay than is normal for that DRG, the PSRO is notified and that patient's case requires a review.[8]

The reader can see the potential. The MRD holds the key. Not only must the fundamental functional operations be performed effectively, but all data analysis activities and programs will require careful monitoring. As in other research activities, a determination must be made regarding the extent of the quantity and level of functions that the MRD will undertake. Understanding the potential support available through computer assistance is a key element in arriving at that determination. Assertive employment of that potential, once it is identified, is another.

Computerized Word Processing and Transcription Management

"Word processing is the transformation of ideas and information into a readable form of communication through the management of procedures, equipment, and personnel." This is the formal definition established by the X4A12 American National Standards Committee of Washington, D.C.[9] Another definition is, "Word processing is the use of a computer–typewriter interface to prepare documents in such a way that they can be extensively edited, rewritten, combined with other documents, and rearranged internally, all with a minimum of effort."[10]

A typical word processing microprocessor-based unit consists of a microprocessor and a video screen incorporated in a cabinet, a keyboard included in the cabinet or attached to it; a small printer separate from but attached to the cabinet and one or more diskettes or floppy-disk read–write units, either incorporated into the cabinet or separate but attached, like the printer.[11]

Documents are generally composed on and edited directly on the CRT screen, although they may be produced by the printer as "hard copy" at any point in their preparation. The most important feature of word processors is that no document need ever be retyped in its entirety; only those parts needing changes have to be altered. Whole paragraphs or pages can be moved within the document or added from other documents previously typed into the system and saved on magnetic disks. After extensive editing and reediting on the CRT, the printer may produce a hard copy much more rapidly and accurately than a typist ever could. The advantages to word processing are increased output and improved turnaround time; greater accuracy with ease of correction; storage capabilities to reproduce a report that has been lost or misplaced; standardization of all report formats; high-speed printout; and ease of operator training.

Word processing can be viewed as one of the two basic computer functions relevant to management needs; the other is data processing. See Table 3–2 for a comparison between data and word processing.

Table 3–2 Data Processing and Word Processing Compared

Data Processors	Word Processors
Short entries, individual field; length generally limited	Variable and virtually unlimited text length
Output in rigidly programmed format	Output format easily and extensively modifiable
Extensive content analysis possible	No computer analysis of text contents possible
Data input strictly ordered	Text entry flexible, word and paragraph order freely changeable

A basic problem solved by the designers of word processing systems is the conflict between the physical units of storage, usually on diskettes or floppy disks, and the logical units of data that the user recognizes and manipulates. As with computer systems, word processing systems have a hierarchy of logical units of data. Each has a unique name and meaning. Table 3–3 illustrates this logical hierarchy of data. The diskettes and floppy disks have physical and data handling system limitations on the way in which data is stored, not corresponding to the logical hierarchy. The solution used by most word processing systems is always to commence a logical unit of data at the beginning of a physical unit boundary, regardless of the wastage of physical storage space involved. That is, a document, page, paragraph, or line (in some cases) will always commence at the beginning of a sector or track of a diskette or floppy disk. This solution is convenient, and the wastage of space caused by the mismatch between the length of logical units of data and the physical space used to store them is made workable by the low cost of diskettes and floppy disks, the fact that most documents are not long enough to make the wastage of storage space significant, and the greatly simplified physical handling of data by the system and the user.

Word processing systems on the market today offer a wide variety of capabilities and are not necessarily standardized from one manufacturer to another. Table 3–4 describes the sequence and priority that an institution or manager can consult before making an initial decision on word processing equipment.

Once a decision is made seriously to study word processors on the market, there are other considerations for the manager. First word processing is still a developmental technology. The terminology, definitions, and characteristics of word processing are not standardized. General areas to be considered in the evaluation and analysis of word processing applications, systems, and equipment are as follows:[12]

- Typing and data entry functions
- Word processing functions
- Text processing functions
- Data processing functions
- Operational and training aids

Terminology, definitions, and characteristics of word processing are not yet standardized between vendors or users, so that even the general terms may be used differently elsewhere. To define the terms and processes further, Table 3–5 lists the detailed functions and operations of word processors.

Table 3–3 Logical Hierarchy of Data and Terminology in Word Processing

Document
The largest logical unit of data. The document has rigidly protected boundaries, but no initially fixed size. The document may be indexed and recalled, by number and content. The document is composed of pages, paragraphs, lines, words, and characters.

Page
The second largest logical unit of data, but one that does not have a permanently fixed data content and is indexable by page number alone, not content, as a result. The page has a size fixed by the size of the page that is to be printed. Automatic repagination in combination with the addition or deletion of data will change the data content. The page is composed of one or more paragraphs including lines of words and characters. One or more pages constitute a document.

Paragraph
The third largest logical unit of data, with rigidly protected boundaries, but no initially fixed size. The paragraph may be indexed and recalled by number, content, and key words in the text. The paragraph contains one or more lines of words and characters. One or more paragraphs is in a page or document.

Line
The fourth largest logical unit of data, but, like the page, one without a permanently fixed data content or data identification, and as a result indexable by the line number alone, not the content. The line has a size that is fixed by the width of the line to be printed on the physical page. Renumbering of the lines due to the insertion or deletion of data changes the data content of a line. The line is composed of words and characters. One or more lines appears in a paragraph, a page, and a document.

Word
The next-to-smallest logical unit of data, with no initially fixed size but with rigidly protected and observed boundaries. The word is the logical unit most frequently used as a key for retrieval of other logical units of data. Several special handling operations, like hyphenation, deal solely with the word. The word may be defined as a group of characters bounded on either side by spaces or punctuation. The word is not frequently used as an independent unit of data. The word contains one or more characters, and one or more words is in a line, a paragraph, a page, or a document.

Character
The smallest logical unit of data, which has both a rigidly protected content and a fixed physical size, occupying just one printed character on a line or page. This physical size may vary in relation to the line or page, as in the use of 10- versus 12-pitch characters (like point size in typographical terminology), or in the use of proportional spacing. There is one or more characters in a word, a line, a page, a paragraph, or a document. Note: According to standard conventions, a "space" is a character, not the lack of a character.

Source: Adapted from "Word Processing—The Changing Image," by Gene Thompson in *Computers in Hospitals* (November-December, 1980).

Table 3–4 Recommended Sequence and Priority of Word
Processing Applications

1. Applications where word processing is currently being used, but on older, less efficient systems and equipment.
2. Applications where the computer is being used as a printer or reproducer with a small data file that changes less than 20 percent during a year.
3. New applications in areas where three to six typists prepare form letters, fill out standard forms, type long reports that must be proofread and modified in several drafts (as with budgets, planning documents, research reports, certificate of need applications, and so on).
4. New applications that collect several long documents requiring correction on an annual or semiannual basis, such as telephone books, staff listings, and procedure or policy manuals, to be produced at a single location.
5. New applications that attempt to centralize the routine typing of an entire organizational unit. (Note: Care should be taken in this type of application to avoid the "word processing pool" concept that was current in the late 1970s, as the training for the personnel required and the management pressure required to make the "pool" concept work have not frequently paid dividends.)

Examples of Word Processing Applications

Hospitals
Case abstract narrative preparation
Preparation of letters requesting information on patient records

Laboratory
Preparation of formal reports and consulting opinions, especially in anatomical pathology and cytology
Preparation of autopsy reports

Medical
Preparation of surgical dictated reports
Preparation of radiology consulting and interpretation reports
Preparation of dictated case summaries

Administration
Preparation of policy and procedure manuals, plus updating
Budget preparation and updating
Plan preparation and updating
Certificate of need and grant proposal preparation
Preparation and maintenance of telephone books or lists and address listings for contacts, as with the blood bank or the volunteers. Preparation and maintenance of staff lists.

Source: Adapted from "Word Processing—The Changing Image," by Gene Thompson in *Computers in Hospitals* (November-December), 1980.

A manager who is considering purchasing word processing equipment should not only watch demonstrations of available systems and equipment, but try them out. This is very important, since the comparison one might make by simply reading specifications from manufacturers is not sufficient because of the unstandardized terminology just previously described, which

Table 3–5 Word Processor Functions To Be Evaluated When Choosing a Word Processing System

Typing and Data Entry Functions
Pitch and spacing control: 10 or 12 pitch and proportional spacing
Margin and tab set and reset control and indicators
Automatic repeating keys and functions
Special characters and type faces available
Mode and mount of page display
Keyboard layout, feel and use for typing versus control functions
Continuous typing

Word Processing Functions
Methods of entry for characters, words, lines, paragraphs, and documents
Methods of adding underscores, punctuation, decimals, super- and subscripts
Methods of inserting characters, words, lines, and paragraphs into documents
Methods of deleting characters, words, lines, and paragraphs from documents
Methods of changing or deleting underscores, punctuation, decimals, super- and subscripts
Methods of preformat for variable entry, as in production of form letters
Methods of special format and data extraction, as in address lists

Text Processing Functions
Copying paragraphs and pages from one part of a document to another, or from one document to another
Identifying and recalling paragraphs, pages, and documents from storage
Inserting or deleting paragraphs, pages, or documents
Merging two sets of text into one paragraph, page, or document
Moving columns on a page
Automatic pagination
Creating, copying, or deleting physical diskettes

Data Processing Functions
Performance of calculations or numeric data in the text
Interpretation and expansion of codes entered or present in the text
Selection of alternative modes or routes of operation based on text contents
Sorting or merging of text by keys in the text
Storage, retrieval, or conversion of data from computer to word processing files and formats

Operational and Training Aids
Instructions and operational aids are presented on the screen, through the system, either automatically or on request
Manuals and written instructions, delivered before or with the system and equipment
Training availability on and off the hospital site
Opportunities to learn from experience, yours or others'

Source: Adapted from "Word Processing—The Changing Image," by Gene Thompson in *Computers in Hospitals* (November-December), 1980.

makes it hard to know what the system really can do without seeing it in action. The user must actually see the screen, feel the touch and position of the keyboard, and work with the printer in order to evaluate properly the workability of the system in a particular setting. Another reason to test the equipment, and to do so as a demonstration in the manager's facility, is the following. A study by the National Institute of Occupational Safety and Health Administration (OSHA) found stress levels among women using the terminals the highest of any group of workers it had ever studied. And a national survey of office workers reported such physical complaints as eyestrain, back or neck pain, and fatigue associated with video display terminal use. Readers may note that Dataspan, a system designed by Management Design of Cincinnati has developed a 12-hour program to elevate the visual capacity of those who work with CRTs. The program reportedly enhances worker perception to such a degree that errors and fatigue are reduced to a minimum.[13]

Recent technological advances in hardware and software have developed to a point where word processing systems can be connected to larger systems. This includes a large hospital computer system mainframe. This capacity of newly developed word processors enables the managers to integrate information and expand the capability of the word processor to do more than traditional word processing functions. It will move word processing closer to data processing. (The capacity of microcomputers in this expanded function is described in Chapter 9.)

The computer has had a major impact on transcription systems. For some practitioners, planning and executing an effective word processing program can provide an excellent showcase of the potential benefit of computers to the medical staff. In addition, the often dramatic efficiencies afforded by computerized word processing enables medical record departments to extend services consistent with improved use of patient data and health information.

As we discussed when describing the current state of the art in the terminology associated with word processing, individual word processing unit features may be described in several ways. The following is a partial list of current terms, including vendor's unique terms. These definitions are provided to illustrate the range of features available on the market today. Be aware that rapid technological advances will have already changed this list.

- A *stand-alone* system is a terminal that contains a microcomputer, sometimes referred to as an "intelligent terminal." It can function independently of a main computer.
- *Shared resources* are intelligent terminals that share storage.

- In a *cluster system* a word processor user can begin with a stand-alone word processor using the systems' diskette storage and character-printing and text-editing capabilities. These may then be combined with a central processing unit capable of connecting to shared printers and other compatible functions. Each terminal can continue to operate in stand-alone mode or in a shared mode and can communicate with other stations.

- The features called "document assemble" (by the vendors Lexitron, CPT, and Wang), "cut and paste" (Lexitron), and "building blocks" (Lanier) all accomplish the same tasks. They bring words, sentences, and paragraphs from memory and assemble them on the video screen of the terminal. These words, sentences, and paragraphs may then be inserted and adjusted to the document format. The terms *document, text,* and *block* are used interchangeably by vendors explaining this feature.

- *Search and replace* is a feature that allows word or phrase replacement without retyping an entire document. In general, the operator specifies what word or phrase is to be changed and what the new phrase is. Different vendors accomplish the task differently, as follows:

  CPT search and replace can read an entire document, automatically changing a word or phrase of up to 33 characters.

  Dictaphone Global's search and replace can either accomplish the desired change automatically at each occurrence or, at operator's option, stop at each occurrence and await operator confirmation. Lanier's search and replace also allows the operator to replace words selectively or automatically.

  Lexitron's search and replace allows the operator to search through a document to a specific text and 1) stop at each occurrence of a particular word or phrase and decide whether to replace it, 2) count each occurrence, or 3) replace each occurrence with specified text.

  Wang's global search and replace searches and highlights every instance of a special character sequence in a document. The operator has the option of automatically replacing a defined character sequence with another throughout the document.[13]

The reader can see that a myriad of opportunities are available in word processing. Computer technology offers extensive efficiency in information handling and enables medical record practitioners to extend patient data uses within the organization.

Questions and Problems for Discussion

1. Define and describe typical functional operations in health record systems.
2. Why is the MPI an excellent candidate for computerization?
3. Prepare a report for the hospital administrator indicating your support for computerizing the hospital ADT (census) function. Include a brief list of benefits.
4. What is the relationship between effective record control functions and medical record department efficiency?
5. Prepare a list for a medical staff meeting that explains how current computer technology has improved medical record functional operations.
6. Describe three ways the MPI could be computerized.
7. Comment on the following statement: "An effectively designed computerized chart completion system can provide timely, financial management data for hospital administration."
8. How can computerized coding offer significant improvements in data quality?
9. Describe some key elements in a computerized discharge abstracting system. How would an in-house computer system differ from a discharge abstracting service program? Illustrate your answer with examples.
10. Explain how computer technology has improved traditional functional operations in MRDs.
11. Prepare a report to a clinic administrator recommending installation of a word processing system. Include description of typical features of word processing and an explanation of the need to investigate alternative systems on the market.

NOTES

1. Edna Huffman, *Medical Record Management* (Berwyn, Ill.: Physicians' Record Co., 1981), p. 531.
2. Joyce Currie Little and Richard H. Austerig, "Some Guidelines for the Analysis and Design of a Computerized Utilization Review and Reporting System," *Medical Record News* (August, 1977): 7.
3. Rosemarie T. Dunn, "Computer Assistance for Utilization Review," *Medical Record News* (August, 1977): 62.
4. "Computer System Streamlines Variance Reporting for Iowa Hospitals," *Hospital Risk Management* 2, no. 1 (January, 1980).
5. James B. Martin, "Towards Effective Computer-Aided Medical Care Evaluation Studies," *Inquiry* 15 (June, 1978).
6. Wanda W. Young; Robert Blane Swinkola; and Marsha A. Hutton, "Assessment of the Autogroup," *Medical Record News* (June, 1980): 72.
7. American Medical Record Assoc. "DRGs—Too Much Too Soon?" *Counterpoint* 17, no. 5 (September, 1980): 8.
8. Jane Fedorowica and Stephen Veazie, "Automated DRG Systems: Unanswered Questions," *Hospital Progress* (January, 1981): 54.
9. Arnold Rosen and Rosemary Fielden, *Word Processing* (Inglewood Cliffs, N.J.: Prentice Hall, 1977), p. 9.
10. Gene Thompson, "Word Processing—The Changing Image," *Computers in Hospitals* (November/December, 1980): 42.
11. R.J. Spinard, "Office Automation," *Science* 215, no. 4534 (February, 1982): 810.
12. Thompson, "Word Processing—The Changing Image."
13. Betty Burnett and Bonnie Mellor, "Word Processing in a Health Care Facility," *Computers in Hospitals* (November/December, 1980): 18–20.

Chapter 4

How to Perform Systems Analysis in Medical Record Settings

Objectives

1. Define and describe the process of systems analysis.
2. Describe the relationship between the process of systems analysis and the development of computer applications in health information.
3. Define and describe a systems life cycle model in the development and implementation of computer-based systems.
4. Define and describe the activities and products of the analysis phase of the systems life cycle.
5. Describe the actual and potential role of health record practitioners in planning and developing computer-based systems through systems life cycle activities.
6. Identify effective tools for systems analysis activities.
7. Present feasibility analysis as a method for selecting proposed system alternatives.
8. Examine cost evaluation methods used in measuring proposed computer applications.
9. Present analysis phase models, including documentation of present systems, feasibility analysis and proposed system requirements, and analysis phase final report.

The purpose of this chapter is to prepare readers to participate in and perform systems analysis for health information settings. Systems analysis is a process. A process, according to Webster's dictionary, is a series of actions or operations conducing to an end.[1] Systems analysis is a process that can be used to examine an activity, procedure, method, technique, or business. Its structured order of inquiries can be used to determine what must be accomplished to achieve change and to select the best method for achieving a goal, and, finally, in the case of computerization of medical records provides direction and evaluation for implementing change or installing new methods or equipment. Included in the structured process of systems analysis is the application of investigative tools. The tools are used to examine and document the activities in question.

Systems analysis should be viewed as a tool. With it managers can investigate and study current and proposed situations conveniently and logically. It is a method for analyzing a great quantity of objective data in a logical, documented format, but it is not a panacea always assuring successful and cost effective solutions to problems. The choice of solutions is still a human one, not the result of the application of a tool.

Another definition of systems analysis is "a scientific method of problem solving, utilizing the tools of description, investigation, research, creativity and judgment. Each skill is applied through a highly structured format of documentation."[2] This definition emphasizes the problem-solving aspect of systems analysis. It also highlights the role of documentation in the analysis process. Documentation can take the form of flowcharts, comparative tables, problem statements, or whatever is useful. The object is to study existing operations, procedures, information, or paper flow in an orderly manner.

Systems analysis is the critical component needed to develop an effective computer application. By computer application, we mean the identification, selection, installation, and evaluation of a software or hardware component. It could be all of the actions identified in the previous statement or some part or combination of those actions. Systems analysis is used to determine what currently exists in any given operation or series of procedures. As such, it demonstrates its characteristic of problem solving. And until the problem solving step is taken, the manager does not know what will be discovered from a given situation. When the manager has uncovered or located a function that needs improvement or solution, the next step is to select alternatives and then to make a choice among them. The choice is based on the alternative that seems most likely to solve the problem or resolve the difficulty. Then, going back to our previous statement, the manager uses either a software or hardware component in resolving the issue. So, systems analysis is the best method for making the choice. It is also the method of choosing in many other business and scientific applications, particularly settings that focus on information flow.

There are two reasons why the MRA can play an important part in the structured process known as systems analysis. The first is that the MRD in a health care facility has a unique relationship with other departments in the organization. It has direct and almost daily contact with the admitting department, the business office, administration, the medical staff as individuals and as groups or committees and other organizational units, depending on the particular facility. In most cases, these departments both rely on and give information to the MRD. Among the documents that regularly flow between the MRD and the other departments are admission lists; discharge lists; corrections, deletions, and additions to various reports and lists; schedules; cost analyses; agendas; statistical reports of services provided to patients; and individual patient information. These relationships show that the MRD is a subsystem in the organizational structure, a subunit of a larger, more complex group of units or subsystems. Therefore, the activities of the MRD are a prime subject for the formally structured analysis process.

The second reason for the importance of the MRA in systems analysis regarding medical records is that the patient record has many unique characteristics. Perhaps one of its most unique characteristics as a document is its intense concurrent and subsequent use by a wide variety of users and providers. During the many stages of development, the record demands and achieves heavy use. After it is fully developed and completed, it continues to have many uses and users. It is during the record's developmental stages that one can readily identify the complex interdependence of people and organizational departments and professional disciplines. For example, admitting, medical records, and the business office all must coordinate procedures to achieve optimum data capture and documentation. Physicians, nurses, and laboratory technicians must all enter and retrieve data in an organized manner in order to communicate about patient care. This brief example could be expanded to further describe the many subsystems inherent in the processing of patient information.

To provide a clear understanding of the knowledge and skills needed to carry out systems analysis, this chapter will address the topic in four ways. (1) A rationale will be presented that describes the reasons for the MRP to learn about the systems analysis process. (2) The purpose and function of the systems life cycle model in the development of computerized health information systems will be explained. The life cycle model illustrates how health information procedures, methods, and processes are transformed from manual to computerized systems. (3) The elements, activities, and products of the analysis phase of the life cycle will be examined. In describing this phase, analysis tools are defined and examined. Examples of products that are the outcome of analysis tools are featured. They will enable readers to apply the principles of the tools in alternate situations. (4) A suggested procedure for accomplishing the analysis phase tasks is provided. Examples and exhibits are included as case studies. Using the information presented in this chapter, readers should be able to select appropriate elements from the analysis phase, identify a strategy for employing them, and prepare a functional requirements document model. The model will support a plan for computerization in an individual medical record department or health facility application.

THE ROLE OF MEDICAL RECORD PRACTITIONERS IN HEALTH INFORMATION SYSTEMS

Understanding Systems within Hospitals Is the Beginning

Patient information systems are derived from patient and service data that are routinely collected and processed from the medical record. Recall that a system can be defined as "related elements that are coordinated to form a unified result; or, specifically, people, activities, equipment, materials, plans and controls working together to achieve a unified objective or whole."[3]

Just as the hospital or health care facility is considered a system, so are the individual departments within it. When considering the whole facility as a system, one may hear the individual departments referred to as subsystems. But since departments have autonomous objectives, budgets, and personnel, they are also correctly referred to as systems. Accordingly medical record practitioners are expected to plan and develop health record systems appropriate for varying sizes and types of health care facilities, organizations, and agencies.[4]

As computer technology enables managers to extend and expand the departmental services they offer, health information managers in medical record departments must plan to extend and expand patient information access. This must be done while maintaining privacy and guaranteeing appropriate access for clinical, analytical, and administrative users. In addition, computer applications must be carefully integrated with the overall information needs of the institution as well as individual departments. One way this can be accomplished is by rigorous adherence to the practice of systems analysis.

The Rationale for Using Systems Analysis As a Problem-Solving Tool

The previous chapters reviewed the architecture of computer hardware, traced information flow from the patient record through the computer, and examined computer applications in medical record departments. In this chapter, specific tools that identify and document the information necessary to select appropriate computer systems applications will be examined.

As computer technology expands and becomes more flexible, more affordable, and more appropriate for patient information systems, the demand for individuals trained in the selection of appropriate technology will grow. The demand for skilled use of methods designed to plan and complement the technology grows accordingly. This means health information managers must be proactive participants in systems planning and development. They must also be able to carry out the various processes of systems analysis.

Analysis tools can be employed as instruments to derive the most complete examination and documentation possible. If done properly, this produces a thorough, organized picture of a selected operation so that problems can be identified and recommendations can be made. In particular, if computer alternatives are among the recommendations for improvement, the precise information that is required to complete explanation of the existing operation or system can be determined. As indicated in Chapter 2, computers operate under programmed instructions that must be written to accomplish the function and satisfy the particular application requirements. To do this, the needs and requirements must be clearly identified.

The Role of Medical Record Practitioners in Systems Analysis

There are four general roles appropriate for medical record practitioners (MRP) in the development of computer applications. Selection of the most effective role will depend upon each individual case and the practitioner's judgment. An MRP may do the following:

- Use specific individual tools for analysis of departmental operations.
- Engage in formal, highly structured systems analysis projects such as those designed to provide a computer solution for major functional operations within their department or health facility.
- Participate as a team member in comprehensive or interdepartmental projects within their health facility.
- Perform systems evaluation to measure effects of computerization or to determine the necessity for the review of existing computer applications.

In order to select one of these roles, the MRP must first have a clear understanding of the elements and activities that make up a systems analysis process. Second, those components of the systems analysis process that are most readily utilized for MRD applications and can be carried out to their optimum extent by the MRP as an individual user-manager should be easily identified by the MRP.

Accordingly, an overview of the life cycle phases and their key elements will be provided. As each phase is described, a suggested list of systems analysis activities for MRPs will be identified. In order of natural occurrence the strategies will identify the steps needed to support effective completion of or participation in the activities carried out in that phase.

The reader will observe that a series of work products will be developed as the process moves along.

OVERVIEW OF SYSTEMS LIFE CYCLE

The life cycle of a computer system has four general phases:

- Analysis
- Design
- Development
- Implementation

A life cycle model is provided in Table 4–1. Note that the phases and tasks of the model are interdependent and do not in reality evolve in sequential order. Many activities are parallel efforts requiring dynamic interaction by participants. The table points out the complexity of planning and implementing computer systems. To some degree this process is carried out in every computerized application that is developed. The level of participation by the user-manager or MRP in the phases and various steps listed depends upon the care setting, the role of that manager, and the policies of the organization.

To supplement Table 4–1, Table 4–2 is provided. A detailed expansion of the life cycle process, it illustrates a breakdown of tasks categorized according to the major phases. It identifies areas for the health record managers' contribution and participation during each phase. If the tasks of Table 4–2 are compared with the headings in the life cycle model of Table 4–1, the reader will see how the corresponding columns indicate the medical record professional's responsibility for direct or delegated action in many of the key activities in each phase. Today's practitioners are involved in these activities according to their unique settings. This means the range of MRP involvement extends from one or two activities carried out in the analysis phase to active participation in three or more activities for all phases.

An effective manager reaches out to become active in as many phases of the development as is possible.

- The analysis phase is dependent on the user-manager for problem identification and performance definition for activities within a given department or area and participation in feasibility analysis.
- The design phase is used to refine a requirements document that will be followed to put together the computer application. User-managers

Table 4–1 Model of Life Cycle of a Computer System

Analysis Phase
I. Problem identification
   A. Plan present system investigation—identify project leader.
   B. Identify constraints.
   C. Survey users of the present system.
   D. Use investigative tools; document activities.
II. Performance definition
   A. Determine goals and objectives.
   B. Develop system flowcharts.
   C. Describe outputs.
   D. Prepare functional requirements.
III. Feasibility analysis
   A. Plan systems analysis activities.
   B. Determine operational feasibility.
   C. Determine technical feasibility.
   D. Determine cost/benefit analysis.
IV. Analysis phase report

Design Phase
I. Functional requirements confirmation
   A. Review design phase issues.
   B. Reconfirm the problem.
   C. Secure management review and approval.
II. Detailed requirements determination
   A. Examine detailed outputs.
   B. Confirm detailed inputs.
   C. Prepare for file design activities.
   D. Review program designs.
III. Documentation needs identification
   A. Review system documentation requirements.
   B. Review programmer documentation.
   C. Plan user documentation.
IV. Design phase report

Development Phase
I. Project planning and estimating
   A. Make formal project schedule.
   B. Establish plans for monitoring projects.
   C. Convert information, procedures, and equipment.
II. Program development and testing
   A. Establish system test plans.
   B. Provide user testing.
   C. Secure effective procedures development.
III. Personnel training
   A. Develop positive attitudes.
   B. Select effective training methods.
IV. Development phase report

Table 4–1 continued

Implementation (Operation Phase)
I. System Changeover
   A. Schedule changeover activities.
   B. Provide back-up support programs.
   C. Work with vendors.
II. Postinstallation audit
   A. Apply performance objectives and standards.
   B. Plan postinstallation evaluation.
   C. Perform technical audit methods.
   D. Integrate computer system performance into work standards.

---

are involved in the design phase to concur with and monitor the requirements.

- The development phase deals with the purchase of equipment, program writing, training staff, and preparation for change. Here, user involvement shifts from design into operational focus.
- The operation or implementation phase includes system changeover and postinstallation audit. Careful recognition of these activities helps to identify cost, performance, and effectiveness of the project. Each phase has responsibilities for the user-manager.

After the analysis phase has been examined in detail, the overview of the life cycle model will be complete. We will then describe how and where the 12 items under the heading *analysis phase* in Table 4–2 can be applied.

The Analysis Phase

In this first phase of the life cycle of a system, a problem area is identified and substantiated. Solutions are proposed and their feasibility validated. Functional requirements of the proposed solution are defined. Beginning with a thorough understanding of the present system, the analysis phase documents and verifies not only current activities but also proposed solutions. It does this in a highly scientific way that offers both measurement and evaluation. The analysis phase contains all primary systems analysis activities and, as such, requires effective leadership and participation by the MRP as user-manager. The analysis phase activities culminate in a formal report. The report documents the results of all investigations and proposed solutions that have been developed during this period. This report should contain the functional requirements of the proposed computer solution as specified by the user. When an MRD operation is under consideration, the MRP should be the primary author in preparing this document.

Table 4–2 Health Information Manager's Role in Computer Systems Development

Requirements Analysis Phase	Design Phase	Development Phase	Implementation Phase
Develop problem identification. Investigate present system. Identify and communicate constraints. Survey users. Determine general and specific objectives. Prepare or review information-oriented flowcharts. Prepare or review system output descriptions. Prepare or review system input descriptions. Determine proposed system evaluation criteria through feasibility analysis. Evaluate potential systems candidates. Prepare evaluation conclusions and recommendations. Prepare analysis phase report.	Review analysis phase report to validate problem identification. Help resolve design issues. Review problem reconfirmation. Review detailed requirements for the proposed system. Review planned documentation for proposed system. Review equipment requirements and functions from user performance perspective. Provide user manager feedback for design phase report. Review design phase report.	Review and approve appropriate implementation plans for: • overall project plan, • conversion plan, • system test plans, • personnel training plans. Assess documentation developments. Approve changeover plan. Provide information for development phase report. Review development phase report.	Coordinate system changeover activity in department. Perform postinstallation audit activities. • Compare actual system output with requirements. • Interview operations personnel about equipment use and procedures. (Does system meet their expected needs as users?) • Assess cost data against projected system costs budgeted for the user department concern. • Assess error rate. Prepare performance evaluation reports. Coordinate system audit activities. Participate as team member in performance review and follow-up recommendation actions.

The Design Phase

The second phase of the life cycle model is the design phase. The design phase is initiated after decisions on the proposed computer solutions have been made. It is based on the activities completed in the analysis phase. Design phase activities include confirming performance requirements that have been specified by the user-managers and developing the functional requirements so that computer system outputs, inputs, file design structures, and programming needs can be mapped out.

In addition, documentation needs are determined. Operational aspects of the computer system and appropriate software documentation are specified, as is the user documentation. User documentation delineates what will be prepared as a resource or manual for users' reference in performing the computerized operation.

A critical component of the design phase is the selection of a project manager or coordinator. This individual oversees the continued planning and development of the application. In many cases this important assignment is made during the analysis phase as the proposed computer solutions become viable and the organizations' management commits itself to support of the project. Project management focuses on planning and estimating activities. These activities are concerned with coordinating and monitoring task progress and estimating anticipatory task completion. They are fully identified in the development phase activities as the need to coordinate the projects many components becomes necessary. In reality many of the design phase and development phase activities occur in parallel. From a descriptive perspective, however, the activities can be discussed separately.

The Development Phase

The third phase of the life cycle model is the development phase, in which the requirements confirmed and addressed during the design phase are translated into actions. In this phase those responsible to carry out the conversion of the old to the new continue to estimate and plan the project's many components.

During this phase programs are written, hardware is purchased and installed, and information is converted from manual files to computer-stored files. It is a period of close collaboration with users to coordinate all people and procedures during the transition from old to new. Here, the proposed strategy for MRPs is directed to assertive management of change in their departments. Notice the recommended activities for design and development in Table 4–2. If practitioners assume responsibility for these activities, more effective coordination can occur. There is also ongoing opportunity for continued integration of health information principles and ideas that reflect the concerns of medical record professionals. As medical record user-managers often have the most complete picture of health information systems' activities and access requirements, they can provide valuable assistance to computer developments in this area.

The Implementation Phase

A look at the implementation phase concludes this overview of the life cycle model. Implementation takes place following final review of the development phase activities. During this period, conversion and system changeover activities occur. For example, medical record department personnel may actually begin working with an on-line discharge abstract system.

Along with the changeover activities, postinstallation audit activities are performed. Notice in the recommended strategy in Table 4–2 that immediate system evaluation techniques are projected. A detailed examination of these activities will be made in Chapter 6.

It is important to relate the need for systems follow-up to the overall life cycle model. This is so the information and performance requirements can be established during the initial system development activities in such a way that effective evaluation can be performed. Health information managers will recognize follow-up evaluation on projected performance objectives as one component in such evaluations.

ANALYSIS PHASE ACTIVITIES

Initiating Analysis Phase Activities through Problem Identification

Problem identification is the result of investigative studies. Problems can be identified by examining existing operations. Operations that do not meet expected objectives or are not achieving new or expanded objectives are targets of the investigative process that will result in the identification of a particular problem. Medical record practitioners will recognize problems appropriate for investigation in the following scenarios:

- Outpatient clinics file a request for duplicate copies of the patient record because they are unable to obtain the central record for scheduled patient appointments.
- The coronary care unit requests central dictation to absorb transcription of all outpatient coronary care follow-up programs.

- The hospital administrator has charged the medical record department to cooperate in the establishment of a computerized risk management program.
- Hospital accreditation is in jeopardy because more than 4,000 patient records are incomplete.

These and similar situations may result in consideration of computer-based alternatives. At this point, systems analysis may begin. The initial thrust in the analysis phase is problem identification through systems investigation.

How to Perform Problem Identification

Identifying the problem(s) requires a firm understanding of current operations. The identification process can only be achieved by someone who is familiar with the jobs, procedures, methods, personnel, and work flow of the work unit under consideration. The investigation must be planned and carried out in an organized manner. The organized approach provides a comprehensive picture. It enables the investigator to acquire a more thorough understanding of the objectives and outcomes of the existing work unit. Only with such familiarity can the problem be identified accurately.

In many computer development projects a project leader is appointed in the beginning of the systems analysis project. Effective project management calls for this step. This is especially true for the reasons cited in the identification process. User-managers start out in the systems analysis process with an understanding of their department that is exclusive to them. They are essential participants in the key step of systems analysis—problem identification. Therefore the MRP should be prepared to enact a leadership role in a computer project and serve as project manager if asked. If not asked but interested in carrying out this role, the MRP should make it known to the administrative staff responsible for the project that he or she desires to be the project manager.

Not all computer projects select a project manager during this initial step. However, user-managers and others who will be collaborating on the project will need to work together in a team configuration. Therefore individual leadership in collaborative activities is a quality the MRP must demonstrate.

We will now describe the three major elements in the problem identification step.

Identifying Constraints: The First Step

The first element in problem identification is identification of constraints. Constraints are the boundaries or limitations, rules and regulations, both legal and organizational, that affect a particular operation. For the MRP this includes Joint Commission on Accreditation of Hospitals (JCAH) standards; national and state legislation dealing with reimbursement for federal patients; organizational boundaries (the activities that are under the control of the MRD); and cost, or budgetary limitations specified for the particular organization. Such constraints must be carefully listed so that the problem can be identified and discussed in their light. For instance, a computerized discharge analysis program must incorporate the UHDDS, PSRO, and JCAH reporting requirements. Certain conditions such as JCAH standards, may constitute constraints that exist for all discharge analysis operations.

A constraints list begins with a description of goals and objectives of the present system. The purpose of the existing operation should be explained. Standards and regulations that affect the operation should be listed. Copies of procedures and job descriptions will also yield information to be included on a developing constraints list. Careful assembling of these and other facts that will necessarily place limits on potential system alternatives helps establish an understanding of the framework within which a new system must be established. Other examples of constraints that affect MRD operations are

Constraints	*Operation Affected*
Medical staff bylaws	Computerized chart completion system
Budget restrictions	Computerized risk management program
State laws	Computerized risk management program
JCAH Standards	Computerized risk management program
MPI information requests of admitting and outpatient departments	Computerized MPI with number control in the MRD
Staffing allocations	Computerized number control in the MRD

Assembling these facts provides needed information in the problem area under consideration. Consideration of users is the second step.

Surveying Users: The Second Step

The second component in problem identification is the employment of user survey instruments. The systems analysis team plans for an investigation by identifying the boundaries or limits of the present system. Included in this is the identification of current users of the operations involved. Investigating a record control program for potential computerization requires a clear description of the record control activities from the point the record is created throughout subsequent record handling activities. In addition a precise list of all users of patient records via the existing control program should be compiled. These facts are then documented in the initial paperwork of the systems analysis. Once these are completed, the systems analysis team is ready to outline events and schedules for problem identification. As the present system is then investigated to uncover all aspects of the problem, one of the first tasks is to contact identified users to collect information about the operation. Often this contact results in completion of user surveys.

User surveys are organized methods for communicating with people. They are used to find out how things operate now and how the users might like to see things operate in the future. They are carried out through interviews and questionnaires. Key personnel may be interviewed to determine their perceptions of the operation. Effective interviewing requires communication skills to listen and hear the information presented. It also affords opportunities to observe attitudes and personalities. It can assist in establishing rapport between the investigator and key personnel. A review of interviewing techniques and a planned approach to each interview are important prerequisites when initiating user interviews.

Questionnaires can also be used to solicit user feedback. They may be general or detailed and can be used when interviews are not feasible or to supplement interview results.

When successfully employed, these instruments answer such questions as: Who uses the information? Who uses the products of the operation? Do the users need the information in summary form at one predetermined time or do they need to know on demand? For instance, it may be important for the administration of a health facility to know what the census is and what its configuration appears to be in the hospital on demand. An effective real-time census system will provide up-to-the-minute status for facility administrators to use for decision making.

User surveys can also provide information about volume, activity, and the operations procedures that will assist in accurately describing the present system. In many cases users report on problems they find in the day-to-day operations.

One method for surveying user needs for a monthly statistical report is to circulate a current report with an attached concurrence sheet asking for each user to review and specify that the attached report is necessary and useful in their area. Such documents are often supplemented through additional user surveys prepared and administered by the systems team. Examples of user surveys for the investigation of a discharge abstracting operation are presented in Exhibits 4–1 and 4–2.

When working with a variety of users, the systems team will likely incorporate all these methods to secure user experiences and ideas to aid in problem identification.

Applying Investigative Tools: The Third Step

The third element in problem identification requires a clear understanding of the information flow and associated procedures that make up a particular application. A number of management and investigative tools

---

Exhibit 4–1 Sample User Questionnaire

TO:
FROM:
SUBJECT: Interview scheduled for _____
                at _____

At the scheduled interview we are planning to discuss with you some aspects of the on-line master patient index (MPI) system now under development. We are interested in exploring two major areas with you at that time.

The first is concerned with the *type and quality* of information you are presently obtaining, the extent to which it meets your needs, and the degree to which it supports your decision making. At the time of the interview, it would also be helpful if you could have a set of samples of those documents (computer reports, forms, etc.) on which you currently receive information.

The second area involves a determination of the ways in which an automated MPI system can be of greatest value to you and your department. We would appreciate you giving some thought to *significant objectives* that you would like your department to achieve *as a result* of an MPI system becoming operational. In identifying these objectives, think in terms of practical levels of achievement, the parts of the department that would be most directly affected, those that would have the responsibility of achieving the objectives, and the relationships of these objectives to long range plans.

Attached is a short questionnaire that we would like you to complete as frankly and completely as possible and return to us in advance of the interview. We appreciate your cooperation and look forward to meeting with you.

Exhibit 4–1 continued

SURVEY   DEPARTMENT:

TITLE OF OPERATION UNDER REVIEW _____

Please answer the following questions to describe the operation named above currently performed in your department. Use additional pages as necessary. Attach examples of all requested items to the survey. Key all examples to the appropriate question.

1. What is the major activity performed in the operation?
2. How many people are involved in performing the activity?
3. When and how is input received, and in what format? (Attach a copy to the survey. Key the copy to this question.)
4. From where and from whom is it received?
5. Is the input used as is, or is it checked, reworked, or just passed on?
6. What is the ultimate use, retention time, filing procedure, and disposition of all copies?
7. What are the controls that ensure accuracy and completeness of the inputs?
8. What information is transmitted and what is the destination of the outputs currently produced by the operation?
9. What is the purpose of each copy and the routing of each copy of a form?
10. How is the output compiled, filled out, sorted, reproduced, checked, and what are the various times required for each?
11. What errors and omissions are referred back to the area under study, and what are the corrective action procedures for these errors?
12. How is the current output controlled to ensure that it is accurate, complete, timely, and delivered only to appropriate personnel?

and methods can be used to ascertain this data. The systems team needs to select from these resources and coordinate the results with the user survey data and the identified constraints to formulate a precise definition of the problem.

Twelve tools can be used to assist in the problem identification phase of systems analysis:

- Organization chart
- Organization function list
- Systems flowchart
- Operation flowcharts
- Decision tables
- Input/output chart
- Forms analysis chart
- Forms distribution chart

Exhibit 4–2 Sample Preliminary User Interviews

Interview with the MRD director:

Q. What is the current type of discharge abstracting system used in your department?

A. We contract out to a shared computer service bureau called Medical Discharge Services (MDS).

Q. How do your employees perform this task?

A. We employ three coder/abstracters who manually fill out abstracts for each discharged patient. A batch of these are sent to MDS every two weeks.

Q. What is the turnaround time for the return of processed information?

A. It usually takes between two and three months to receive the reports and indexes.

Q. How accurate do you think the input and output data are?

A. We seem to be having a definite problem with the input performed by MDS into the computer. Sometimes numbers are misinterpreted. If this occurs in patient numbers or the diagnosis or procedure codes we can lose the information.

Q. How frequently are the MDS reports actually used?

A. They are limited in use because we don't receive the information in a timely manner. Our department provides routine reports to medical staff committees as well as administration. If necessary, these must be manually prepared in order to meet deadlines.

Q. Can you estimate the frequency of this practice?

A. Eight out of twelve reports required were manually prepared during the last six months.

Q. Where are these reports used?

A. We supply information to most departments in the hospital. In addition to the medical staff committees and administration, most hospital departments receive a copy of a general statistical report on hospital services each month. Other departments depend on this report to validate their own data. We provide general tracking data from month to month.

Q. How satisfied are you with the current system?

A. We are dissatisfied. Not only do we lose valuable staff time, we increase the risk of error rate due to the necessity to transfer information by hand. We are actively seeking alternatives.

Interview with the hospital administrator:

Q. Do you believe the current MDS system used in medical records is able to meet the needs of the hospital?

A. I feel it doesn't meet the current needs at all. Furthermore, I don't believe the system is flexible enough to handle variations in data collection that we anticipate will occur.

Exhibit 4–3 continued

Q. What types of variation do you expect?

A. The current demand to contain health care costs will require more sophisticated data manipulation than we have commonly practiced. We will need the resources to accommodate the data manipulation required in working with Diagnosis Related Groups (DRGs), case mix analysis, or another method for linking and evaluating patient services with costs.

Interview with the director of the hospital business office:

Q. What types of information do you receive from the Medical Record Department?

A. We receive patient diagnosis and operative procedures for all discharges.

Q. Does this information flow into your department in a timely manner?

A. Usually. There are problems when backlogs occur. Medical Record staff have to locate the discharge data. If, for instance, the information is not available from the record, they contact the physician and ask for the information. Of course, any delays in acquiring the information can result in delays for our department to secure payment for patient bills. This can delay revenue and cost money.

- Process flowcharts
- Work distribution chart
- Layout flowchart
- Data element analysis

These tools are used by the systems team or by an MRP engaged in analyzing a current operation. They provide methods for describing and documenting information and activities.

Systems teams select the tools that will help complete a thorough investigation of the present system. They can help analyze and document existing information uses and functional operations involved in the problem area.

The organization chart is a graphic depiction of the functions performed in a department or an organization. Through vertical or horizontal lines it demonstrates the authority and responsibility relationships of workers and administrators. It depicts line and staff relationships. It identifies the titles of jobs in accordance with job descriptions.

Review the organization chart in Figure 4–1 to determine the organizational structure of a given health facility. The chart identifies the organizational elements of the operation and displays the areas of responsibility where departments or information structure flow naturally. Understanding information problems must begin with a clear understanding of the information structure of the organization.

Figure 4–1 Organization Chart for a Community Hospital MRD

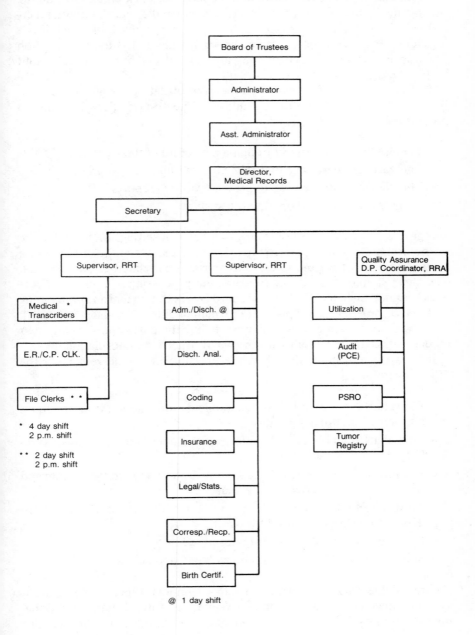

Secondary to use of the organization chart is the use of an *organization function list*. It is a document that is prepared for each unit shown on the organization chart in order to describe the specific major activities performed by that unit. It is keyed to the organization chart, of which it is a logical extension. In the example shown in Exhibit 4–3 the organization function list specifies what actions are carried out by a particular unit within the organization.

The *systems flowchart* portrays the flow of data through the system, files being used, and programs required.[5] The systems flowchart should properly show

- The flow of information through a part or all of the system
- The various operations that take place within the system
- The files that will be used to produce various reports
- Programs required to operate the system

Figure 4–2 is a systems flowchart for a hospital setting. In the MRD one of the first and most valuable steps to be carried out is the creation of a systems flowchart for the information structures related to patient data. The chart should map out the fundamental processing of discharged records beginning when a discharged record enters the system and concluding with the completed record placed in permanent filing. The users who work with the information can be asked to review the chart and indicate their agreement or disagreement with the way it traces the information flow.

In creating a systems flowchart the practitioner should remember the following points:

- Flowcharts are intended to communicate information to users about a specific functional operation.
- Flowchart symbols should conform to standard systems analysis symbols.
- The information described should flow from the top to the bottom and from the left to the right of the chart.
- All descriptions should be written within the symbol, and one-verb descriptors should be used for clarity.
- Each activity should be clearly described.
- Start and stop activity should be included and so noted.

Figure 4–3 identifies the correct flowchart symbols. If properly prepared, systems flowcharts can provide an excellent picture of major functional operations in medical record departments. They can also provide an excel-

Exhibit 4–3 Organization Function List

DISCHARGE ANALYSIS

1. Maintains all incomplete records of recently discharged patients until completion.
2. Codes all diseases and procedures.
3. Completes data abstracts.
4. Edits transcribed work received from transcription service.
5. Obtains records for hospital personnel when requested.
6. Maintains computer data base of all medical record deficiencies and generates deficient record reports.

---

lent illustration of systems that coordinate an interdependent system among several departments. Patient registrations and MPI operations are an example. Figure 4–4 is another example of a systems flowchart.

Operation (programming) flowcharts are the fourth tool. The operation flowchart is a graphic depiction of the sequence of activities and identification of decision points that are encountered in carrying out a function. It does this by displaying the synthesized steps that take place. It specifies *how* a function is performed, not what is performed. Whenever possible, a written explanation should be placed within the symbols used in this chart.

Figure 4–5 is an operation (programming) flowchart. Notice the sequence detail in the illustration. This flowchart is an excellent communication tool between users and programmers.

A *decision table* can be defined as an analytical tool that provides a tabular technique for describing logical rules. In analyzing an operation, the decision table can be used to communicate the logical sequence of that operation. An example is listed in Exhibit 4–4, which demonstrates the use of the PSRO preadmission certification process in decision table format. It lists the steps used in completing the process. Describing a particular operation in an MRD in this format provides another communication tool to systems analysts and programmers in developing computer-based operations.

The next tool is the *input/output chart*. It is used to analyze source documents and to identify their relationship to output forms or reports that contain data elements derived from the source document. In the example in Exhibit 4–5, an analysis of information that is entered into a manual census reporting system specifies the source documents that are used in the preparation of the census reports. By charting out the source documents and the associated census reports, the origin of the information

Figure 4–2 Systems Flowchart of MPI and Visit Log in a Hospital Outpatient Department

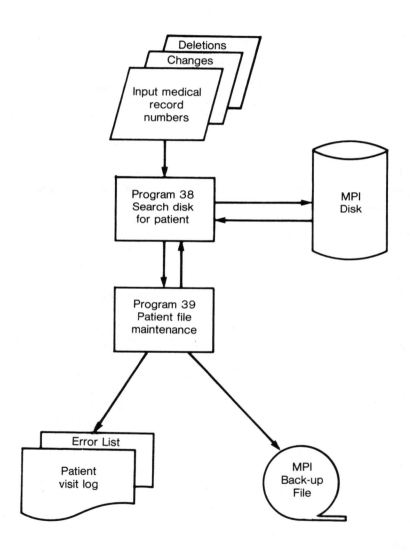

The inquiry mode of the system utilizes the medical record number, which then searches the disk for the appropriate patient file. This then displays the patient data, which is then utilized to achieve a hard copy display incorporated on an outpatient encounter form and an automatic entry on the patient visit log.

Figure 4–3 Standard Flowchart Symbols (These symbols conform to the International Organization for Standardization (ISO) draft recommendations on flowchart symbols for information processing.)

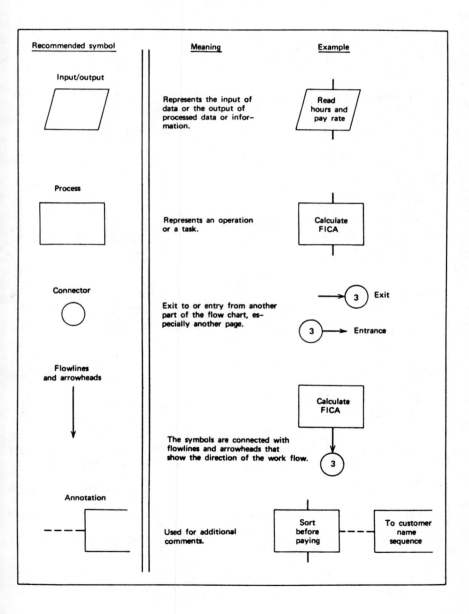

Figure 4–3 continued

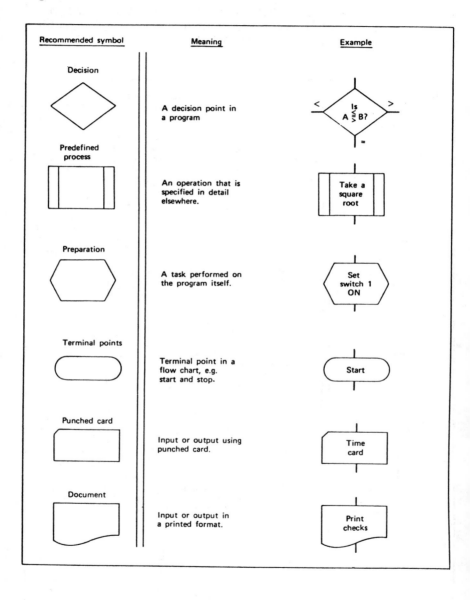

Figure 4–3 continued

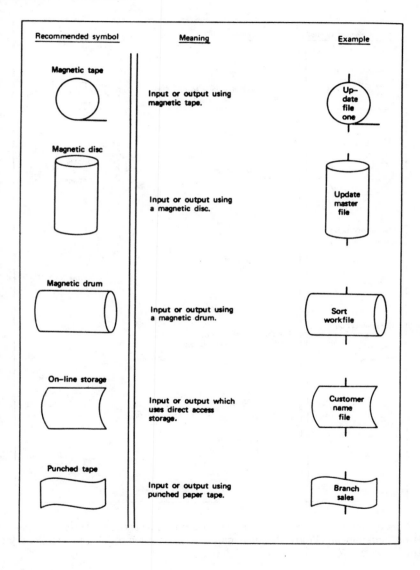

Figure 4–3 continued

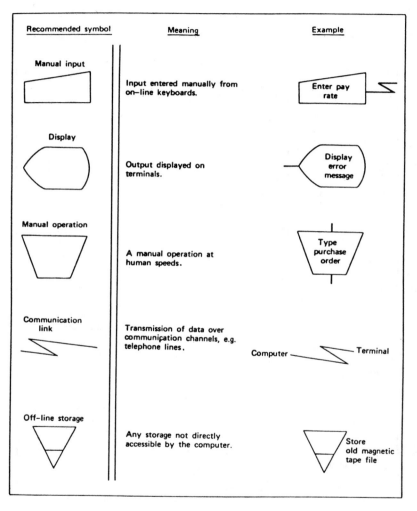

Figure 4–4 Systems Flowchart

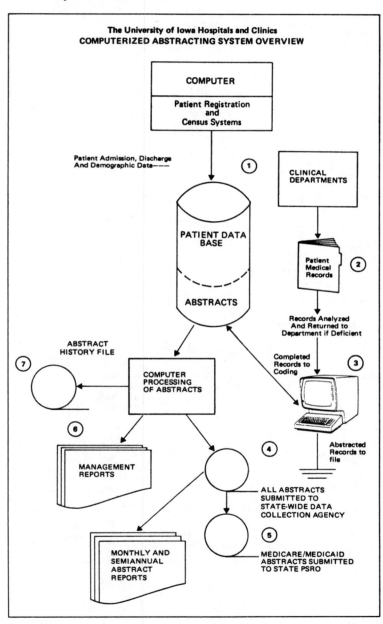

The University of Iowa Hospitals and Clinics
COMPUTERIZED ABSTRACTING SYSTEM OVERVIEW

Source: Reprinted from *Medical Record News* by permission, © April 1980.

Figure 4–5 Operation Flow Chart of Transcription Operation

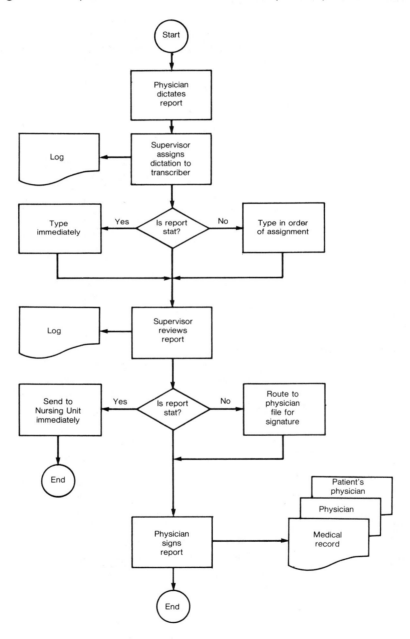

Exhibit 4–4 Decision Table for PSRO Admission Certification

R O W		Rule Number									
		1	2	3	4	5	6	7	8	9	10
Conditions	1 1st admission review conducted	Y	Y	Y	Y	Y	Y	Y	Y	Y	Y
	2 2nd admission review conducted					Y	Y	Y	Y	Y	Y
	3 3rd admission review conducted								Y	Y	Y
	4 Patient meets admission criteria	Y	Y	N	N	Y	Y	N	Y	Y	N
	5 Patient meets standard criteria	Y	N	Y	N	Y	N	Y	Y	N	Y
Actions	6 Approve admission	X	X	X		X	X	X	X		
	7 Refer for physician review				X					X	X
	8 Conduct 2nd admission review		X	X							
	9 Conduct 3rd admission review						X	X			
	10 Conduct continued-stay review within 3 days	X				X			X		

Decision Table Form—Guide for Completion

1. Header information: program name, program number, decision table name, decision table number, date of completion, page numbers, and by whom prepared, are to be completed in the appropriate space.
2. Condition statements should be assigned row numbers and entered on the form. A heavy line should then be drawn separating the condition statements from the action statements.
3. Action statements should be assigned row numbers and entered below the heavy line.
4. Rule numbers should be assigned horizontally across the top of the form. The spacing used can be variable and heavy vertical lines drawn down the form if required.
5. The condition entries, in limited or extended form, are entered opposite applicable condition statements.
6. The action entries are entered opposite the applicable action statements.

Exhibit 4–5 Input/Output Chart

Input Source Documents	Census by Floor	Census by Service	Patient Admission List	Medical Record Admission Log	OB List	Preliminary Op. Schedule	Room List for Return Op	Medical Record Discharge Log	Discharge List
Preadmission registration						X			
Hospital clearance and admit order					X	X			
Admission and discharge record	X	X	X	X	X		X		
Record of ER admissions	X	X	X	X					
Admission log from admitting			X	X	X		X		
Room bed transfer slip	X	X			X		X	X	X
Discharge statement	X	X						X	X
Discharge patient clearance card								X	X
Outpatient referral form	X	X	X	X	X	X			
Discharge transfer form	X	X						X	X
Statistical ledger								X	

contained in the census reports can be identified. This helps ensure that the design for a computerized census will include all necessary points of information origin. In addition, a careful review of this example suggests a logical consolidation of the manual census output reports prior to the initiation of a computerized system. Accurate and thorough information analysis by means of this and the other tools described in this chapter must be performed if good systems design is to occur.

A *forms analysis chart* identifies which fields of information are common to a group of forms. This allows the practitioner who is analyzing an operation to determine the use of the information. Forms analysis is also used to provide data element analysis. Exhibit 4–6 is provided to demonstrate how the discharge abstract form can be analyzed to identify the items of information routinely collected there that are subsequently used to produce required output reports. This information is needed for a programmer to prepare programs that collect and store the data elements and automatically generate the required output reports. In hospitals where "tape-to-tape" transfer is used for billing third party payers, the programmer specifies that the stored data elements required for billing be transferred to magnetic tape, which is then sent to another computer center and entered directly into another system. Such a procedure is used by the PSROs in collecting discharge information. Medical record managers can expect to see this transfer procedure more frequently in the future.

The *forms distribution chart* is used to determine the number of copies of a particular report to be distributed. It enumerates departments that input or use the copies.

Like the input/output tool, the forms distribution chart helps prevent duplication of effort. It often improves distribution procedures in manual systems even before a computer-based system is installed.

Detailed employment of this tool, while time consuming, can streamline information flow within an organization. Investigation often demonstrates that personnel request copies of forms because information access is inhibited at some point in the paper record system, usually because of other users' demands on the record. Computers have the potential to correct this. CRTs can display information in several locations at once if necessary. Thus, the computer display far surpasses paper copies in the ready availability of information, since paper copies, even when available, are accessed through manual files and therefore subject to the delays inherent in manual processing. Users may be well satisfied with a form displayed on a CRT screen rather than a paper copy, if the screen display is available as they need it.

The flowchart in Figure 4–6 is a forms distribution chart of the census. The recipient individuals or groups are listed by job function for each

Exhibit 4–6 A Forms Analysis Chart

Forms Analysis Chart of Recurring Data	Disease Index	Operation and Procedure Index	Physicians' Index	Death Registry	LOS Analysis	Discharge Statistics	PSRO	Committee Reports	Demographic Breakdown	Billing
Date of Analysis page 1 of 1 pages										
Activity Systems Invest.										
Discharge Abstract-VAH										
Patient name										X
Patient number	X	X	X	X	X		X	X	X	X
Date of Birth	X	X		X	X		X			
Age (computed)	X	X			X			X		
Sex	X	X		X	X		X			
Race	X						X			
Marital status							X			
Zip							X		X	
Method of payment										

Admission date	X	X	X		X	X				
Discharge date	X	X	X		X	X				X
Length of stay (computed)	X	X	X		X	X				X
Special unit days		X			X	X				
Hospital service	X	X	X		X	X			X	
Attending physician	X	X	X		X		X	X		
Operating physician	X	X	X		X		X	X		
Principal diagnosis disc.				X				X		X
Other diagnoses disc.				X				X		X
Diagnoses codes	X	X	X		X		X		X	X
Principal procedure disc.				X				X		X
Other procedure disc.				X				X		X
Procedure codes	X	X	X		X		X		X	X
Date of procedure		X					X			
Complications	X							X		
Nosocomial infections	X							X		

Exhibit 4–6 continued

Forms Analysis Chart of Recurring Data	Disease Index	Operation and Procedure Index	Physicians' Index	Death Registry	LOS Analysis	Discharge Statistics	PSRO	Committee Reports	Demographic Breakdown	Billing
Date of Analysis — page 1 of 1 pages										
Activity — Systems Invest. — Discharge Abstract-VAH										
Tissue pathology								X		
Transfusions								X		
Discharge disposition						X				
Death within 48 hours				X		X				
Death after 48 hours				X		X				
Anesthesia death				X		X		X		
OR death				X		X		X		
Post-op death				X		X		X		

Figure 4–6 Forms Distribution Chart

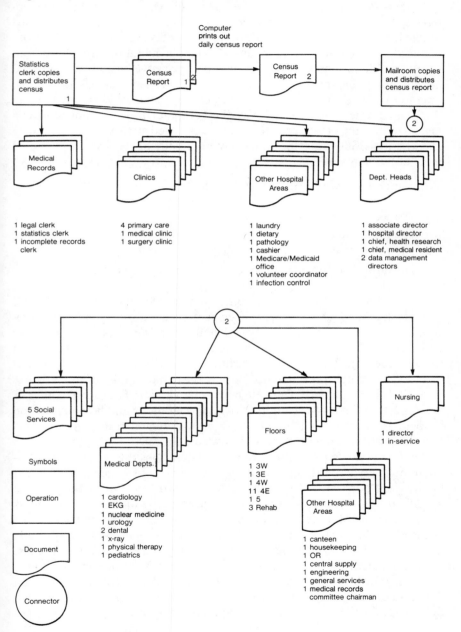

department, with the number of copies of the form per recipient specified.

Process flowcharts (PFC) are the ninth tool. The PFC is a graphic step-by-step display of the detailed series of activities that take place from start to finish in a particular process or in the work of one subject only (person, material, or paper form). By means of symbols operation, transportation, inspection, delay, and storage as well as distance covered and time involved are all noted on the chart. The PFC is helpful in analysis of manual operations. It can verify the sequence of steps and activities involved in the manual operation. In this way the computer programming staff can visualize the steps that apply to a computer-based structure. The example in Exhibit 4–7 illustrates processing of appointment request slips handled in an outpatient setting.

A *work distribution chart* (WDC) identifies

- The work activities performed
- The time it takes to perform them
- The individuals who are working on these activities
- The amount of time spent by each person on each activity

Two preliminary lists must be completed in order to compile the WDC: 1) the task list, which states what each person actually does, not what they should be doing, and the number of hours per week spent doing these tasks; and 2) the activity list, the major activities that should be performed to fulfill the objectives of the department.[6] Task lists should be prepared by individual employees; activity lists are the responsibility of supervisors. The WDC is particularly valuable because it shows what each member of the department does in relation to overall operations and how much time is spent accomplishing the tasks. Exhibit 4–8 demonstrates chart review functions within a department as recorded on a WDC.

The *layout flowchart* (LFC) graphically describes the physical environment in which activities or operations are carried out within a department or unit. Its purpose is to chart the flow of work activities through the specific desks and pieces of equipment used in the functional operations of the department. (See, for example, Figure 4–7.) Layout flowcharts can be used in the analysis of manual as well as computer-based systems. LFCs help reinforce the other tools and enhance the investigative process.

Data element analysis is a tool used to identify all the data elements necessary for a given product. It points out the data elements to be used for output and specifies the format and size of the information that must be stored. Exhibit 4–9 indicates how a computerized discharge abstract screen display might be analyzed on a data element analysis sheet.

Exhibit 4–7 Process Flowchart for a Process Needing Improvement

PROCESS FLOWCHART-Appointment Request Slips

Details of current activity proposed	Operation	Move	Inspect	Delay	Store	Distance (feet)	Quantity	Time (minutes)	Notes
1. Request info. completed, fill in (M.D. office assistant)	●	⇨	□	▽	◇			3	(initial request info)
2. Slip taken to floor desk clerk (M.D. office assistant)	○	▶	□	▽	◇	200		4	
3. Slip sent to records dept. via lift (floor desk clerk)	○	▶	□	▽	◇	3 floors		5	2-3 days in advance of appt.
4. Slip received by medical records dept. (M.R.D. lift/mail clerk)	●	⇨	□	▽	◇			1	
5. Slip placed in proper bin for chart search (routing clerk)	●	⇨	□	▽	◇	2		.5	bins listed, filed by patient #
6. Slip awaits removal from bin (M.R.D. file clerk)	○	⇨	■	▶	◇			120	variable time
7. Slip removed from bin (MRD file clerk)	●	⇨	□	▽	◇			.5	
8. Slip carried to appropriate file area (MRD file clerk)	○	▶	□	▽	◇	120		3	
9. Chart sign-out confirmed, indicated on slip (MRD file clerk)	●	⇨	■	▽	◇			3	inspection process done first
10. Slip request dated (MRD file clerk)	●	⇨	□	▽	◇			.6	
11. Slips separated; copies stamped for routing (MRD file clerk)	●	⇨	□	▽	◇			3	
12. White slip copy placed in box for future search (MRD file clerk)	●	⇨	□	▽	◇			.1	
13. Yellow copy sent via lift to requestor (MRD lift/mail clerk)	○	▶	□	▽	◇	3 floors		5	
14. Requestor acknowledges slip info; discards (MD office ass't)	●	⇨	■	▽	◇			2	involves short inspection
15. Pink copy sent to area chart is currently signed to (MRD lift/mail clerk)	○	▶	□	▽	◇	2 floors		5	
16. Pink copy held for location of chart (office ass't of location)	○	⇨	■	▶	◇			5	short delay
17. Pink copy sent with chart to MRD via lift	○	▶	□	▽	◇	2 floors		5	
18. Pink copy received MRD via lift with chart (MRD lift/mail clerk)	●	⇨	□	▽	◇			.3	
19. Pink copy removed from chart and discarded (MRD file clerk)	●	⇨	□	▽	◇			.2	
20. White copy found and stapled to chart (MRD file clerk)	●	⇨	□	▽	◇			1	
21. White copy and chart sent via lift to requestor (MRD lift/mail clerk)	○	▶	□	▽	◇	3 floors		5	
22. White copy and chart sent to requestor's office ass't. (floor desk clerk)	●	▶						.2	
23. White copy and chart sent to office assistant (messenger from floor desk)	○	▶				200		2	
24. Chart received; white copy discarded (MD office ass't.)	●							.2	end of appt. slip process

Exhibit 4–8 Work Distribution Chart

ACTIVITY	Total Hours (all employees)	Agusta Bernard Supervisor	Hours	Laura Case Clerk II	Hours	Martha Cossian Clerk I	Hours
Telephone Requests	8		1		3		4
In-Person Requests	14		11		3		
Subpoena/ Depositions	12	• Prepare charts • Attend court sessions	12				
Obtaining Patient ID for requests	6					• Checking patient master file	6
Chart Retrieval	11			• Complete chargeout cards • Retrieve charts	1		10
Abstracting Reply	42		10		26		6
Photocopying	7					• Daily Photocopying	7
Miscellaneous	20	• Supervisor's Meeting • Session with trainees	4 2	• Emergency coverage in files 5/15	7	• File area coverage 5/16	7
TOTAL	120		40		40		40

Summary of System Analysis Tools

The tools previously described were presented to provide a resource list for the analysis process. The tools presented in Table 4–3 could be used in a health information application. By employment of one or more tools, coupled with the constraints list prepared in the first step and feedback from user survey questionnaires, the manager can accurately formulate the problem identification and complete the first portion of the analysis phase. In addition, many of these tools will assist the system team to complete other aspects of the analysis phase such as those that occur during performance definition.

The following items may have been assembled to document an actual problem identification:

1. Problem statement
2. Narrative description of present system operation (optional)
3. List of constraints
4. Notification letter to users
5. Summary or results of user interviews
6. Organization chart and/or function list
7. System flowchart describing the activity
8. Input/output charts
9. Forms analysis charts
10. Data element analysis document or other documented tools used to complete the analysis activities

Figure 4–7 Layout Flowcharts

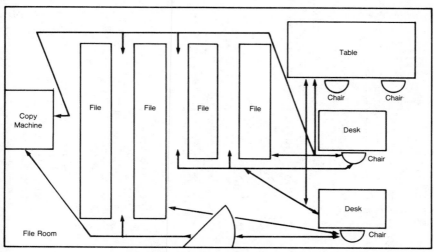

Layout Flowchart—Current Arrangement

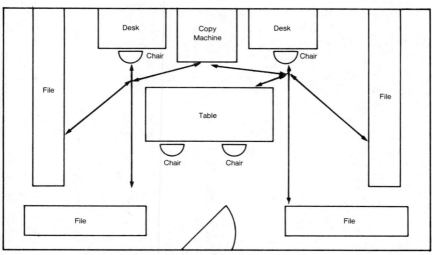

Layout Flowchart—Proposed Arrangement

Exhibit 4–9 Data Element Analysis Sheet

Discharge Abstract Screen Display Format

Field*	Item	Size**	Source	Format
01	Patient number	6	Admissions	
02	Name	35	Admissions	Last name, first name, and initial
03	Zip code	5	Admissions	
10	Age	3	Admissions	Age in years as given. Age of infants admitted will be designated in months from 1 to 11, followed by an M. Newborns are classified as NB.
11	Sex	1	Admissions	1-Male, 2-Female
12	Race	1	Admissions	1-White, 2-Black, 3-Other
13	Birth date	8	Admissions	e.g., 01/22/55
14	Financial code	4	Admissions	
20	Admission date	8	Admissions	Same format as birth date
21	Time	4	Admissions	Given as military time
22	Admission type	1	Admissions	1-elective, 2-emergency
30	Discharge date	8	Nursing staff	Same format as birth date
31	Length of stay	3	Computer action	
32	Disposition	2	Nursing staff	
33	Service	2	Medical records	Two-digit code
40	Attending MD	4	Admissions	Four-digit MD user number
50	Consultations	2	Medical records	Two-digit service code for service of consultation
51–54	Other consultations	2	Medical records	As above
60	Transfusions	4	Central service	1-whole blood, 2-packed cells (number of units), e.g., 1-02
61	ICU-CCU days	3	Census system	Total number of days
70	Principal diagnosis	6	Medical records	Use ICD-9CM codes
71–76	Add'l. diagnoses	6	Medical records	Same as above

80	Date of operation	Surgery	8	Same format as birth date
81	Surgeon	Surgery	4	Four-digit MD user number
82	Assistant surgeon	Surgery	4	Four-digit MD user number
83	Primary operation	Medical records	5	ICD-9CM procedure coding
84–87	Other operations	Medical records	5	ICD-9CM procedure coding
90	Date of operation	Surgery	8	Same format as birth date

Field is the numbered location of the information element as it is shown on the form or screen.

**Size* is the maximum number of item characters per field. It includes alphabetical and numerical characters.

Table 4–3 Systems Analysis Tools For Common MRD Applications

Title	Function	Example
Organization chart	Depicts departmental functions and the authority/responsibility relationships of workers and administrators	Describes an organizational picture of an MRD
Organization function list	Describes the major activities performed by each unit identified on an organization chart	Identifies a concise description of an individual functional operation in an MRD
Systems flowchart	Displays flow of data	Displays information flow for Registration, Admission, Discharge, Transfer
Operations flowchart	Depicts the sequence of activities and identifies decision points involved in completing an operation	Documents the function of chart deficiency system
Decision table	Provides a tabular display of logical rules followed in performing an operation	Demonstrates the admission screening process for utilization review
I/O chart	Describes relationships of input documents to output reports	Illustrates relationships of admission documents to discharge abstracting operation
Forms analysis chart	Identifies which fields of information are common to a group of forms	Demonstrates repetitive notice of admission documents in hospital admission operation

Forms distribution chart	Identifies who receives specified reports	Illustrates census report distribution throughout the hospital
Process flowchart	Traces the steps of one subject in a process	Traces the chart requisition process for an outpatient clinic
Work distribution chart	Describes work activities and the time it takes to do them	Demonstrates the work activity involved in abstracting and chart deficiency review
Layout flowchart	Displays the movement of functional operation through the physical environment	Analyzes floor space problem in record control
Data element analysis sheet	Identifies and describes the characteristics of each data element required for a given output	Describes the required and optional data fields planned for a computerized discharge abstracting system

Exhibit 4–10 gives samples of the first three of these components of problem identification that would result from carrying out some of the activities listed so far in this chapter. Figure 4–8 illustrates item 7 on the list; it is a generalized system flowchart for the same problem identification.

Performance Definition Begins with Goals and Objectives

The second part of the analysis phase is the *performance definition*. This entails the precise description of what functional or systems requirements are necessary to solve the problem, through design of a computer system. It provides a foundation for the design phase of the computer-based solution as well as criteria for postimplementation audit activities. During this phase, goals and objectives of the proposed system are identified. Departmental objectives are included in the performance definition to help specify the purpose of the process being reviewed. In many applications the goals and objectives of the present system are refined and updated to reflect the expectations of the computerized operation. The revised goals and objectives provide the initial definition of the system requirements. Table 4–4 lists performance objectives for a proposed computerized chart deficiency system. Once goals and objectives are established, verification of a formal system flowchart is carried out by user operations personnel. Model outputs are prepared to satisfy previously identified constraints and system requirements. System flowcharts and model output documents are part of the functional requirements package that is necessary in order to solve the problem.

System flowcharts are reviewed as a follow-up to the activities carried out in the problem identification section. Once the systems flowcharts of the proposed system and corresponding narratives are complete, an initial description of the functional requirements of the proposed application can be prepared. This is a list describing how the system is to work. A preliminary description, it will be supplemented in more precise detail as the systems analysis process continues. The performance definition concludes with a description of the outputs that the computer is required to produce.

The *functional requirements document* is also initiated by the systems team during the second part of the analysis phase of project development. The functional requirements document is the foundation for the functional description of the completed system. One of the most important documents prepared during the development of a project, it presents the design of the system together with costs, benefits, and implementation plans such that both users and designers may understand, evaluate, and formally agree on the proposed system. It identifies how the proposed system is to operate. It details purpose, information flow, identified input, expected oper-

Exhibit 4–10 Sample Components of a Problem Identification

---

1. Problem Statement
   The hospital's accreditation is in jeopardy because of over 4,000 incomplete patient records.
2. Narrative Description of Present System Operation
   Incomplete records are monitored by a manual deficiency tracking system that is currently performed in the MRD. All patients records are processed through the system immediately following discharge. Incomplete patient record data are communicated to the medical staff by the MRD on a routine basis.
3. Constraints List
   a. The incomplete chart system must satisfy JCAH standards as defined in the hospital bylaws.
   b. The system must monitor record completion deficiencies for nurses' and physicians' data entries.
   c. Individual incomplete records must interface with record control system to allow quick retrieval of desired records.
   d. Any proposed automated system must use the present in-house computer.
   e. The development costs must not exceed $20,000.
   f. The development costs must be paid back through reduced operational costs within five years of operation.

---

ation, points of communication with other operations, narrative procedures when available, and proposed outputs. These are assembled to specify how the proposed system is to work.

Initiated in the analysis phase through as detailed a description as possible, functional requirements are under continual development and refinement throughout subsequent phases. Examine Table 4–5, the complete outline for a functional requirements document. Consider how systems analysis for a computerized discharge abstracting system might be detailed to provide key elements for this document.

It is important to note that the analysis phase is preliminary to more detailed system design that includes inputs, outputs, file requirements, and so on. However, when system functions are precisely described in the analysis phase, the consequent detailed design activities are facilitated.

Medical record practitioners should request a copy of the functional or system requirements outline used by the computer services department in their organization. A functional requirements outline should identify all the components needed to prepare a complete description of the application. Analogous to preparation of architectural drawings and mechanical specifications for a new building, this document will evolve and develop into the specific information requirements needed to build the system. Refer again to Table 4–5. Many items listed can be identified according to a primary source, by user, technical application designer, systems designer,

Figure 4–8 Generalized Systems Flowchart for a Manual System of Correcting Physicians' Incomplete Data on Charts

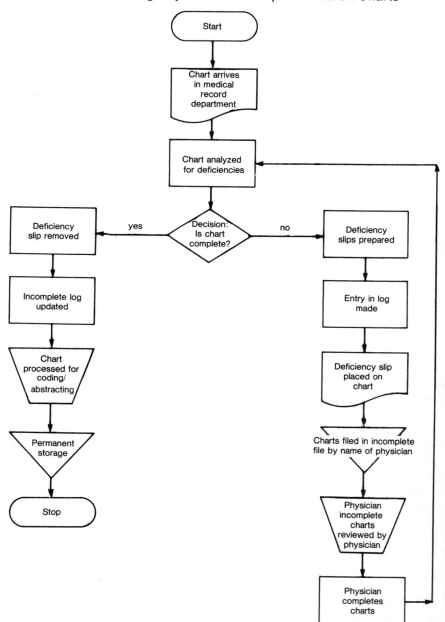

Table 4–4 Measurable Performance Objectives for Proposed
Computerized Chart Deficiency System

1. To decrease the clerical time needed to monitor the incomplete system by 95 percent.
2. To decrease the search time needed to locate incomplete charts to less than 30 seconds 90 percent of the time.
3. To decrease the total number of incomplete charts to 5 percent of the total number of monthly discharges within 48 hours of discharge.
4. To provide weekly status reports to physicians of deficiency types on their patients' charts and the exact incomplete data status for each one.
5. To provide weekly status reports to department heads on the length of time open and the number and types of chart deficiencies for physicians in their service area.
6. To provide administration with biweekly status reports of the patients' financial codes, the total number and length of time records have been open, by service, and a decimal proportion comparison of each service stating their number of delinquent records per number of monthly discharges.
7. To provide daily reports of discharge information concerning patients and their diagnoses to the billing department.
8. To provide a screen display on demand to the MRD listing the charts with unfinished data abstracts due to incomplete status.

---

and so on. Other items would be mutually derived by various members of the systems team. Still other items can be seen to be documented products of the analysis phase activities. Item 2.2 on the outline is an example of this.

OUTPUT DESCRIPTIONS—SELECTING COMMERCIAL PACKAGES

Outputs are defined as the end products of the computer-based system. This can mean either the improved operations expected for a given task or reports and screen displays. Unless outputs are clearly described in precise detail in the first phase so that the design phase blueprint may be drawn to accommodate them, there will be distinct holes in the application. Figures 4–9 to 4–11 are examples of outputs for three computer-based systems in medical record departments.

Determination of outputs is not limited to development of in-house systems. Medical record practitioners must communicate the same information to vendors if their careful analysis is to result in effective package selection. Most practitioners are familiar with statistical report outputs offered by the various commercial discharge abstracting services. Many have already made decisions and recommendations based on analysis and assessment of these and other outputs of such systems.

Table 4–5 Functional Requirements Document Outline

1. Introduction
   1.1 Background description of present system
   1.2 References
2. Proposed system description
   2.1 Purpose and function
   2.2 Goals and objectives
   2.3 System design approach
   2.4 Users involved, by department
   2.5 Project guidelines
   2.6 Future expansion
   2.7 System evaluation
3. System flowchart
   3.1 Master system flowchart
   3.2 Transaction flowcharts
   3.3 Narrative procedures outlines
4. System requirements
   4.1 Computer interfaces
   4.2 Input definitions (data entry terminal screens)
       4.2.1   Data input forms samples
   4.3 Inquiry definitions
       4.3.1   Inquiry request masks (e.g., patient ID look-up)
       4.3.2   Inquiry summary response masks (i.e., display summary census)
       4.3.3   Display/modify inquiry response masks (e.g., name change)
5. Output definitons (reports)
   5.1 Scheduled reports
   5.2 Special reports
   5.3 Output models
6. Data field definitions and record descriptions
   6.1 Master file record description (record layouts)
   6.2 Master file data element description (data element definition)
   6.3 Data element editing and conversion requirements
   6.4 Data element reference documentation (See analysis tools)
7. Operating requirements
   7.1 System environment
       7.1.1   Machine
       7.1.2   Software
       7.1.3   Resources
                7.1.3.1 Core
                7.1.3.2 Disk
                7.1.3.3 On-line
       7.1.4   Terminals
   7.2 Performance requirements
       7.2.1   System availability
       7.2.2   Responsiveness
       7.2.3   Frequency of operation
       7.2.4   Data accessibility
       7.2.5   Record retention duration

Table 4–5 continued

Whether designing new outputs or selecting from vendors, health information managers will develop stronger data use and system acceptance if they analyze these items against the output design principles in this text. Output design is featured in detail in Chapter 5.

Summarizing and Integrating Analysis Phase Activities

How are the various activities coming together in the systems developments? If project coordination has been established at some point prior to or during these activities, schedules may have been identified and progress monitored. If the analysis work accomplished thus far is preliminary investigation, a project coordinator or formal systems team may be assigned at this point. This can be as a result of the user-manager presenting these findings to administration. Readers will note the variety of approaches possible for initiating computer development projects. The most important factor is *communication* among all users and managers involved in the

Figure 4–9 Sample Disease Index Output

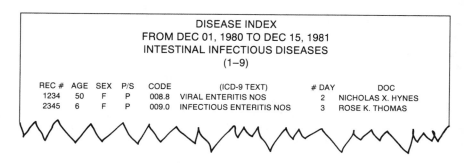

Figure 4–10 Sample Operative Index Output

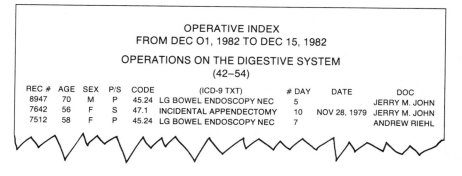

Figure 4–11 Sample Chart Request Output for File Folder

```
                                 MEDICAL CHART
        PATIENT NO: U-2-29-3i-96           REQUESTED BY: KEV
        ORIGINATING TERMINAL: JK2Pi        OSFEJJi    1500
        PATIENT:                                      DOB:  i9SEPi936
        OTHER ID: SSN: 50i-0i-6983         RACE: C    SEX: M

        NOTE:
        **********SEND CHART PLEASE

        SEND CHART TO DEPT:  ER            RM: NE 207     NAME:  KEV
        CHART DELIVERED (  )   DATE: _____    TIME: _____
        CHART OUT          (  )   TO DEPT: _____    ROOM: _____
                                          NAME: _____
```

operation or system being reviewed. The following section illustrates how the activities might be organized by users and a systems team.

Reviewing the Role of Users and Systems Teams

The sequence of steps identified so far in the systems analysis process can be related to the systems teams' responsibilities. Each organization confers its own unique type and amount of authority, autonomy, direction control, and so on, to the systems team. Users also have a specified and limited range of responsibility. But users and the systems team are the key people in the development of a computer project. Their various tasks and activities are the basis for compiling a draft functional requirements document. The drafting of this document can be initiated when results and documentation of problem identification have been freely evaluated; users asked to review a description of the problem as augmented by additional information from analysis tools, and performance definition described. The following sequence illustrates how users or systems team members can carry out the systems analysis functions described so far.

Step 1   Develop a written problem identification, which should include a clear explanation of the problem elements. This explanation will usually include present system description.

Step 2   Prepare a written constraints list so that required factors can be included in the evolving problem identification process.

Step 3   Summarize user surveys, which report on performance in the problem area.

Step 4   Summarize results of other selected investigative tools.

Step 5   Initiate performance definition by review or revision of department objectives. Specify operational objectives that apply to the problem. This delineates the role of the MRP because both departmental and specific functional operations are identified. The organization function list is a good tool to identify the functional operations that warrant corresponding objectives.

Step 6   Write a description of criteria for the objectives so that the expected performance for each objective is specified and documented.

Step 7   Prepare a flowchart of the proposed system with any necessary narrative explanations.

Step 8   Prepare examples of system outputs and include them in the documentation as part of the performance specifications.

Step 9   Write a preliminary description of the known functional requirements of the anticipated computerized application. This enables the designers to identify a framework from which to work. (See Figure 4–11 for the kind of information that would be included here.)

FEASIBILITY ANALYSIS AND THE SELECTION OF ALTERNATIVES

Feasibility analysis is the comparative assessment of alternative solutions to determine which is the most acceptable. Feasibility analysis can occur at any point where alternatives must be evaluated. It can and probably should occur at the very beginning of computer systems development. It can then include all the tasks and activities carried out in the analysis phase. This is particularly appropriate when a significant investment is to be made in the analysis activities. The feasibility study team may recommend further analysis of the present system for computerization or recommend against the use of computers. Here it is critical that the team include experienced information management personnel so that the study can accurately identify the use to which a computer will be put and then identify the best computer for the job. Four questions are the focus of a feasibility analysis:

- Is it feasible to continue a comprehensive systems analysis to identify and select appropriate alternative solutions?
- Is the proposed alternative operationally appropriate within the organization?
- Is the proposed solution technically appropriate for the computing resources available to the organization?
- Is the proposed solution economically feasible for the organization?

Let us examine these questions.

Generally, the feasibility analysis affords the first opportunity for thorough description of projected computer solutions. The steps that have gone before should have resulted in a complete description of the project. Alternative computer solutions are usually projected so that management can select the most effective one. Along with the proposed computer solutions, the feasibility analysis should incorporate a realistic solution using the present resources. This will enable the decision makers to consider a noncomputer solution among the alternatives. Practitioners may find it useful to consider the benefits of applying feasibility principles at the onset of large projects. Alternately, in other MRD applications the feasibility process may be effectively employed at the conclusion of the investigation of the present system when problem identification and performance definitions have been completed.

Feasibility analysis is initiated with a careful look at the systems analysis cost itself. This includes the investigation of personnel costs and the cost of proposed alternatives. What is the cost of released time of the RRA manager to engage in the systems analysis process? What is the cost of the released time for the user-manager to engage in feasibility activities? What is the cost of the collaborative effort needed with the data processing professional and other staff within the agency or health facility? This needs to be a planned budgetary issue. A serious commitment to change will build in analysis time for key personnel. Without such planned costs, project funds may be eaten up by these activities before system design work is actually underway. It will cost money to carry out a comprehensive analysis and this expense should be calculated into the overall expenditures.

Why is systems analysis included in the feasibility process if it has already been carried out beforehand? In many cases, the analysis activities and performance definition completed prior to feasibility analysis are just preliminary. Hospital administration may consent to a more formal process that will require review of the completed activities. This is to make sure that all users affected by the problem under investigation are adequately involved in the investigative process. This could mean supplementing the analysis results through the employment of additional systems analysis tools to be sure the process is comprehensive. Identifying completed activities and incorporating all appropriate users will help make the performance definition as precise as possible. Because this is the point at which the organization determines whether or not to continue in the systems analysis activities, it is important that all factors be identified. This means that management needs to be aware that systems analysis itself is a cost item. Feasibility analysis should always be conducted prior to any commitment to a solution that requires a large investment.

Operational feasibility can be defined as the determination of the capability of a given computer-based system to perform the intended functional operation. This step is a practical investigation. What procedures will data entry clerks follow in order to retrieve information regarding census activity? What changes will be made in the manual procedures to operate the computerized system? Operational feasibility considers staff allocation and placement of hardware such as terminals for access and use. It should report on operating features such as CRT terminal response time so that the performance of a particular alternative can be ascertained. Terminals may replace typewriters in hospital admitting departments when a computerized admission system is introduced. If patient admitting time is increased because the response time from one terminal display to another is too slow, then the operational feasibility of the system should be seriously questioned.

Technical feasibility evaluates the ability of the proposed computer system to meet the technical performance definitions that have been prepared. The technical issues must be identified by data processing and information management personnel and must include consideration of the hardware and system requirements needed to create or select a particular computer system. A data base system for patient registration, admission and outpatient scheduling may be technically feasible. But if the hospital is planning extensive hospitalwide computer developments, then that data base design may be modified or deferred until hospitalwide implementation plans are better defined. Technical feasibility deals with such questions as these:

- Can the proposed system admit the patient through the Admitting Department and automatically create an MPI available to the MRD?
- Is it technically possible to store the disease and operation index on magnetic tape and have it available on 24-hour notice for research and medical audit projects?
- Is it technically feasible to use a microcomputer in the MRD for MPI and connect it to other computer installations in the hospital so that other departments may access the MPI as needed?
- Will the proposed computer system be able to maintain the applications currently used and also allow for expansion?

The answers will depend on the particular hardware and software capabilities of the proposed system and may differ considerably from one setting to another.

Economic feasibility analysis considers four types of cost: systems analysis, design, development, and operation and maintenance costs. The cost of systems analysis comes primarily from personnel costs for analysts and for clerical staff who provide support. Consultants' fees would be included here.

Design costs include personnel costs for data processing staff required to work on the project; consultant fees, if consultants are used; and prorated computer costs where existing equipment is used for design activities. For instance, if the existing computer is used for writing programs, costs are incurred for the operation of the CPU, storage devices, input/output equipment, and so on.

Development costs are incurred with the development of the system. These include continued program development as the system is con-

structed, programmers' salaries, computer operations personnel's salaries, and, again, any consultant fees. The cost of purchased computer
hardware is also included.

Operation and maintenance costs include the costs of computer system
conversion and changeover, maintenance, and salaries of system personnel and users. Unlike the analysis, design, and development costs, these
costs can be used to project the continuing costs for the installed system.

Cost/Benefit Analysis

The cost data are collected or projected for systems analysis, design,
development, and operation and maintenance. They are organized to answer
the following general questions.

- What are the major cost categories of the overall system?
- What methods of analysis are available and appropriate?
- Can all costs be accurately identified and estimated?
- Which benefits cannot be estimated in dollars?
- What criteria will management follow in evaluating these factors?

There are several methods used for performing cost/benefit analysis. We
will describe two of these and demonstrate examples.

The payback period is used to judge the benefits of a system. It does so
by determining how long it will take for the system to accumulate savings
or earnings sufficient to cover its costs. Table 4–6 illustrates how this
might be displayed. Project B would probably be preferred because a
larger amount of the investment would be recovered during the first year.
This fact must be combined with additional accounting data to complete
a thorough payback analysis.

Another method used is comparison of the current system with the new
system. This is performed in the following manner.

1. Estimate the projected useful life of proposed system.
2. Calculate the proposed system's operating costs.
3. Calculate the present system's operating costs.
4. Compare the two operating costs. Fixed investment costs, including
   one-time implementation, should be calculated. *Implementation costs*
   are one-time costs to create new capabilities. *Investment costs* are
   one-time costs to acquire new equipment. *Operating costs* are ongoing outlays required for operating the system.

Table 4–6 Computerized Word Processing: Annual Return

Project	Year 1	Year 2	Year 3	Year 4	Total
A	10,000	10,000	10,000	10,000	40,000
B	25,000	5,000	5,000	5,000	40,000

The cost elements related to salaries, equipment use, and supplies discussed earlier would be incorporated into this method. Tables 4–7 and 4–8 illustrate how this could be displayed.

As a further example, see Table 4–9, which lists cost/benefit analysis factors.

Appendix 4–A illustrates the complexity of a thorough cost analysis process. It points out the need for critical review of cost justification data in determining realistic costs for any new system. The cost checklist, the benefits checklist, and the cost analysis strategies listed in the tables should be reviewed. The critical analysis featured in the case study offers insight to MRPs in their role as user-manager. Better understanding of the cost analysis process results in more effective cost data preparation at all levels. Interested readers may also wish to review return on investment methods, present value methods, and marginal efficiency of investment (MEI), which can be found in the references listed in this text. More detailed accounting procedures are beyond the scope of this book.

Steps in Feasibility

All the components of feasibility analysis described are brought together in a process that builds on the initial activities of the analysis phase. In developing computerized systems, the feasibility analysis activities provide the method for selecting a particular course of action. In the following list of steps in the feasibility analysis, notice that specific products of the systems analysis carried out thus far are incorporated into this decision-making step. Feedback from earlier steps is continual. The reason for this is to see that the decisions made about implementing computerized systems are based on precise and thorough analysis with checkpoints along the way.

1. Charge the system team with the assigned project.
2. Prepare summary of the problem identification.
3. Develop existing general systems flowcharts from performance definitions in study phase.

Table 4–7 Comparison of Current PAS Abstracting Costs (for 5,500 Discharges) with Costs for PAS+ System (for 96 Beds)

	YEAR 1	YEAR 2	YEAR 3	YEAR 4	YEAR 5
PAS	$ 4,950.00	$ 4,950.00	$ 4,950.00	$ 4,950.00	$ 4,950.00
Postage	200.00	200.00	200.00	200.00	200.00
Personnel	25,000.00	25,000.00	25,000.00	25,000.00	25,000.00
Total	$ 30,150.00	$30,150.00	$30,150.00	$30,150.00	$30,150.00
PAS+	$ 5,200.00	$ 5,200.00	$ 5,200.00	$ 5,200.00	$ 5,200.00
Install PAS+	7,500.00				
Postage	12.00	12.00	12.00	12.00	12.00
Personnel	12,500.00	12,500.00	12,500.00	12,500.00	12,500.00
Computer	9,900.00				
Maintenance	1,700.00	1,700.00	1,700.00	1,700.00	1,700.00
Shipping	200.00	—	—	—	—
Total	$ 37,012.00	$19,412.00	$19,412.00	$19,412.00	$19,412.00
Yearly Savings	$ −6,862.00	$10,738.00	$10,738.00	$10,738.00	$10,738.00
Cum. Savings	$ −6,862.00	$ 3,876.00	$14,614.00	$25,352.00	$36,090.00

Table 4–8 Cost Analysis of Hospital Transcription Services: Five-Year Forecast

Current System	1978	1979	1980	1981	1982	1983
Labor (salary and fringe benefits)	$72,200 5.5 FTE	$78,697 5.5 FTE	$85,779 5.5 FTE	$107,780 6.5 FTE (+ 1.0 FTE)	$117,480 6.5 FTE	$128,054 6.5 FTE
Equipment (includes service and maintenance agreements)	$ 5,050	$ 6,610 (+ 2 typewriters)	$ 5,300 (Increased maintenance)	$ 6,080 (+ 1 typewriter)	$ 5,550 (Increased maintenance)	$ 5,550 (Increased maintenance)
Supplies	$ 2,600	$ 2,834	$ 3,089	$ 3,367	$ 3,670	$ 4,000
Renovation of space		$ 4,400				
Total	$79,850	$92,541	$94,168	$117,227	$126,700	$137,604
Proposed System						
Labor (salary and fringe benefits)	$59,448 4.5 FTE	$64,798 4.5 FTE	$70,630 4.5 FTE	$ 76,987 4.5 FTE	$ 99,481 5.5 FTE (+ 1.0 FTE)	$108,433 5.5 FTE

Equipment (includes service and maintenance agreements)	$14,718	$14,718	$14,718	$ 14,718	$ 14,718	$ 14,718
Supplies	$ 2,076	$ 2,263	$ 2,467	$ 2,689	$ 2,931	$ 3,195
Salvage of typewriters	-$ 200					
Shipment of mag-cards	+$ 100					
Annual total	$76,142	$81,779	$87,815	$ 94,394	$117,130	$126,346
Annual savings	$ 3,708	$10,702	$ 6,353	$ 22,833	$ 9,570	$ 11,258
Total five-year savings	$64,484					

Table 4–9 Cost/Benefit Analysis Factors

Costs	Benefits
Direct costs	Direct and indirect cost reductions
Computer equipment	Elimination of clerical personnel and/or manual operations
Communications equipment	
Common carrier line charges	Reduction of inventories, manufacturing, sales, operations, and management costs
Software	
Operations personnel costs	
File conversion costs	Effective cost reduction, for example, less spoilage or waste, elimination of obsolete materials, and less pilferage
Facilities costs (space, power, air conditioning storage space, offices, etc.)	
Spare parts costs	Distribution of resources across demand for service
Hardware maintenance costs	
Software maintenance costs	Revenue increases
Interaction with vendor and/or development group	Increased sales due to better responsiveness
Development and performance of acceptance test procedures and parallel operation	Improved services
	Faster processing of operations
Development of documentation	Intangible benefits
Costs for backup of system in case of failure	Smoothing of operational flows
Costs of manually performing tests during a system outage	Reduced volume of paper produced and handled
Indirect costs	Rise in level of service quality and performance
Personnel training	Expansion capability
Transformation of operational procedures	Improved decision process by providing faster access to information
Development of support software	Ability to meet the competition
Disruption of normal activities	Future cost avoidance
Increased system outage rate during initial operation period	Positive effect on other classes of investments or resources such as better utilization of money, more efficient use of floor space or personnel, and so forth
Increase in the number of vendors (impacts fault detection and correction due to "finger pointing")	
	Improved employee morale

Source: Reprinted from *Fundamentals of Systems Analysis* by Ardra F. Fitzgerald, Jerry Fitzgerald, and Warren D. Stallings, Jr., by permission of John Wiley & Sons, Inc., New York, © 1980.

4. Develop system candidates (for example, alternative commercial systems for discharge abstracting).
5. Prepare detailed description of candidates; include special features.
6. Target meaningful system characteristics. How well do the characteristics meet the needs?
7. Identify cost elements for each candidate system.
8. Weight projected candidates' performance and cost elements using simple rank scale (1–10) and display.
9. Recommend the best candidate.

Once sufficient information is collected to describe and display the alternative computer candidates, a review group evaluates and ranks the alternatives. A candidate matrix chart is used to summarize the functions in the alternative systems. Table 4–10 is a matrix chart. When the information has been accumulated, the chart can be completed and used as a summary document for the evaluation process. The systems analysis project team, management staff, and other users engage in this process. Usually the operational, technical, and economic feasibility questions are incorporated into the feasibility report, and resource persons who are able to answer questions in these areas are included in the review process. All members of the review group rank all alternatives to determine which has the most benefit. Medical record practitioners may recognize the ranking process from equipment selection procedures and other decision-making situations. The evaluation factors are listed on the table and weights are assigned. By definition, some factors such as cost can be determined exactly. Others will be evaluated more subjectively, depending on the individual judgment of each reviewer. One tool commonly used for recording the results of the evaluation is a candidate evaluation matrix. Table 4–11 is an example of this kind of chart. Evaluators can individually rank each candidate and the results provided in the feasibility report.

CONCLUDING THE ANALYSIS PHASE

The final tabulation and feasibility results are incorporated into the analysis phase report. This report is presented to the chief executive officer of the organization at the end of the analysis phase. It should include all the facts necessary for management to make an informed decision whether or not to commit to the proposed computer system. The analysis phase report provides supportive background data and the results of key staff judgments about the total topic of discussion—the identified problem.

The analysis phase report is a critical document. For medical record practitioners participating in large systems analysis operations, this report

Table 4–10 Candidate Matrix Chart

Candidate System / Functions	Present System	Candidate 1	Candidate 2	Candidate 3
Output Screen Hardcopy				
Input Terminal Entry Documents				
Storage On-Line On-Line Interactive Back-up Storage				
Processing Screen Response Time Turn-around- time				
Data Security				

will contain their input and analysis results. Practitioners working with single modules, such as record control systems, discharge abstracting systems and others, will need to prepare analysis phase reports themselves. Appropriate report formats would follow the models displayed in this chapter and might look like Exhibit 4–11, which conforms to the systems analysis process identified in this chapter. Notice the elements included. Remember that the purpose of this report is to assist administration in the decision-making process. The written expression of the analysis tasks and functional documentation elements completed should be carefully edited to ensure the most effective communication possible. Exhibit 4–12 provides a suggested format for verbal presentation, which requires careful planning and preparation to earn the desired result.[7]

Table 4–11 Matrix Comparison of System Candidates for Physician's Incomplete System

Candidate System Evaluation Criteria		Candidate 1 Manual		Candidate 2 Key-Punch		Candidate 3 On-Line CRT System	
Performance:	*Weight*	rate	score	rate	score	rate	score
Accuracy	(5)	5	25	7	35	9	45
Security/ Confidentiality	(2)	8	16	5	10	9	18
Flexibility/ Additional interfacing ability	(4)	3	12	5	20	7	28
Reliability	(5)	2	10	6	30	8	40
Speed	(4)	3	12	5	20	9	36
Storage capacity	(5)	1	5	8	40	8	40
Total Maximum	250		80		155		207

Performance criteria were each assigned a numbered weight ranging from a high of 5 to a low of 1. Each candidate was rated on a scale of 1 to 10, with 10 the maximum attainable. Rates were multiplied by the assigned numbered weight resulting in a numerical score.

Management Evaluation

The end of the analysis phase is a decision time for management. A go-ahead decision based on a firm investigative foundation should be possible. Commitment to the next phase moves the project into the design phase of systems development or into selection of software packages to meet the projected performance.

This text will examine the design phase activities in Chapter 5 and the selection of software in Chapter 6. It is important to note that many design phase activities refine the analysis phase products and, even when packages are purchased, software modifications and implementation must be planned.

Questions and Problems for Discussion

1. Define systems analysis. What is its relationship to problem solving?
2. What are the major phases of a systems life cycle? What activities are carried out in each phase?

Exhibit 4–11 Analysis Phase Report Document

I. Management Summary
II. Problem Identification Statement
  1. Summary of present system operations—defined boundaries
  2. Constraints List
  3. List of users
  4. User survey results
  5. Summary of other investigative tools
                = Summary report of investigation of the present system
III. Performance Definition
  1. Written goals and objectives of required system performance including criteria (These will serve as a basis for evaluation)
  2. Systems flowcharts
  3. Functional requirements description (input, proposed procedures)
  4. System output models
                = Proposed performance requirements necessary to ascertain if a feasibility analysis is appropriate—requires a *Decision To Proceed*
IV. Feasibility Analysis
  1. Operational feasibility summary of proposed computer application
  2. Technical feasibility summary
  3. Economic feasibility summary
  4. Value Analysis charts
                = Answers to questions of appropriateness of computer solutions
V. Conclusions and Recommendations
                —requires a *Decision To Proceed*
VI. Appendix (All support documentation keyed to analysis report category)

---

3. Prepare a report recommending a team role for the health record practitioner in designing and developing a computerized discharge abstracting system. Include a projected list of tasks.
4. What is the role of feasibility analysis in planning for computer applications? What are the specific kinds of feasibility analysis that are performed?
5. What are the elements involved in problem identification?
6. Prepare an organization function list for a transcriptionist.
7. Prepare a decision table for use in determining whether or not to issue a new patient number.
8. How can the MRP participate in user surveys? Give examples.
9. Prepare a list of performance objectives for a computerized record control application for use in a large MRD that serves an outpatient clinic of 43,000 visits per year.
10. How can the MRA prepare for cost/benefit analysis prior to initiating formal systems analysis activities?
11. Prepare a sample cost/benefit analysis report. Use the illustration in this chapter as a guide.

Exhibit 4–12 Format for Verbal Presentation

Verbal presentation
I.   Introduction (¹⁄₁₂ of time allowed).
   A.   State the problem.
   B.   Explain how the systems study progressed.
   C.   Explain the division of the lecture, for example, what follows the introduction.
II. The body of the presentation (½ of total time).
   A.   Describe, very briefly, the existing system.
   B.   Describe the new proposed system.
   C.   Present economic cost comparisons.
      1.   Be careful not to make unsupported claims.
      2.   Be conservative.
   D.   Make recommendations.
III. Summary of the body of the presentation (¹⁄₁₂ of total time).
IV. Open discussion (¼ of total time).
   A.   Answer management's questions.
   B.   Keep lively discussions going; it is more profitable not to cut off a lively question session just because the allotted time has elapsed.
V.   Conclusion (¹⁄₁₂ of total time).
   A.   Summarize the points of agreement that were brought out during the open discussion.
   B.   Make a positive statement as to what you are going to do next, such as get more data, install the system, start programming, or conclude the project.

---

12. Develop an outline for an analysis phase report. Design a poster presentation of the results for an administrative committee meeting.

NOTES

1. *Webster's New Collegiate Dictionary* (Springfield, Mass.: G. & C. Merriam, 1973).

2. Ralph R. Grams, *Problem Solving, Systems Analysis, and Medicine* (Springfield, Ill.: Charles C. Thomas, 1972), p. 3.

3. Kathleen Waters and Gretchen Murphy, *Medical Records in Health Information* (Rockville, Md.: Aspen Systems Corp., 1979), p. 492.

4. Essentials of an Accredited Educational Program for Medical Record Administrators, American Medical Record Association, Chicago, Ill., 1975.

5. Jack D. Harpool, *Business Data Systems, A Practical Guide* (Dubuque, Iowa: Wm. C. Brown, 1978), p. 59.

6. Arda Fitzgerald; F. Fitzgerald; and Warren D. Stallings, Jr., *Fundamentals of Systems Analysis* (New York: John Wiley & Sons, 1980), p. 241.

7. *Ibid.*

Appendix 4–A

Cost Analysis Case Study

A model hospital's "Cost Justification Analysis" is reproduced in Figure 4A–1. Case analysis will be based on these data, bringing in other data from the application or making assumptions where no data is provided.

Initially, it may be noted that savings exceed costs in every year—the return on investment is infinite! Similarly, if a target return of 10% is used, calculation from the seven year stream of "total cost less savings" results in a present value of $432,014. Since there is no initial cost, this investment exceeds the 10% investment criteria (or any other rate we might have set). Indeed, if enough investment opportunities like this were available to the hospital, patients would no longer have to be charged for health care!

To begin a critical analysis, it is first necessary to make explicit the key assumptions:

- Useful life—seven years is used in Figure 4A-1 and it is agreed that this is reasonable.

- Inflation rate—6% is assumed in note #1 in Figure 4A-1. This seems somewhat low but it will be accepted. Other rates are used in notes #2, #3 and #4.

- Installation Schedule—The application indicates five months from start to equipment installation and four more months to "bring up" the system throughout the hospital.

- Benefit Realization Schedule—None is stated although Figure 4A-1 suggests benefits are realized simultaneously with installation. It will be assumed to require three months after completion of installation. Thus, benefits will commence in the second year (nine months installation plus three month benefit realization periods).

- Financing Method—The application suggests "software and hardware" and "installation" are funded by the vendor over the seven

Source: Case study taken from "Health Planning Review of Medical Information Systems" by Melvile H. Hodge, National Center for Health Services Research Report Series, DHHS Publication No. (PHS 81-3303), May 1981, U.S. Gov't. Printing Office, Wash. D.C., pp. 52–56.

year period at a rate in excess of the hospital's local bank credit lines. An informed "guess" is that the hospital's local rate is one point over prime and financing rate is three points over prime. Since the prime was 10% at the time of this application, it is reasonable to assume 11% and 13% respectively.

The analysis method calls for separating out the effect of financing so the financing of "software and hardware" are "backed out" at 13%, leading to a calculation of the purchase cost of $732,628. Similarly, the "installation costs" are $43,028.

Now, it is necessary to consider any omitted installation costs. In the absence of specific information, the following informed "guesses" will be made using Table 4A–1 as a checklist:

Equipment Shipping	$ 1,500
Site Preparation	12,000
Electrical	5,000
Cabling	7,000
Applications development (data tables), (2 man-years @$15,000/m-y, plus 25% fringe)	37,500
Electrical Utilities	2,000
Sales Tax—5%	36,181
Casualty Insurance—1%/year	7,236
Training and installation support (5 man-years @$15,000/m-y, plus 25% fringe)	93,750
Supplies	10,000
Industrial Engineering (1 man-year @$25,000/m-y, plus 25% fringe)	31,250

Next, it is necessary to consider operating costs in years two through seven. "Equipment maintenance cost" will be used as shown. In addition, using the Table 1 checklist, the following operating costs for the second year will be projected. These will be inflated at 6% per year for years three through seven.

System Coordinator (data table mainte-nance), (1 man-year @$20,000/m-y, plus 25% fringe)	$25,000
Electrical Utilities	2,000
Casualty Insurance—1%/year	7,236

Figure 4A–1 Cost Justification Analysis

	Year One	Year Two	Year Three	Year Four	Year Five	Year Six	Year Seven	Total
Software and Hardware	$ 163,620	$ 163,620	$ 163,620	$ 163,620	$ 163,620	$ 163,620	$ 163,620	$1,145,340
Installation Costs	9,729	9,729	9,729	9,729	9,729	9,729	9,729	68,103
Equipment Maintenance[1]	31,332	33,212	35,204	37,316	39,554	41,927	44,442	262,987
TOTAL COSTS	$ 204,681	$ 206,561	$ 208,553	$ 210,665	$ 212,903	$ 215,276	$ 217,791	$1,476,430
Recovery of Lost Charges[2]	$ 109,000	$ 119,900	$ 131,890	$ 145,079	$ 159,586	$ 175,545	$ 193,100	$1,034,100
Reduction of Waste Meals[3]	13,000	14,170	15,445	16,835	18,350	20,001	21,801	119,602
Forms Cost Reduction[4]	9,900	10,890	11,979	13,177	14,495	15,944	17,538	93,923
Discount	6,000	6,000	6,000	6,000	6,000	6,000	6,000	42,000
Reduction of EDP Equip.[5]	2,600	2,600	2,600	2,600	2,600	2,600	2,600	2,600
Increased Cash Flow[6]	20,000	21,800	23,762	25,900	28,231	30,772	33,541	184,006
Personnel Savings[7]	75,000	81,000	87,480	94,478	102,036	110,198	119,014	669,206
TOTAL SAVINGS	$ 235,500	$ 256,360	$ 279,156	$ 304,069	$ 331,298	$ 361,060	$ 393,594	$2,161,037
Total Cost Less Savings	($ 30,819)	($ 49,799)	($ 70,603)	($ 93,404)	($ 118,395)	($ 145,784)	($ 175,803)	($ 684,607)
Cost Per Patient Day	—0—	—0—	—0—	—0—	—0—	—0—	—0—	—0—

Notes

1. Equipment Maintenance is the only variable cost to the system during the seven year contract. The increase is tied to the Consumer Price Index and for the purpose of this comparison is averaged to a 6% per year increase.

2. Recovery of lost charges is based on a conservative 2.5% of ancillary charges with a 10% per year increase. Using national averages of 3–5% for lost charges would result in significantly more savings than those reflected above.

3. Reduction to wasted meal savings represents a projected 9 meals per day at $3.75 each which are currently lost due to transfers, surgery, diet changes, etc. Those will be eliminated as a result of the instant patient diet status communications. An annual increase of 9% was projected.

4. Form costs have averaged and can reasonably be expected to continue to average a 10% increase for the next seven years.

5. This reflects the elimination of the census system which presently costs $.68 per patient day.

6. The system will result in the patient bill being generated four days sooner than is now possible. This will result in an annual cash flow increase of $222,000, invested at 9% per year.

7. Personnel savings represent 6 FTEs with fringe benefits.

Table 4A–1 Cost Checklist

1.0 *Equipment*
    1.1   Computers and peripherals
    1.2   Terminals and printers
    1.3   Communications
    1.4   Interface devices
    1.5   Shipping
    1.6   Storage racks
2.0 *Facilities*
    2.1   Floor space
    2.2   Site preparation
    2.3   Air conditioning
    2.4   Electrical
    2.5   Cabling
    2.6   Sub-flooring
    2.7   Controls, monitors, alarms
3.0 *Software*
    3.1   Operating systems
        3.11 Rental
        3.12 Development/conversion
        3.13 Maintenance
    3.2   Application programs
        3.21 Rental
        3.22 Developmental/conversion
        3.23 Maintenance
4.0 *Maintenance* (labor, equipment, parts and supplies)
    4.1   Computers and peripherals
    4.2   Terminals and printers
    4.3   Communication equipment
    4.4   Facilities (air conditioning, etc.)
5.0 *Utilities* (installation and usage)
    5.1   Electrical
    5.2   Telephone
    5.3   Air conditioning
6.0 *Taxes and Insurance*
    6.1   Sales tax
    6.2   Property tax
    6.3   Casualty insurance
7.0 *Training*
    7.1   Initial training and installation support
    7.2   Inservice training
    7.3   Documentation
8.0 *Supplies*
    8.1   Tapes and disks
    8.2   Printer paper/ribbons
    8.3   Forms and labels
    8.4   Punched cards

Table 4A–1 continued

9.0 *Management*
    9.1 Hospital
    9.2 Facilities management
    9.3 Consultants
10.0 *Industrial Engineering*
    10.1 Implementation
    10.2 Benefit realization
11.0 *Labor Fringe Benefits*
    11.1 Group insurance
    11.2 Retirement
    11.3 Payroll taxes
    11.4 Vacation and holidays

Training (included in hospital training staff duties without staff addition)	-0-
Supplies	10,000
Industrial Engineering (½ man-year @$25,000/m-y, plus 25% fringe)	15,625

The next step is to examine projected savings. The first adjustment will be to exclude all savings in the first year as savings cannot logically accrue until installation and benefit realization occur.

Next, savings from "Recovery of Lost Charges" are deleted. A substantial portion of hospital costs is reimbursed on a cost rather than a charge basis. Lost charges have no bearing on Medicare, Medicaid, and other third-party cost-reimbursement collections. Procuring a medical information system to improve charging precision will increase the cost of health care to the community with no associated increase in health care delivery productivity.

All other projected savings are accepted as presented. It is constructive, also, to examine the "Personnel Savings." Elsewhere in the application these are identified as:

Department	F.T.E. Personnel Reductions
Admitting	2
E.D.P.	2.5
O.P. Registration	1
Radiology	0.5
Total Reductions	6 F.T.E.

These appear quite reasonable for the system proposed. Reductions in admitting and registration may be expected. The radiology reduction undoubtedly is related to more efficient handling of outpatients. The E.D.P. reduction presumably reflects a reduction in key punch operators since charge collection is now automatic and undoubtedly relates to "reduction of E.D.P. equipment," which is probably key punch machines.

Fractional F.T.E. savings are questionable but the assumption will be made that part-time people are being removed from the payroll. Ordinarily, fractional F.T.E. savings are not realizable.

Note that there are no savings in Nursing. Since nursing is by far the largest department in any hospital, typically representing half or more of the work force, major productivity improvement must include major nursing department staff reductions. This is possible only with more comprehensive systems. While no employee data is presented, the subject hospital probably has a work force of one thousand or more. Thus, the productivity impact appears to be less than 1%.

Review of the Benefits Checklist presented in Table 4A–2 suggests that all relevant savings were claimed. It is now possible to adjust the Cost Justification Analysis presented in Figure 4A–1. The adjusted Cost Justification Analysis is presented in Figure 4A–2. Ignoring financing costs, it may be seen that the initial investment of $1,038,405 is not recovered over the seven year useful life but is only reduced to $644,259. Thus, the project has a negative return; that is, it costs more than it saves. Therefore, a review decision must be made on the basis of whether the nonfinancial benefits (e.g., patient care benefits) are worth the increased cost.

To establish this cost, we must calculate the net present value (or cost) using an interest rate approximating the hospital's weighted average cost of capital. It was estimated that the hospital had access to bank credit at 11% and long term lease credit at 13%. Therefore, a reasonable figure for the hospital's weighted average cost of capital is 12%.

Using a financial calculator, the net present cost of this project is calculated to be $804,234. (This calculation discounts the stream of costs or savings back to the present at the chosen interest rate.) The hospital board must thus decide if the patient care benefits are worth approximately $800,000.

This example illustrates how an apparently cost-effective project may not be cost-effective at all. Of course, the estimates made here may be open to debate. Nevertheless, this methodology may prove to be useful to those who must make real decisions.

Risk was not considered in our example. The proposed system is a widely installed system from a reputable vendor. Therefore, it is unlikely that results will differ significantly from our adjusted analysis. The major

Table 4A–2 Benefits Checklist

1.0 *Realizable Labor Savings*
    1.1 Labor
    1.2 Fringe benefits
    1.3 Supervision and management
2.0 *Consumables*
    2.1 Forms
    2.2 Medications and supplies
    2.3 Meals
3.0 *Previous System Costs*
    3.1 Labor and fringe benefits
    3.2 Equipment and maintenance
    3.3 Supplies
    3.4 Services
4.0 *Interest Costs*
    4.1 Accelerated billing
    4.2 Reduced receivables aging
    4.3 More accurate third party claims
5.0 *Capital Facility Costs*
    5.1 Reduced length of stay
    5.2 Improved scheduling
    5.3 Shared facilities

Cost Analysis Strategies

- Identify each hospital function affected by the system (e.g., "admit a patient").

- Flow chart each function using pre-system manual procedures and again, using post-system automated procedures.

- Assign standard times to each element by skill category in each flow chart.

- Subtract the times with the automated procedure from those with the manual procedure.

- Estimate the frequency each function will be performed (e.g., "admissions per month").

- Estimate the cost of labor involved (e.g., "admitting clerk monthly wage and fringe").

- Examine fractional savings in a given department (e.g., "Admitting") resulting from different functions and establish a strategy for consolidating these fractional savings into full time positions which can be eliminated (realizable benefits).

- Establish a time phased schedule of positions to be eliminated, identifying *specific positions*.

- Secure written concurrence of the affected manager (e.g., "Admitting Manager") and the hospital administrator.

- Convert eliminated positions into realizable dollar savings by applying the cost of labor to each identified position and aggregating by month.

Figure 4A–2 Cost Justification Analysis—Adjusted

	Year One	Year Two	Year Three	Year Four	Year Five	Year Six	Year Seven
Costs							
"Software and Hardware"	$723,628	—	—	—	—	—	—
"Installation Costs"	40,028	—	—	—	—	—	—
"Equipment Maintenance"	31,332	$33,212	$35,204	$37,316	$39,554	$41,927	$44,442
Equipment Shipping	1,500	—	—	—	—	—	—
Site Preparation	12,000	—	—	—	—	—	—
Electrical	5,000	—	—	—	—	—	—
Cabling	7,000	—	—	—	—	—	—
Applications Development	37,500	25,000	26,500	27,825	29,495	31,264	33,140
Electric Utilities	2,000	2,000	2,120	2,247	2,382	2,525	2,676
Sales Tax	36,181	—	—	—	—	—	—
Casualty Insurance	7,236	7,236	7,236	7,236	7,236	7,236	7,236
Training and Install Support	93,758	—	—	—	—	—	—
Supplies—forms	10,000	10,000	11,000	12,100	13,310	14,641	16,105
Industrial Engineering	31,250	15,625	16,563	17,556	18,610	19,726	20,910
Total Costs	$1,038,405	$ 93,073	$ 96,623	$ 104,280	$ 110,587	$ 117,319	$ 124,509
Cumulative Cost	$1,038,405	$1,131,478	$1,230,101	$1,334,381	$1,444,968	$1,562,287	$1,686,796
Savings							

Systems Analysis in Medical Record Settings   191

"Recovery of Lost Charges"	—	$14,170	$15,545	$16,835	$18,350	$20,001	$21,801
"Reduction of Waste Meals"	—	10,890	11,979	13,177	14,495	15,944	17,538
"Forms Cost Reduction"	—	6,000	6,000	6,000	6,000	6,000	6,000
"Discount"	—	2,600	2,600	2,600	2,600	2,600	2,600
"Reduction of EDP Equipment"	—	2,600	2,600	2,600	2,600	2,600	2,600
Increased Cash Flow	—	21,800	23,762	25,900	28,231	30,772	33,541
"Personnel Savings"	—	81,000	87,480	94,478	102,036	110,198	119,014
Total Savings	—0—	$136,460	$147,366	$158,990	$171,712	$185,515	$200,494
Cumulative Savings	—0—	$136,460	$283,826	$442,816	$614,528	$800,043	$1,000,537
Total Costs Less Savings	$1,038,405	($43,387)	($50,743)	($54,710)	($61,125)	($68,196)	($75,985)
Cumulative Cost Less Savings	$1,038,405	$955,018	$904,275	$849,565	$788,440	$720,244	$644,259

Cost and savings in quotation marks are items contained in the original analysis (Figure 4A–1). The other costs and savings items are added as described in the text.

risk is likely to be failure to realize the 6.0 F.T.E. personnel reduction which accounts for a majority of the projected saving. Therefore, as part of a review of this project, a written commitment by the cognizant department managers and the administrator to eliminate the targeted positions on a stated schedule should be required.

In examining a less proven system, it would be necessary to examine the effect of, say, a one year slip. Installation labor costs would continue for year two and savings would not commence until year three. Using the figures from this example, one would conclude the "downside" was perhaps a quarter of a million dollars associated with a one-year schedule slip.

REPLACING INEFFECTIVE SYSTEMS

The method suggested above requires comparison of costs and savings projected for a proposed new system with hospital costs prior to its installation.

To illustrate, consider the extreme case where a hospital has installed a system costing $1 million per year and producing no benefits. The hospital now prepares to replace that system with a new system costing $500,000 per year and producing no benefits. Using the suggested methodology blindly will lead to the result that the new system will save $500,000, and hence, is cost effective. Common sense suggests that an even better plan would be to throw out the old system and reject the new one—and save $1 million!

If circumstances are encountered where this problem may be present, it can be easily avoided by a second economic analysis using the hospital *without* the existing medical information system as the baseline for comparison with the proposed new system.

PARTIAL SYSTEMS

Occasionally, the reviewer must consider an application to review replacement of a portion of a system or addition to an existing system rather than a totally new system. The procedure is identical to that which we have used for a complete system.

Keep in mind, however, that *marginal* costs and *marginal* savings should be used; that is, only the *changes* in costs and savings. These marginal costs and savings are then related by a return on investment or net present value analysis.

CONCLUDING NOTE

The reviewer analyst must remind himself that he is attempting to estimate future results. Despite use of six or seven significant figure numbers and sophisticated analytical techniques and calculations, the precision of the analyst's results are still largely limited by the validity of estimates and assumptions.

The real issue before the analyst is whether the project is going to reduce or increase health care costs for the community. If they will increase, some estimate must be made of how much, which can then be related to the nonfinancial benefits. In the case study, it was concluded that the decision makers must decide if the patient care benefits over seven years are worth $800,000 to the community. Failure to focus on this central question often results in the analysis being ignored when the decision is made.

Chapter 5

How To Participate in Design Phase Activities

Objectives

1. Define and describe the design phase of a systems life cycle.
2. Relate the purpose of the design phase components to the overall development of computer applications in health information settings.
3. Describe the actual and potential role of the MRA and other health record practitioners in design phase activities.
4. Define and describe the major elements of the design phase and their relationship to appropriate activities coordinated by the MRA.
5. Define and describe design phase products and activities and illustrate their use by health record practitioners.
6. Present the potential impact of alternative design techniques in health information systems development.
7. Define and describe documentation requirements in computer applications ranging from computer system documentation to user procedures documentation.
8. Explain the role of the design phase report and its relationship to the MRA's responsibilities as a user-manager.

The purpose of this chapter is to prepare readers to continue to take an active part in systems design and development throughout the design phase. Design is the second major phase in the systems life cycle. During this phase, a design team validates the information and application requirements that were identified in the analysis phase. Team members confirm the initial performance requirements that have been specified by the user-managers. They refine these requirements to develop additional, more detailed functional requirements so that systems outputs, inputs, file structures, and programming needs can be mapped out.

Additional actions that take place in the design phase involve allocation of resources, determination of hardware and software requirements, and

195

identification and assignment of personnel. Documentation needs are also determined in this phase. Documentation describes the operational aspects of a computer system and provides detailed descriptions of the software. It includes user instruction guides. The guides delineate what will be prepared for the people who will be performing the newly computerized operation on a day-to-day basis. Like an architect's drawings, documentation should provide a comprehensive picture of the computer system.

If not yet selected, a project manager is assigned in this phase. The project manager assumes responsibility for the remaining project planning and activities.

Health record practitioners have an opportunity to continue an active user-manager role through strong teamwork in the design phase. Table 4–1 and Table 4–2 illustrated the continued participation in this phase. This chapter will explore these activities as they apply to MRDs and other related patient information systems in health facilities.

Using the information contained in this chapter, the reader should be able to prepare a written strategy that outlines tasks to be performed by health record practitioners and other user-managers in maintaining a strong contributory role in systems development during the design phase. (See "Design Phase" portions of Tables 4–1 and 4–2.)

DESIGN PHASE OVERVIEW

The design phase is initiated after the feasibility analysis has been completed and a particular system description outlined. This phase of systems analysis and development is made up of three major steps:

1. Confirming performance requirements
2. Preparing detailed design requirements
3. Identifying documentation needs for system performance and maintenance

Confirming Performance Requirements

Performance requirements are confirmed by reexamining and validating the problem statement and resolving the design issues that have been uncovered in the analysis phase. Recall that the analysis phase results are compiled in a formal analysis report. To ensure that the original problem as identified and documented in the initial system investigation is still a valid expression of user needs, a formal review is performed by the systems design team and the health facility management. This important review process includes a critical look at the proposed application's performance objectives and desired outputs. The outputs are actually information products that must be designed for each unique application. Each product's characteristics must be evaluated before the product can be created. Similarly, input forms and formats also include information characteristics that will need to be considered. These are termed design issues. Design issues can be defined as unresolved questions about the form and use of information or data elements that apply to the application under consideration. In general, design issues can be identified through review of the performance objectives established during the analysis phase. Recall that performance objectives are the specific planning method that outlines and describes the purposes of particular departments and individual operations. Well-written objectives clearly identify what is expected of the operation. In the following example the objective clearly states what the output should be:

> *Objectives of a Medical Record Tracking System*
> The system should provide the ability to input a medical record request on a CRT; have the current location of the record checked automatically against the computer file; and have the request and location displayed on the screen and simultaneously reproduced on a computer-generated requisition form.[1]

This objective informs the design team that screen displays will be required to display record requests and record locations. It also states the need for designing a computer-generated requisition form.

A design team may review each performance objective with user-managers to specify the required outputs more clearly. The characteristics of these outputs will be evaluated as design issues. It will be seen that this evaluation results in more exact expressions of the application requirements. Consequently, more precise system requirements can gradually be developed so that the technical design tasks can be performed as efficiently as possible.

During the first step of the design phase, medical record practitioners as user-managers are integral members of the systems design team. They serve as advocates for the application itself. Unless this responsibility is assumed, decisions about the project will be made by people who are not trained in patient record system principles and theory. Further, the failure to represent the user can seriously impair the project's design efforts at the outset. It is a good policy for practitioners to plan to participate whenever patient information systems are under development.

Outlining Detailed Design Requirements

The second major step in the design phase is the preparation of detailed design requirements. Building on information gathered early in the project, the design staff typically refines and expands on a written, detailed outline of the outputs, inputs, file structure, and programs that are necessary to develop the application. This is a technical activity. It translates the needs identified in the analysis phase into more specific detail that is needed by the system developers. In this activity the internal processing, data flow elements, and computer operations necessary for the proposed application are laid out. The analysis phase was directed to investigating and understanding the user environment to best describe the proposed application in user terms. Efforts were made to express the information needs and the functions of the desired application in general user language. Recall Table 4–4. It lists tools that were used to explain and document data and information flow. A primary goal was to ensure that all participants understood functions and processes in nontechnical terms and that the proposed application was fully described at a general, operation level. In this phase specific technical approaches are determined. Here again, systems flowcharts will be prepared and additional data relationship charts will be used. However, these will be employed so that the internal, technical components of the system can be identified and plans for developing or acquiring them can be formulated.

Identifying Documentation Needs

The third step in this phase is identification of documentation needs for the project, including

- System documentation, which is a written, technical description of the operational aspects of the overall computer system architecture and operation
- Program documentation, which is a record of all the programs that will be written to run the application
- User documentation, which is the procedure guide for the application users to follow in the daily operation of the application tasks

Documentation is prepared in manuals or stored in automated text files. It should contain clear descriptions of the system and discussions of its functional components. It should be available for all concerned personnel when the system is completed. Like a schematic chart of a piece of machinery or a detailed map of a geographical location, documentation is essential for adequate understanding, operation, and maintenance of a computer system. An illustration of documentation is seen in the excerpt of the PROMIS system in Exhibit 5–1.[2] Notice how the information explains some of the very structure and components of the systems features. Given this documentation, would users be more or less prepared to work with system features?

Design Phase Philosophy

Design phase activities are primarily directed by the scope of the project under consideration in one of two ways. The activities may be integrated with a master plan for overall computer systems development for the organization. This approach promotes coordination of general systems development tasks with any additional applications planned or under development. In an alternative approach, the tasks may be performed for a stand-alone application that requires less coordination with other projects in the organization. Decisions about these and other design phase matters will usually be made by the project manager and the design team. This text supports the philosophy of project integration for the institution as a whole. This philosophy will be explained in greater detail in Chapter 7. Meanwhile, a more detailed look at the design phase elements will help the reader understand the complexity and potential of this phase in the computer systems life cycle.

EXAMINING DESIGN PHASE ELEMENTS

Design Issues

The design issues relevant here are unresolved questions about the form and use of the information to be processed in a particular computer application. There are general and specific information issues related to input, output, file, and program design. As general issues are often easily addressed by users, medical record practitioners need to become familiar with them. Specific issues of technical system design are dealt with in the references. However, some key information and data issues are discussed here. These are[3]

- Data volume reduction
- Computer-stored tables
- Program calculations

Exhibit 5–1 PROMIS System Design

III. THE SYSTEM DESIGN FOR PROMIS

A . Technologic characteristics:

A PROMIS system designed to help solve four major problems facing medicine today (lack of coordination, failure to record logic, dependence upon human memory and lack of meaningful feedback loops on the everyday practice of medicine) must have three technologic characteristics:

1) Responsiveness: It must be very responsive and allow the information originator to be directly interfaced using a touch sensitive CRT computer terminal;

2) Reliability: It must be very reliable so that it is continuously accessible and usable as an information utility; and

3) Access to large files: It must allow access from any terminal tied to the system (if confidentiality allows) of all patient records kept within the system and of a large, structured, medical information data file used for the generation of the patients record and to provide the medical guidance in decision making.

These three characteristics are the major architectural elements of a PROMIS system.

B. Design characteristics:

The characteristics of the design approach followed by PROMIS Laboratory to implement this architecture are:

1) Build the system so it is modular and scalable (i.e. expandable or contractible).

Source: Reprinted from *Medical Records in Health Information* by Kathleen A. Waters and Gretchen Frederick Murphy by permission of Aspen Systems Corp., Rockville, Md., © 1979.

Exhibit 5–1 continued

2) Build the system so it has minimum redundancy yet under various types of failure can still operate.

3) Build the system out of reasonably inexpensive components (i.e. mini-computers) that are technologically updatable without having to modify the application software.

4) Build the system so that the difficult systems problems can be solved (i.e. the access to the distributed POMR file) while the system is still reasonably small and well understood.

5) Build the system out of "off-the-shelf" components or at least "mostly-off-the-shelf" components. (If anything must be built that is not now a standard product, get it from a manufacturer who will make it a standard product.)

6) Build the system using recognized software development approaches which increase programmer productivity, decrease debugging time and allow all software to be maintained by a team instead of the individual who originally wrote it. These approaches include language processors that allow "structured" (go-to-less) programming and the use of decision tables as an integral part of the program specification process.

7) Build the system so that all application and medical content programming is done in a higher order language that has machine independent characteristics. This will help maintain the software base if the hardware base is changed.

8) Build the system so that all software tools required to compile, edit and debug all application software are available within the same hardware and operating system base as the health care delivery systems base. This is a corollary to "top-down" program development (where commonly used pathways are exercised very early in the software development cycle) since the commonly used pathways in the total system are used to get the application software developed. This helps increase the reliability of the system very quickly.

Many of the above characteristics should enhance the exportability of the final system.

Exhibit 5–1 continued

C. The major elements of the new system design—the PROMIS technology.

1. general information

The new technologic approaches at the highest levels involve multiple nodes tied together in a (limited) network. The network is limited in the sense that the functions to be performed are very well defined. Each free standing node (VARIAN-75 CPU with associated peripherals) will handle the frames (i.e. the medical content to be displayed) and medical record files for the segment of the health care facility(ies) served by its terminals. Internode communication will facilitate retrieval and storage of patient information from the other segments of the health care facility(ies) served by other nodes. In the event of a node failure another node will be able to pick up the failed nodes functions (i.e. by switching terminals and moving the patient record files).

2. increased capacity

The technologic approaches employed using the latest hardware technology allow much greater capacity than the CDC 1700 system.

a) The central memory of the V-75 (VARIAN-75) can be increased up to 1 million words (maximum on CDC was 64,000).

b) The secondary memory (disk storage) can be increased up to thirty-two 46 million character disks (maximum on CDC was 45 million characters).

c) The network approach allows for modular growth up to several hundred terminals (CDC limit was 28).

3. increased flexibility

The design of the new PROMIS Programming Language (PPL) and supporting software environments allow much more flexibility in the programming of applications. For example:

Exhibit 5–1 continued

> a) Limits on size of patient record on the CDC system, of order and problem list files are virtually eliminated.
>
> b) "Memory buffer pool" design allows utilizing more central memory without reprogramming for increasing responsiveness or permitting more terminals to be added.
>
> c) PPL frees programmer from most of the concerns about physical characteristics of data, allowing concentration on the medical logic.

4. distributed data base improvements

    The distributed patient POMR record design allows tailoring of the system capacity to handle a wide range of loads, while maintaining reliability and continuous access. The frames and programs are duplicated on each node with this approach and the medical records are located where most heavily used across the network.

5. CATV transmission system improvements

    The PROMIS terminal communications system allows installation of terminal taps in many locations and terminals can be plugged into any tap. This permits great flexibility in system usage—including portable terminals for "rounds." The communications medium is a standard two-way CATV network which can be used for entertainment TV, closed circuit TV, paging and other communications simultaneously with the computer-to-terminal signals and the computer-to-computer signals. See Appendix 10.

6. PROMIS terminal improvements

    The PROMIS terminal allows a more flexible touch-screen format (no longer a rigid two column screen), greater screen capacity (1,920 characters instead of 1,000 as on the CDC) and more levels of enhancement of the displayed text (upper and lower case plus intensified video and inverse video instead of only upper case on the CDC). (See Appendix 9)

- Streamlining input
- Validation and edit checking techniques
- Scheduled versus on-demand outputs

Data Volume Reduction

One common design issue is the need to reduce the volume of information handled. In the analysis phase user-managers identify expected output. This output can be evaluated and consideration given to ways to reduce the data volume. This includes reviewing source documents and data entry documents to identify methods for streamlining data. For example, codes should be used to represent information whenever possible.

Discharge codes in a discharge abstracting system can represent the discharge category, as in the following list.

Discharge Disposition Code Plan

01 Home self-care
02 Home health care
03 Skilled nursing facility (SNF)
04 Intermediate care facility (ICF)
05 Other hospital
06 Left against advice
07 Expired/no autopsy
08 Expired/autopsy

Consider other data fields on a Uniform Hospital Discharge Data Set (UHDDS) or a risk management screening abstract. How can other individual fields be designed to capture data in a similarly reduced fashion? We will see how other design issues can help answer this question.

Computer-Stored Tables

Another design issue is the need to identify information that can be produced by using a computer-stored table. For example, a physician index can be produced by using computer-stored physician identification numbers that users will see as names displayed on a CRT screen or as listings on printouts. These can also be translated to specified reports as needed. Computer-stored physician ID numbers can be used for references in chart locater systems and for physician identifier codes in risk management and quality assurance screening systems. In the same fashion, in a computerized chart deficiency system, numbers can be assigned to record deficiency categories such as signatures or complications. These can be stored in the computer for more efficient use, where they would be listed as follows:

Chart Analysis Code Plan
01 Signature missing
02 Report missing
03 Report undictated
04 Cosignature required
05 Audit screen flag

Additional examples of computer-stored tables appropriate for medical record applications readily suggest themselves. The more experience practitioners gain in working with computer applications, the more creative they will become in applying these design principles.

Program Calculations

Another design issue deals with identifying calculations where information can be produced through a specific program action by the computer. The character of such a "data-driven" calculation is determined by the type and amount of the values in the data. Consider a hospital census system for example. When all admissions have been entered and the computer is notified that the nth entry is the final admission for the time period in question, this signal will initiate the automatic calculation of the census data. This calculation can be a complex function: the attributes of each admission will produce the admission list and appropriate notifications to departments. Until the computer has received this unique data item, the specific calculation cannot be performed.

Streamlining Input

Another design issue is a critical need to evaluate streamlined methods for entering data elements into the computer. Operations that are currently performed manually are evaluated to determine how they might be streamlined to provide efficient integration into a new automated operation. For example, a clinic appointment scheduling system may use a copy of the patients' appointment slip to notify the medical record file room of charts to be pulled for clinics. The sequence of the information listed and the actual data elements on that slip may be streamlined for more efficient data entry into a computerized scheduling system. (See Figure 5–1.) By changing the manual system appointment slip to a more streamlined version, system requirements are more efficiently met. The simplified data

Figure 5–1 Streamlining the Appointment Slip

```
                      number
        patient name _____
        clinic name _____
        appointment date _____

        This is to notify you that

        _____ has an appointment in the above
        clinic at the date and time noted
```

a. Appointment slip used in manual system

```
        ┌─┬─┬─┬─┬─┬─┬─┐
        └─┴─┴─┴─┴─┴─┴─┘

        patient number (terminal digit)
        patient name _____
        clinic number ☐☐      time          date

        01 medical        etc. . . .
        02 surgical
        03 end
        04 eye
```

b. Appointment slip used in automated system

elements are organized to match a coded, data entry format. Such a format may correspond with a punchcard design subsequently used to keypunch the data.

When an on-line scheduling system is established, these data elements could become part of a screen format in which the appointment information is entered. In this case, the applications may include a printout notification to the patient, the clinic, and to the file keeper in the medical record department. How could data input be streamlined in building a computer application for an MPI?

Validation and Edit Checking

Validation and edit checking are also design issues. Validation must be controlled by users who are responsible for the quality of the data. A hospital may have the admitting department personnel register patient data into the hospital MPI. The staff in this department do not routinely work with the MPI for identification inquiries. They may not have access to previous information or be allotted time to investigate and validate the data. Consequently, a common problem in such a setting is duplicate number assignment. Control and validation of the information that enters into the system is a very important design issue. Controlling and validating disease and operation coding procedures in acute, ambulatory, and long-term care settings is another major concern for medical record professionals. Automated coding systems will have significant impact on data accuracy and use and will become major features for MRD considerations. Other general design issues are

- Determining retention requirements of the data processed by the computer system. As incidence of computerized patient records increases, this issue will become increasingly important.
- Identifying access frequency for providers and other users. Who needs to access the new computer system? We will explore this issue in a discussion of output demands in the following section.
- Determining permanent data storage versus variable data storage. A disease index may be stored for five years, whereas computer-stored statistical reports may be purged at the end of the fiscal period. Each application will have different needs.
- Specifying edit checking procedures to be carried out by the computer. Examine Table 5–1 to see what this entails.

Consider how these issues would be reviewed in designing an in-house chart deficiency system. What recommendations could be made? Each of the general design issues requires consideration by both user-managers and systems analysts in order to ensure comprehensive construction of a project document. The list of data checking in Table 5–1 clearly illustrates how individual data elements can be analyzed to identify effective methods to maintain accuracy and completeness.

Scheduled versus On-Demand Outputs

The last general design issue to be discussed here is determination of the access and production of the application outputs. The users must indicate the access needs for each case. One of the design team's tasks is to review the output schedules with various users to be sure the expectations are necessary. One feature of computer applications and users is the phenomenon of creative expectation in which "wish" lists are not distinguishable from essential output needs. Table 5–2 compares scheduled and on-demand output categories. As responsible managers of patient information systems, MRAs will be constantly involved in resolving application outputs according to these categories. Skills in human relations and com-

Table 5–1 Data Checking Methods

Sequence Test

This operation determines whether the input records are arranged according to a predetermined sequence. In such cases, the program will check each record to ensure that all are ordered by a particular sequence. If one record is missing, then normally all records that follow will be considered as errors. For example, in a radiology system, a certain set of numbers, ascension numbers for instance, can be assigned for use on one day's x-rays. The numbers would be sequential and inclusive—from 150,888–160,000, for example, so that on a given day, numbers 150,888 and all numbers up to and including 160,000 should be used. At the end of the day the computer can provide the billing department with a list that verifies numbers assigned. It can also display billing data, but no numbers, thereby raising the question as to whether the number was used or not.

Completeness Test

This operation determines whether a record has the proper number of fields contained on the input form. For example, on an MPI entry display, a validation routine checks to see whether all five fields are completed. A value must be indicated for each defined field. This kind of testing could be carried one step further to ensure that each field contains only valid characters as indicated in the format below:

Field	Record Positions	Remarks
Patient number	5–11	Must be numeric

If an invalidity were detected, an error message could be displayed on the screen.

Existence Test

This test makes certain that a certain field does exist in the input record. For example, if admission transactions are being processed with discharge transactions, the unique transaction code should be checked for that type of record. In this case, D records (discharge) could then determine which fields in the census file master are to be updated. A records (admits) would be likely to update the census file master somewhat differently. Other existence tests are to check certain fields for alphabetic characters, such as a name field.

Automatic Correction

A program can be designed to generate error free data automatically. Assume that a predetermined field is blank. The field position may be filled with blanks or zeros. Such automatic correction procedures can be utilized by designing the program to emit constants onto a valid output record. A report message showing such action should be produced as output.

Reasonableness Test

This test checks a field value against a predetermined criterion. In a utilization review system, an "upper limits" test of a predetermined maximum number of days could be considered a reasonableness check for admission reviewed.

Table 5–1 continued

Combination Test
This determines whether a known relationship exists between two or more fields. If one value exists in the transaction, then the other value must also exist in the record. For example, for codes designating an abdominal hysterectomy procedure in one field, there must be a field value of 2 (female) in another specified field.

Limit or Range Check
This is applied to both the input data and the output results of the computer. Each transaction field to be tested under this criterion must fall within a predefined limit of minimum and maximum values. For example, hours worked must be no less than 20 but no more than 59.

Reformat Option
When the source document cannot be conveniently redesigned, the validation program may reformat the input to conform to multiuser output needs. In this way the output record will better meet format requirements of other programs using it.

Expansion Option
Input data records must often be enlarged. A typical record may be expanded to a larger record length to accommodate additional fields that may be required in further processing. Most key-to-tape and key-to-disk devices have this capability.

Valid Key-Check Digit Test
Check digits are used to detect errors that occur when numbers are transcribed from one document or medium to another. This technique assures that field values are valid by applying a check digit to the base number. This check digit is a single digit that is initially computed, assigned to the number and then checked in the validation program.

Source: Jack D. Harpool, *Business Data Systems: A Practical Guide* (Dubuque, Iowa: Wm. C. Brown, 1978).

---

munications are valuable assets in these matters. An awareness of effective data use is another.

Continuing Problem Reconfirmation

When these and other design issues have been reviewed and suggestions made for dealing with them, the results provide another look at the problem. It is important for ongoing review to be performed so that designers and users can have the opportunity to communicate. All participants need to understand clearly the developments of the system. Is the original problem still the target area? In too many projects, individual team members or alternate system users have tried to incorporate additional or alternative components that have resulted in failure to solve the original

Table 5–2 Output Alternatives

Scheduled outputs
Outputs that are scheduled for production at regular intervals. The daily census report and daily admission and discharge lists are examples.

Unscheduled (on-demand) outputs
Outputs that are produced as needed. They are of three types:
1. Outputs with predesigned formats but not routinely scheduled for production, like audit screening reports.
2. Outputs without predesigned formats that may require programming to produce.
3. Outputs that may be produced by user–computer interface, such as report generator programs that work off a patient data base. Other examples include statistical packages that are frequently available on time-share systems and administrative report programs.

---

problem. Periodic problem review helps protect the project's integrity. Critical design efforts and findings are added to the analysis phase findings so that a revised product is then available for user and management review. This is a critical time for medical record practitioners. They must take this opportunity to validate their own needs as identified in the analysis and design process and to document unresolved problem areas. In this way they may begin to keep a log of potential design issues for use in future systems revisions and developments.

Constructing Detailed Design Requirements

Constructing detailed design requirements is the second major element of the design phase. Like blueprints for builders, these are the maps for programmers. They are used in writing the programs and constructing the unique features of the system. Although change should be regarded a constant, these requirements will likely be established as a baseline from which general project direction will be initiated. Also included would be interim record outputs expected to be incorporated in an ongoing portion of the patient's record.

Begin Detailed Design Specifications with Output Design

The project manager must see to it that the form of the final output and details of the input form have been specified formally. System designers will map out the appropriate file design and project the program design requirements as well. User-managers must examine the finalized version of output and input to determine if it satisfies the objectives originally identified in the project. Illustrations of outputs for several medical record applications were featured in Chapter 4.

The more specific users can be in stating what the system outputs should be the more likely they will be satisfied with the end products of the system. Outputs, the visible end products such as reports, displays, and so on, can be further broken out into external, internal, and user categories. External outputs are any routine products made by the computer system plus any special inquiry products. In a computerized clinical lab system, patient lab report summaries generated by the system would be an example of external reports.

Internal outputs are reports on the activity of the system itself. These are primarily used for the benefit of the data processing department. However, some internal reports will provide feedback on the use of the particular application programs and, as such, are valuable to managers in various application areas. The third kind of output is user output. This serves as a clerical or auditing process and deals with information specifically for the user of the system. It can be limited to those at the operations level such as the data entry clerk. The clerk could get a monitor report on the error rate for data entry. In a word processing system, this might be a monitoring report provided for each terminal operator on the speed or volume handled by their terminal. In the medical field it is useful to extend this concept. It can include the notion that user reports should also be extended to include auditing features for clinical users and the managerial users. Such reports might report on the activity of the application to monitor how well it appears to meet the needs of physicians, nurses, MRAs, clinical lab managers, and others.

Input Design

Many of the input design considerations have already been discussed in the look at design issues. It is important, however, to identify and focus on input so that we coordinate all system input needs. The detailed design specifications for input prescribe the precise methods by which computer systems input will be accomplished. This may be by establishing source documents for admissions staff to fill out. It may be by designing screen displays requiring users to enter appropriate information into the computer for subsequent use. Let us pause here and examine the steps in designing CRT screen displays.

Steps in Designing Video Display Screen Formats. Screen displays will be used for both input and output design. Specific guidelines for designing screen displays are featured here to assist practitioners in preparing screen formats. The CRT screen is the window to the computer and is the medium by which people communicate with the computer about their unique information. Consequently, medical record practitioners often have to design screen formats. The following steps are required to design screen display:[4]

1. Determine exact screen capacity by outlining its representation on a grid or printer spacing chart. Consider full- or split-screen options.
2. Determine the primary purpose of the display. Identify and write out portions of the display that can be handled through a second page. What do you want the display to look like for the operator?
3. Evaluate the information that will be displayed in terms of exception reporting that focuses on specifics such as abnormal x-ray or record charge-outs outstanding. For example, a lab report displayed on a nursing station terminal may indicate outstanding orders in an order entry system with the message "No return on items X, X."
4. Carefully break down the information elements required on the displays through character analysis. Notice the documentation and identification on the next page. The precise use of this breakdown is necessary in order to plan the most cost effective and efficient use of the computer storage and retrieval capabilities.
5. Just as you carefully break down information elements as to size and form, the screen display designer must also consider the use of coded information to communicate better use of the screen.
   Be sure to consider screen display formats just as you do forms themselves, making sure to review and update them periodically based on users' experience with them. Most computer systems will be considerably rewritten at three-year intervals, at least, due to information changes.
6. Draft samples of your format, which is now converted into screen displays. This description of your end product should be as exact a picture as possible.
7. Circulate your draft or sample among users. Review the response to be sure that essential user needs such as ease of use are met.
8. While the draft is circulating for review among users, the projected display should also be reviewed by the programmer-analyst for technical feedback. What further breakdown will be needed? For example, the programmer may want to know where the information originates.
9. Be sure and include data input staff for consideration of operational impact for both data entry and data updates. Anticipate minor changes to the system soon after implementation.
10. Consider building in two or three sample displays that may be used for training new staff. This can be an excellent practical aid in your

training operations. Design a test system and use it as a training module.

General System Design Requirements

Design phase activities carried out thus far should be resulting in written requirements as each step is accomplished. These written materials should be organized into a systems requirement document. This document should be the combined product of the analysis phase report, resolved design issues, problem reconfirmation, and detailed output, input, file, and program requirements. Table 5–3 is one model of a design specification document.[5] Compare this model with the functional requirements document

---

Table 5–3 Design Specification Document

1. Statement of system objectives
2. Output specifications
   a. Purpose
   b. Content
   c. Format
   d. Distribution and reporting schedule
   e. Estimated volume
3. Input specifications
   a. Sources of data
   b. Forms
   c. Procedures for converting to machine-readable form
   d. Coding systems
   e. Schedule and estimated volumes
4. Master file specifications
   a. Content and format
   b. Access requirements (including data privacy protection)
   c. Estimated volumes and file size
   d. Updating and purging schedules
   e. File security procedures
5. Procedures and data flow
   a. Flowcharts
   b. Narrative description
   c. Computer program specifications
6. Cost/benefit analysis
   a. System development costs
   b. System operating costs
   c. Maintenance costs
   d. Anticipated benefits
7. Management approvals (signatures)
   a. Operating departments involved (users)
   b. Hospital administration

outline, Table 4–5. Notice how the elements compare. Remember this has been carefully constructed throughout the analysis phase and refined in the design phase activities. This model is an outline containing the key items to look for in the design phase review process. The MRA should be working toward developing a document that addresses the content of this model. Before full-scale design requirements can be drawn up, all of the information components in this outline need to be completed.

File Design Requirements

File design requirements relate to the physical storage of data and programs. Determining them is a task of the technical design staff. In particular, the process is concerned with

- Development of a data model
- Cost of the storage medium
- The complexity of programming a designed system
- The volume of information the medium can conveniently hold
- The access methods possible, and
- How the file is to be used in the system

It includes determining data content, reducing field size, and determining the sequence of data. How medical record practitioners translate this in order to provide input for systems designers in developing file design is highly dependent on the user's information needs. The following case example illustrates how users' roles affect file design in computer systems.

Automated Ambulatory Medical Record System
The medical record file is primarily a file where clinical data is stored. Subsets of data are moved from the magnetic tape stored medical record file to an on-line medical data base for processing. Because of its storage function, the automated medical record file will be required to contain a large volume of data. Because of the nature of medical records, the file will be of variable length. (Consequently, the patient information can be stored on a tape and made accessible by reading the data from the tape onto a disk.) This allows the large volume of data on the tape to be maintained and accessed in a less expensive mode than using disk storage alone. By combining tape and disk capability, access is available at all times and a large volume of data can be stored for the fifteen hundred ambulatory patients in the system. The data base files that are stored on disks are used for manipulating data. In one of these files, a stored data description file acts as

an index for the medical record file for random access to the file. This means that the stored data description file, for example, might be maintained on line (accessible to the user through a CRT terminal) while the medical record for an individual person might be kept on a tape.

The characteristics of the system just described and the fact that subsets of data are transferred from the tape medical record file prior to processing can reduce the complexity of programming in this design illustration. The selection of the file structure permits changes in files' contents without concurrent changes in the programs that operate on them. Common utility programs can be used to edit, store, sort, index, report or otherwise manipulate the contents of the data base files or patient records. Cost savings occur because the programs are easier and faster to write. The reader can see how expected use of patient data can affect file design issues. Medical record practitioners should have a general understanding of what is involved in file design. While the technical expertise resides with the data processing professional, practitioners should be able to determine data content appropriate for their unique applications. A method for identifying and projecting data content needs is explained in Chapter 7.

In the example used to illustrate factors in file design, one can also see how program design, which is the fourth area of detailed design requirements, will be affected by the expected use of the systems. This means that user-managers need to understand enough about program design to be aware of the impact of certain kinds of programming techniques on the use of their information.

Program designing is based on the use of information. When health information is in machine-readable form, it can be manipulated in many ways. The same processes used for manipulating noncomputerized data apply to computerized data. Exhibit 5–2 specifies ways of handling information and illustrates the kinds of functions performed by computer programs. It can be said that medical computing is moving toward real time applications in which users can expect to interact with their information on an as-needed basis with immediate updates of information. We commonly expect this capability in an on-line census system. The role of program design and the use of data base methods must therefore become part of the knowledge base for clinical and other health care practitioners.

Documentation Needs

The third major component in the design phase is identification and validation of documentation needs for the system. This means identifying

Exhibit 5–2 Examples of Performing Data Operations with Various Data Processing Methods

Methods \ Operations	Capturing and Verifying	Classifying	Arranging	Summarizing	Calculating	Storing	Retrieving	Reproducing	Disseminating and Communicating
Manual Method	Voice; observation; handwritten records; forms and checklists; writing boards; pegboards	Hand posting; coding; identifying pegboards	Alphabetizing; indexing; filing; edge-notched cards	Hand calculator	Human calculation; pencil and paper; abacus; slide rule	Columnar journals; ledgers; index cards; paper files	File clerks; stock clerks; book-keepers	Hand copying; carbon paper	Handwritten reports; hand carried or mailed
Electro-mechanical Method	Typewriter; cash register; autographic register; time clock	Posting machine; cash register; accounting machine	Semi-automatic (roto-matic)	Adding machine; calculator; cash register; posting machine	Accounting machine; adding machine; calculator; cash register; posting machine		Mechanical files (rotary or tub files); microfilm	Duplicating equipment (carbonization, hectograph, stencil, offset, photocopying, thermograph); addressing equipment	Telephone; teletype; machine prepared reports; message conveyors; hand carried or mailed reports
Punched Card Equipment Method	Keypunch; verifier; marksensed cards; prepunched cards; machine readable tags	Sorter; collator	Sorter; collator		Accounting machine; calculator; summary punch	Card trays	Sorter; collator; hand selection	Reproducer; interpreter	Same as above
Electronic Computer Method	Keypunch; verifier; paper tape punch; magnetic encoder; optical character recognition (OCR) enscriber; cathode ray tube (CRT) terminal; point-of-sale (POS) terminal; sensor; voice recognition (VR)	By systems design	Software	Central processing unit (CPU)	Central processing unit (CPU)	CPU; direct access storage device (DASD); magnetic tape; paper tape; punched cards	Online inquiry into a DASD; report generation; CRT terminals; teletype; other key terminals	Same as above, plus online copies from line printer; computer input/output; microfilm	Same as above, plus online data transmission (telecommunication); visual display; voice output

Source: Reprinted from *Information Systems: Theory and Practice* by John G. Burch, Jr., Felix R. Strater, and Gary Grudnitski by permission of John Wiley & Sons, New York, © 1979.

systems documentation requirements, programmer documentation requirements, and user documentation requirements. Documentation guidelines have been published by the American National Standards Institute and are available through the federal information processing standards put out by the National Bureau of Standards. All computer systems users should be aware that there are standards published for documentation. One of these is Federal Information Processing Standards (FIPS) Publication 38, "Guidelines for Documentation for Computer Programs in Automated Data Systems, Category Software, Subcategory Documentation." According to the introduction in this standards publication, documentation provides information to support the effective management of the Automatic Data Processing (ADP) resources and to facilitate the interchange of information. It serves to

- Provide managers with technical documents to review at significant development milestones, to determine that requirements have been met and that resources should continue to be expanded.
- Record technical information to allow coordination of later development in use/modification of the software.
- Facilitate understanding among managers, developers, programmers, operators and users by providing information about maintenance, training, changes and operation of the software.
- Inform other potential users of the functions and capabilities of the software so that they can determine whether it will serve their needs.[6]

System documentation refers generally to a variety of requirements for specification documents. The list below is drawn from the standards area to show which of these document types would fit under the three categories of documentation needs.

System Documentation

- *Functional requirements document* provides a basis for mutual understanding between users and designers of the initial definition of the software, including the requirements, operating environment, and development plan.
- *Data requirements document* provides, during the definition stage of software development, a data description and technical information about data collection requirements.

- *System/subsystem requirement* specifies for analysts and programmers the requirements, operating environment, design characteristics, and program specifications (if desired) for a system or subsystem.

Program Documentation

- *Program requirements* specifies for programmers the requirements, operating environment, and design characteristics of a computer program.
- *Data base requirements* specify the identification, logical characteristics, and physical characteristics of a particular data base.
- The *program maintenance manual* provides the maintenance programmer with the information necessary to understand the programs, their operating environment, and their maintenance procedures.
- The *test plan* for testing software comprises detailed specifications, descriptions, and procedures for all tests, as well as criteria for data reduction and evaluation.
- The *test analysis report* documents the test analysis results and findings, presents the demonstrated capabilities and deficiencies for review, and provides a basis for preparing a statement of software readiness for implementation.

User Documentation

- User documentation is often provided in a *users' manual*. The purpose of the manual is to describe the functions performed by the software in non-ADP terminology sufficiently clearly that the user organization can determine whether, when, and how to use it. It should serve as a reference document for preparation of input data and parameters and for interpretation of results.
- *Operations manual*. The purpose of the operations manual is to provide computer operation personnel with a description of the software and of the operational environment so that the software can be run.

Further review of documentation requirements would identify that the responsibility for developing appropriate documentation ranges from project managers to participating team leaders. It includes user-managers who assist in targeting the necessary requirements for documentation. Too often a system has been designed, developed, and marketed with user documentation prepared at a technical level too difficult to be immediately used by operations personnel. Consequently user-managers will bypass the system or expend additional effort further translating the user document so that their operations personnel can carry out the particular functions with the computer terminals.

Equipment Review

Another function often carried out in the design phase is equipment specifications review. The review helps determine that the level and expectation of equipment functions and capabilities that will be necessary to support the system are carefully defined. Although many of these are spelled out in the study phase, particular features of hardware are more actively explored and evaluated for their appropriate employment in the overall system development.

Hardware assessment requires technical expertise. However, application users are concerned with the peripheral components that affect the input and output tasks that are frequently managed in the user environment. The most common example of this is the computer terminal. Of course, operation of the terminal in an MRD will be a concern of the staff and management. For this reason, a discussion of computer terminal features is included. How can effective computer terminal selection be made? A major factor is a clear recognition that the effectiveness of this piece of equipment will directly affect the functional operation it supports. Effective selection of the terminal can make the difference between employees' willingness to work with a system and their resistance.

In order to define the CRT terminal requirements, examine several factors. First, the intended use of the terminal must be identified (input or output, data entry, inquiry or graphics display). Second, whether the terminal is to be used by a full-time operator or an occasional one must be determined. Full-time operators can be trained thoroughly for the job and will have plenty of time to practice. The occasional operator, such as the manager, will not be highly trained in terminal use. In this case, special care must be taken to ensure that instructions and responses are simple and clearly understood. As a final step in defining the terminal requirements, the following characteristics need to be considered:[7]

- Typewriter keyboard
- Numeric pad (3 × 4 matrix)
- Special function keys
- Good cursor controls
- End-of-line bell or indicator
- Buffer into which data can be keyed and modified prior to transmission (block mode transmission)

- Skip and tab keys
- Erase or backspace keys
- Page or scroll keys
- Screen angled for viewing ease
- Screen wide enough to display required number of characters
- Characters large enough for easy legibility
- Screen that does not flicker
- Screen color that does not tire operator's eyes
- Image bright enough for operator to see clearly, protected from external glare
- Diffused light intensity to reduce screen glare
- Graphics capability
- Protected/unprotected fields
- Individual fields can be highlighted
- Upper and lower case characters
- Selectable character sets
- Selectable horizontal tabs and other formatting capabilities
- Unique terminal identification by the computer
- Lockable keyboard
- Nonprinting password when keying it in

This technology is rapidly changing. Readers may find additional features available.

Alternative Design Approaches

Ever greater reliance on automation seems inevitable for the foreseeable future. The user community will expand and greater capability will be expected of automation. Generally, people who plan to work with computer applications should develop some understanding of the approaches to application development. This chapter has thus far described key functions carried out in building a particular application as they might occur in the design phase. It is important to point out that alternative design approaches exist. Let us pause and consider this concept. What is meant by alternative design approaches? This expression simply refers to the fact that there are several ways that a particular application can evolve through the design process. In general, the directions that can be selected are turnkey, custom design, or some combination of the two.

The team can decide to select a *turnkey* approach. A turnkey system is one in which a finished product is delivered. This product often includes software, hardware, documentation, and training components. Some are designed to be transported and immediately "turned-on" in the user setting. A word processor is an example of a turnkey system in medical record settings. The major disadvantage of these systems is the limited flexibility offered to users.

The team can decide to develop a *custom-designed* system. This is one designed directly from users' needs and built to the specifications of a particular organization. This approach offers more flexibility to users. A hospital information services department that designs a custom chart deficiency system for the MRD works with this approach. Because there generally are programming resources available within the organization, this approach offers some advantages to the users. However, it is a more expensive approach because the programs have to be written. Software development is the most expensive part of building computer systems.

A design team may also select a combination of the first two paths, in which they identify and purchase particular software packages and use them in some components of the overall system design. In general, many information managers feel they have a responsibility to see what is available to avoid reinventing an application design already developed. It is apparent that this approach has some advantages. Not only can already proven products be used, but custom design costs will be lower.

Another approach has evolved that is derived both from concerns raised by the high costs of continued custom-designing techniques on the one hand and from the frustration of inflexible turnkey systems on the other. In the new approach interactive design functions and data base management resources are combined to provide a sophisticated design process. One illustration of this approach has been developed at the Johns Hopkins University in a program generation system called TEDIUM.[8] The system stores text descriptions and formal specifications on particular application functions. TEDIUM uses the specifications to generate the necessary programs. The text descriptions are written in application or user expressions that enable users to be polled to verify that the descriptions accurately reflect their needs.

Programmers called application designers actually try out the programs that reflect the application need. All programs are generated from a data base of specifications. Because these specifications are maintained as a network, it is possible to access or link associated data elements. Therefore the designer can branch from a requirement written as text expressions such as *appointment schedule processing* or *produce appointment lists* to identify specific functional requirements. Users are then asked to specify processes such as scheduled new appointments, cancelled appointments, rescheduled appointments, and so forth. These processes are linked to programs that execute the system functions in the computer. Designers can access a particular line in a program and a data element within the line. TEDIUM is actually a program generator. It focuses on the analysis phase of the systems life cycle in which the user needs are determined. The program writing is a byproduct of the original text specifications, which in turn are linked to the necessary detailed formal specifications capable of producing the appropriate programs. User-managers could sit down at a terminal with an application designer in this approach and in interactive mode view a series of displays specifying the components and processes for their particular application. They could participate in the evolution of those requirements. Imagine how powerful and effective such a design alternative could be. Although not yet fully available, this kind of design technology will be available in the future. It has been proposed that such design functions will be available in the latter half of the 1980s.[9]

Design Phase Report

At the conclusion of the design phase, a report should be prepared that updates the analysis phase report. The design phase report includes the update of performance definition as prepared earlier. It should incorporate the detailed performance specifications and the design phase requirements that have been identified for building the new system. Figure 5–2 illustrates another model of computer system requirements. In this example, we see how a requirements outline is constructed from the analysis and study phase products and expanded through subsequent developmental stages. Careful attention to dovetailing these systems analysis milestones is a key factor in effective completion of computer-based projects.

The design phase report should provide a clear picture of progress from the analysis phase activities. MRAs should plan to review the design phase report as a fundamental management responsibility.

Figure 5–2 Computer Systems Specifications Evolution

PERFORMANCE SPECIFICATION (Study Phase)	DESIGN SPECIFICATION (Design Phase)	SYSTEM SPECIFICATION (Development Phase)
EXTERNAL PERFORMANCE DESCRIPTION	EXTERNAL DESIGN REQUIREMENTS	EXTERNAL SYSTEM SPECIFICATION
Information-oriented flowchart →	Information-oriented flowchart →	Information-oriented flowchart
System output description →	System output requirements →	System output specification
System input description →	System input requirements →	System input specification
System interface identification →	System interface requirements →	System interface specification
System resource identification →	System test requirements →	System test specification
	Equipment specifications →	Test data
	Personnel and training requirements →	Test results—samples
		Equipment specifications
		Personnel specification and training procedures
		User's reference manual
		Programmer's reference manual
		Operator's reference manual
INTERNAL PERFORMANCE DESCRIPTION	INTERNAL DESIGN REQUIREMENT	INTERNAL SYSTEM SPECIFICATION
System flowchart →	**Computer program design requirement** →	**Computer program specification**
Data processing description	System flowchart	System flowchart
Data storage description	Expanded system flowcharts	Expanded system flowcharts
	Computer program control requirements	Computer program control specification
	Computer program test requirements	Computer program test specification
	Data base requirements	Test data
	Design requirements	Test results—samples
	(for each CPC as required)	Data base specification
	Detailed system flowcharts	**Computer program component (CPC) specification**
	Transaction file requirements	(each CPC as required)
	Control requirements	Detailed system flowcharts
	Interface requirements	Computer program (CPC) logic flowchart
	Test requirements	Transaction file specifications
	Special requirements	Control specifications
		Interface specifications
		Computer program (CPC) listings
		Test specifications
		Test data
		Input-output samples
		Special specifications

Source: Marvin Gore and John Stubbe, *Elements of Systems Analysis for Business Data Processing,* © 1975 by William C. Brown Co. Reprinted by permission.

Questions and Problems for Discussion

1. Describe the major components of the design phase of the systems life cycle.
2. Describe an appropriate role for the health record practitioner in participating in design phase activities.
3. How do the activities in the design phase build on the steps that have been carried out in the analysis phase?
4. What is the purpose of performing problem reconfirmation? Does it require the same process that was followed when the problem was initially identified?
5. Comment on the following statement: "As user managers, the health record practitioners are the most appropriate team members to specify the application outputs for those applications developed in MRDs."
6. Explain the concept "design issues." Give three examples of application design issues likely to occur in patient data computerization.
7. You are the director of patient information services. A hospitalwide chart request and tracking system has just been implemented. There are some problems in the system. You ask to review the documentation. What documentation would you review? Why?
8. What aspects of equipment review should be performed by the MRA?
9. Why are review and sign-off by users so important in the design phase of the systems life cycle?
10. Why is it useful for user-managers to understand the components of design phase activities such as the data-checking methods and input design considerations?
11. Why do user-managers need to understand the processes involved in file design and program design activities?
12. What is meant by alternative design approaches? Why is it important to consider them?

NOTES

1. Elizabeth Capozzoli, "An Automated Approach to Medical Record Tracking," *Topics in Health Record Management* (December 1981).

2. Kathleen A. Waters and Gretchen F. Murphy, *Medical Records in Health Information* (Rockville, Md.: Aspen Systems Corp., 1979).

3. Jack D. Harpool, *Business Data Systems: A Practical Guide* (Dubuque, Iowa: Wm. C. Brown, 1978), p. 220.

4. Ibid.

5. Charles Austin, *Information Systems for Hospital Administration* (Ann Arbor, Mich.: Health Administration Press, 1979), p. 80.

6. FIPS Publication 38; *Guidelines for Documentation for Computer Programs in Automated Data Systems, Category Software, Subcategory Documentation,* U.S. Dept. of Commerce, National Bureau of Standards, U.S. Govt. Printing Office, Washington, D.C., 1976.

7. Ardra F. Fitzgerald, Jerry Fitzgerald, and Warren D. Stallings, Jr., *Fundamentals of Systems Analysis* (New York: John Wiley & Sons, 1981), p. 132.

8. Institute of Electrical and Electronic Engineers, *Proceedings, Fifth Annual Symposium on Computer Applications in Medical Care* (New York: IEEE, 1981).

9. Marvin Gore and John Stubbe. *Elements of Systems Analysis* (Dubuque, Iowa: Wm. C. Brown, 1975), p. 352.

Chapter 6

How To Participate in Development and Implementation of Computer Systems

Objectives

1. Define and describe the development phase of the systems life cycle.
2. Define and describe the relationship of the major elements of the development phase to the management and planning of health information systems.
3. Examine the products and activities of the development phase and identify responsibilities appropriate to the MRA.
4. Present guidelines for the management of computer projects.
5. Describe methods for monitoring individual computer applications.
6. Describe the actual and potential role of medical record practitioners in development phase activities including system testing, training and conversion planning. Examine alternative methods of system conversion and their selected use in MRD applications.
7. Define and describe the implementation phase in the systems life cycle.
8. Explain the use of request for proposal (RFP) in planning and implementing health information computer system applications.
9. Present and explain the significance of performance review in implemented computer systems.
10. Present a model for RFP development in hospital information systems.

This chapter will continue discussion of the systems life cycle in the development phase and conclude with implementation phase activities. In the development phase of the life cycle, the project planning initiated in the analysis phase and reinforced with the establishment of the design team moves into a more visible role. During the development phase, estimation and planning all of the project's components, system installation, staff training, cost control monitoring, and changes in job role are carried out. It is during this time that the new system's methods, processes, and outcomes are tested. Programs are written or purchased, hardware is purchased and installed, and information is converted from manual to computer storage. It is a period of close collaboration with users. This provides a smooth transition into the new system. Alert user-managers use this time to test new procedures, comparing performance of the computer with the requirements that were determined in the design phase.

The implementation phase marks the actual changeover to the new system. It is at this point that the new computer system becomes the new medical record chart location system, the computerized outpatient appointment system becomes operational, and so on. Along with actual system changeover, implementation phase activities include postinstallation review. This is a process by which the performance of the new system is monitored and problem areas are resolved. It is a time for vigorous feedback from operations personnel. Review techniques and responsibilities appropriate to the manager will be featured.

The chapter is divided into three parts. The first explains the development phase process by examining the major elements involved. In the second, a request for proposal (RFP) is defined and described and its use in hospital computer systems planning is explained. The implementation phase is explored in the third part. Throughout the chapter the role and responsibility of medical record practitioners are discussed. Given the experience of participation throughout the analysis and design phase, practitioners will find continued participation in the development and implementation phases to be a logical evolution in participatory systems development.

Think of this as an opportunity to educate both data processing professionals and health record professionals. They can learn about the development, collection, processing, and retrieval aspects of patient record systems. As advances in technology expand opportunities to access patient data in new ways, careful and regular review will be required. It will be required in order to safeguard established management and ethical recordkeeping principles.

THE DEVELOPMENT PHASE

An Overview

The four major steps in the development phase of a computer system life cycle are

1. Project planning and estimating
2. Program development and testing
3. Personnel training
4. Development phase report

In the first step, planning and estimating are coordinated to describe and monitor the project's progress. At this point a transition is made. The system requirements determined in the design phase are translated into tasks to be allocated in the development phase. Here, the project managers, systems design team members, and programmers coordinate their efforts to prepare integrated plans and estimates to map out particular tasks. Assignments and target completion dates are incorporated into formal project schedules. These in turn are routinely used to monitor developmental progress. Inherent in estimation activities is a careful look at plans for conversion, the changing from the old to the new. Conversion alternatives will be compared. Once again the user role becomes highly visible. It is important to remember that building the system involves a number of detailed tasks to be performed by individuals possessing a variety of technical skills. Technical programmers are not the only skilled personnel needed. Also needed are individuals who know that data and can specify data base, file, field, and network terminal particulars. Called "data base administrators" and "data communications administrators," these individuals help define the detailed information necessary for the particular computer application. Medical record practitioners (MRPs) should be prepared to participate in this process. For instance, should development of a physician incomplete system be underway, the MRP is eminently qualified to define and explain the fields, records, and data elements required in the application. The MRP would describe these items in terms of field lengths, type of information (alphabetical, numerical, or coded), quantities, and names. Together with other technical resource people, a practitioner may map out the details on each computerized chart deficiency record that will be set up by patient or physician name.

In the second step, program development and testing, the required programs are written and tested. Application programmers write the programs that perform application functions required by a system. These application programmers interact with various system programmers and with users of the application itself. Systems programmers maintain, support, and enhance current computer system features. They typically work with operating systems, access methods, data base systems, data communications operations, and other such activities. Notice that technical teamwork is as necessary as the teamwork between designers and users called for in the earlier stages of the system creation. Generally MRAs are application users who will interact with the application programmers who are working on a particular project. For example, coordination between an MRA and an application programmer in developing a chart locator system is an illustration of this point.

A word should be said about program writing and software packages. In many applications today programmers are integrating software packages with present system functions. Many software packages are available for use in medical record settings. A hospital may purchase a software package that performs statistical analysis, for example to provide statistical analysis of discharge and financial abstracts. Program planning may include the addition of the package to the existing computer system. An MRP then can access and retrieve pertinent information via a CRT terminal in the MRD. In this example, the use of software packages alters development phase programming activities.

Testing is the validation of completed programs. It should be performed at both the programming and user, or operation, levels. We will examine a variety of testing methods and discuss their significance. Practitioners can then identify reasonable management activities that reinforce effective testing practices. Although testing methods are well defined, they are not always carried out in practice. One reason for this is that other development phase tasks and planned target dates become the focus of the project. It is important for managers to consider the testing activities as a facet of the changes that are about to occur and to participate actively in the testing. Along with the specific program being tested, new or revised procedures that correspond with the program activity are often identified. From a management point of view, if procedure changes are planned in coordination with the application development activities, the stress of change can be minimized and new system implementation is facilitated.

The third step of the development phase is personnel training. Anyone who will need to know about the new system must be identified. Teaching methods are determined and formal training programs are mapped out. Next, actual training sessions take place. Individual instruction, classes, exercises, and even testing are included in many training programs.

Personnel training is well illustrated in the following excerpt from a project in which a master tracking system was developed for an MRD:

> A comprehensive user manual was prepared that covered routine use of the system, report generation, troubleshooting and back-up procedures. The manual was written and illustrated for easy interpretation by employees. After a general overview of the MTS (master tracking system) for all medical record department employees, training was initiated approximately one month prior to implementation. Training was most beneficial on a one-to-one basis, starting with key users, such as the lead file clerk. Time was scheduled for employees to work independently with the MTS computer and the manual. Exercises were written to simulate actual tasks performed by the MTS.[1]

Routine feedback is required to coordinate communication between learners and training staff during this period. As established procedures and job routines are revised, managers need to budget extra time and effort to maintain effective communication. A flexible approach in getting jobs carried out is also necessary. It is valuable to secure documentation of revisions as they occur so that all procedures reflect actual current status of the jobs in question.

The final step of the development phase is preparation of a formal report to the management or administration. The report identifies exactly what has been completed during this phase. Similar to the analysis and design phase reports, it is a benchmark or milestone for management to use in making decisions about the project. Table 6–1 features the activities of the development phase. The following sections describe how the individual development phase activities fit together to form an organized segment in the computer system life cycle.

Table 6–1 Development and Implementation of Systems Life Cycle Model

Development phase

I. Project planning and estimating
   A. Make formal project schedule
   B. Make plans used to monitor projects
   C. Convert information, procedures and equipment
II. Program development and testing
   A. Establish system test plans
   B. Provide user testing
   C. Secure effective procedures development
III. Personnel training
   A. Develop positive attitudes
   B. Select effective training methods
IV. Development phase report

Implementation (operation) phase

I. System changeover
   A. Schedule changeover activities
   B. Provide back-up support programs
   C. Work with vendors
II. Post installation audit
   A. Relate performance objectives and standards
   B. Plan postinstallation evaluation
   C. Perform technical audit methods
   D. Integrate computer system performance into work standards

Project Planning and Estimating Begins with a Formal Project Schedule

Project planning and estimating were initiated in the analysis phase of the computer systems life cycle. They became more defined as the design team was established and as the need to coordinate project efforts emerged. As individual tasks are specified in the development phase, however, project planning must become even more precise. The following items must be produced:

- Formal project schedules
- Plans for monitoring projects
- Plans for conversion of information, procedures, and equipment

The formal schedule for implementation of an automated system includes projected dates for milestones, and achievement of certain activities. This information is most useful when it graphically depicts interaction between the systems team, users, and operations personnel. The project schedule is a guideline for all project personnel to identify their own related tasks. Its availability serves as a visualization of individual or team assignments. The calendar for completion of key tasks serves as a reminder for all to consult when checking on the status of the project. Effective schedules require the cooperation of MRAs and project managers. MRAs need to provide accurate, current descriptions of job-related tasks in their departments. They must also provide factual projections on the training and equipment aspects of the program including details of how conversion of information, procedures, and equipment will be accomplished. One example is the covering of jobs of those who are participating in training programs. Other project plan features are target dates specifying when software and hardware should be ready for the training program. Another feature is specification of what products are tied to payment and acceptance from the vendor. Coordination and cooperation are key elements in successful implementation.

Charts Used To Monitor Projects

During the development phase the data processing staff works from the requirements or prototypes drawn up by the systems team during the design phase. Figure 6–1 illustrates the complexity of this process. Many activities must come together at this point. The effective employment of project planning and monitoring documents is of paramount importance to the user-manager. Two charts that are particularly important are the Gantt chart and the PERT chart.

Figure 6–1 Coordinating, Planning, and Estimating

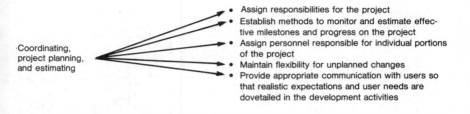

·Coordinating, project planning, and estimating

- Assign responsibilities for the project
- Establish methods to monitor and estimate effective milestones and progress on the project
- Assign personnel responsible for individual portions of the project
- Maintain flexibility for unplanned changes
- Provide appropriate communication with users so that realistic expectations and user needs are dovetailed in the development activities

Gantt Charts in Project Management

The Gantt chart is often used to track the progress of computer-based projects. In this document, activities are scheduled over a period of weeks or months. The length of the open horizontal bar in Exhibit 6–1 corresponds to the expected completion time of the activity. As time progresses, the bar is filled in to depict the portion of the activity actually complete. Notice how the illustration demonstrates this point. Gantt charts provide an excellent view of project status for users and developers. Projects of major scope—of long duration with involvement of many personnel or departments—are especially enhanced by the use of Gantt charting. Charts are useful for monitoring progress on many kinds of projects. Practitioners can incorporate these tools into management planning for a wide variety of projects. Exhibit 6–2 is an adaptation of the Gantt chart.

PERT Charts in Project Management

Another graphic technique for planning is the Program Evaluation and Review Technique (PERT). This depicts the relationship between tasks and schedules. It also indicates what activities can be done independent of other activities and those that are interdependent. When properly executed the PERT chart will allow a comparison of scheduling of tasks in order to show the items that accomplish completion of the project in the least amount of time and at the lowest cost. Some of the advantages of the PERT chart are as follows:

- Project members are forced to consider the sequencing and implementation of various phases of the project far in advance of their occurrence. Frequently this results in the uncovering of tasks not previously recognized as necessary.

Exhibit 6–1 Project Plan and Status Report (Gantt-Type Chart)

Project Plan and Status Report

Project Title
Planning and Designing
Medical Record Department

Planning Committee
Programmer/Analyst

Project Status Symbols
o Satisfactory
□ Caution
△ Critical

Planning Progress Symbols
□ Scheduled Progress
■ Actual Progress

▽ Scheduled Completion
▼ Actual Completion

Committed Date: Nov. 1, 1981
Completed Date: March 30, 1982
Status Date: Feb. 15, 1982

Period Ending (Weeks)

Activity/Document	Percent Complete	Status	1	2	3	4	5	6	7	8	9	10	11	12	13	14
Investigation	100%	C														o
Goals and Objectives	100%	C			o											
Policies	100%	C					o									
Staff Requirements	100%	C						o								
Job Descriptions	82%	I													□	
Procedures	66%	I														□
Equipment, Furniture, Supplies	40%	I														□
Layout	100%	C												o		
Budget	100%	C											o			
Report (Draft)	92%	I														□
Evaluation	86%	I														o
Final Report	50%	II														o

Exhibit 6–2 Gantt Chart Adaptation

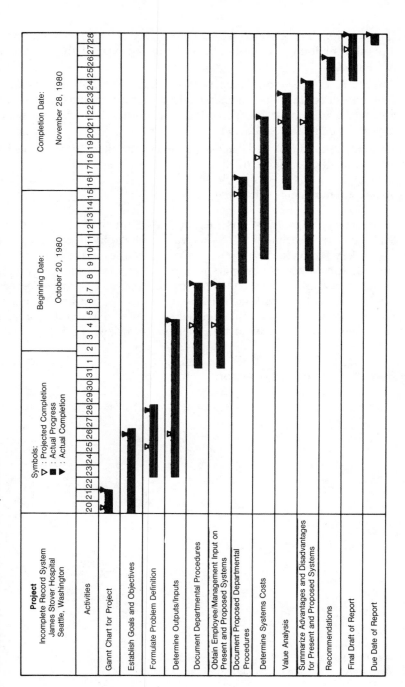

- Interrelationships among tasks are described more clearly than with any other planning technique, thus permitting early analysis of future bottlenecks.
- Attention is focused on the parts of a project most critical to completion on schedule. This focusing takes place early enough so that corrective action can be taken. It also reveals activities from which resources can be diverted.
- A framework is developed that can be used to test alternative approaches to a project.[2] See Figure 6–2.

Ongoing Communication with Users

If a systems design team has included users in the activities of the project, thus far, continuation of the team philosophy is easily accomplished. A key factor in development phase activities is the extension of user interaction to operations personnel. As the formal implementation plan is developed and presented to user-managers, key operations personnel (staff) who will be working with the new system can be brought in. Staff morale depends on careful consideration of all aspects of change in daily activities. Employees who are expected to work with new equipment

Figure 6–2 Critical Path Network

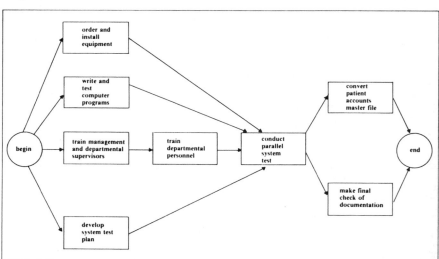

Source: From *Information Systems for Hospital Administration* by Charles J. Austin by permission of Health Administration Press, © 1979.

and procedures need to have an advance picture of the effect on their routine. This sharing process can offset unrealistic fears and expectations about job performance. Honest acknowledgement about unresolved matters will help prevent an us-versus-them attitude among staff. Employees should have some say in review of the implementation plan. Even though careful system investigation may have been carried out during the analysis phase, particular idiosyncrasies that may inhibit planned implementation schedules may not have been uncovered. For example, an increase in work volume or new tasks that have been incorporated into the daily operation since the analysis documentation was completed can create problems at this point. The review process can have a positive effect on MRD personnel. Not only does it inform them of what is to come but it creates an opportunity to respond to changes and offer input to the process. The following excerpt shows what an implementation schedule might look like. This sample is taken from the Master Tracking System (MTS) described earlier in the chapter in which a master tracking system was designed and implemented for an MRD.

The implementation phase started approximately six months before actual use of the system. The Project Committee determined it was necessary to develop an implementation schedule that included the following:

- Hardware delivery and set up
- Software keying, testing and debugging
- User training
- User manual development
- System start-up schedule
- Maintenance and repair procedures
- System troubleshooting procedures

As the equipment and programming coordination progressed, the medical record department management team prepared for the installation of equipment and training of employees for the upcoming changes that would take place because of the MTS. It was necessary for routine department operations to be as smooth and current as possible. Several physical changes, such as rearranging work stations, were made to improve the work flow between processing functions. Information regarding the progress of the MTS was shared with the employees, and this led to an enthusiastic acceptance of the new system.[1]

Planning for Conversion

The final component of project planning involves system conversion. Planning for conversion from manual to automated data processing is a cooperative management task. It requires coordinated efforts from all individuals affected by the change, as shown in the following example.

A Case History

A consortium of physicians and administrators in a Midwestern city was formed to develop a combined tumor registry system. Their premise was that for data to be retrievable, computerization was essential. The combined registry, originally developed for two hospitals, has subsequently expanded to include 10 others. Data collection was designed in the following way.

The primary criterion for selecting data to be collected (besides the data required by the American Cancer Society (ACS) for setting up an accredited tumor registry) was that the data be of significant clinical value in patient care. Some of the data strongly desired by the physicians, such as type of female hormones used in treating the patient, type of birth control pills taken by the patient, number of pregnancies, and smoking habits were not regularly obtainable from the patient's hospital chart. A physicians' committee meeting with the nursing services of the hospitals obtained the latter's cooperation in revising the admitting nurse's notes so that these data could be obtainable on all patients.

Some important data elements are obtained from the physician's office (for example, type of chemotherapy drugs administered to the patient after discharge from the hospital), through the use of a checklist of missing data sent out by the registry. This is required infrequently. The next criterion was that the abstract, the source document on which information is collected, be kept simple and not be a duplication of the medical record. A final criterion was to include epidemiological data available in the medical record (for example, occupation). A reliable, cost effective method could not be found to obtain additional specific epidemiological data such as exposures to selected chemicals and details of past residence.

Three major categories of reports are produced by the system: case summaries, structured reports, and a special report. The case summary is a computer printout of the most pertinent clinical data on one cancer. Structured reports include accession registers, master index, site list, annual report, histology list, reports required by the ACS Commission on Cancer, and so on. The special reports are provided as needed by means of a report generator feature of the computer.

The system was designed to meet mutually established objectives that resulted in a multihospital on-line cooperative system with a shared data base. The first hospital's tumor registry was computerized in 1977, the others converting subsequently.

The computer programs are designed so that data elements can be added without reprogramming the entire registry application. The new elements can and have been added to the tumor registry abstract in their logical place, rather than being added at the end.

A person in the central office updates codes, aids new medical staff members, and so on, without having to go through a computer programmer. These changes are made when decided by the users committee, which is composed of a medical advisor and the tumor registrars from each hospital. This has been critical in keeping an up-to-date registry.

Storing and Retrieving Data. To ensure that all data items could be retrieved for reports, planners took forms, codes, and input into consideration. The design of the abstract, the source document on which information is gathered, is sectional. Each section's contents are numbered data items. To simplify the input of data, sections of information prompting on the CRT screen are identical to the information sections on the abstract. By clearly defining each item to be abstracted and then entering that item into a specific storage location in the memory, retrieval is ensured. After determining the data elements, it was necessary to develop codes for data retrieval. Standard coding systems are used whenever available.

A major function of the tumor registry is to follow all patients with cancer for as long as they are alive. Any patient from whom no information has been received in the past year is due to be followed. Typing letters to request follow-up information consumed a major amount of time in manually operated registries. Though the computer can print these letters rapidly, decisions had to be made on type and frequency of output.

The letter to physicians requesting follow-up information is structured as a checklist that can be responded to quickly, often by the physician's office nurse from the office chart. Yet it is structured so that data can be transferred directly to the computer.

The letters are printed at the end of each month. It was not found to be cost effective or necessary to design computer-generated letters requesting information from other sources if the physician had not seen the patient, since the number of such requests is not large. The letters require special sensitive handling, so are taken care of manually.

Converting an old, large manually operated registry presented unique problems. To maximize immediate use of the computer, an abbreviated data system was designed that expedited entering information on all the living cases into the computer so that the large task of follow-up letters could be computer generated. Each month the cases selected for data entry were those due for follow-up. This offered an organized method to allow the registry to remain current in working its caseload while converting.

To allow valid report calculations of a hospital's cancer experience, it is critical that the data be entered also for patients who have expired. To obtain reports as soon as possible one year's group of expired cases were entered at a time. When reporting on the total data base, special programs sidetrack the converting hospital's data for incomplete years.[3]

Notice the range of users affected by converting the old system. The coordination shown in this case history illustrates the fundamental principles of conversion planning. Let us now take a more detailed look at this issue.

Conversion planning can be broken down into three segments: information conversion, procedures conversion, and equipment conversion. Each segment requires attention to detail in order for the overall conversion to be as smooth as possible.

Information Conversion

In an information system, conversion is the process of changing from one method of collecting, storing, processing, and retrieving information to another. Information conversion includes the transfer of manual or written information presently stored on paper files or earlier versions of computer systems files to a format compatible with the new computer system. The user must help determine what data will be converted. If data is to be condensed prior to entry into the computer, the cost for reviewing and modifying the data must be taken into account. For a hospital converting from a manual MPI, each MPI card must eventually be prepared for automated data entry. The cost of changeover from manual to computer-compatible cards includes personnel time and equipment. Whether the MPI is to be converted all at once, piecemeal, or in a phased-in way as in the tumor registry example, the costs of conversion remain constant. Budgeting for these conversion activities is essential. Cash outflow would be spread over time in the tumor registry conversion example.

Even when data are being converted from an existing computer system to a new one, a careful review of the data elements should be made to be sure that the data are still used or required. Maintaining awareness of information use is a key responsibility of health information managers. Today, a national concern for improved data quality in all health information areas is proof of the need to incorporate data monitoring at all levels of data collection and processing.

Analogous to number system conversion, plans for information conversion must be comprehensive and strategic so as to provide efficient transition with the least disruption of day-to-day operations. Typical examples of conversion in medical records are

- Transferring a manually prepared MPI file to computer output microfiche (COM)
- Transferring disease and operation indexes previously handwritten or typewritten and manually stored and retrieved to magnetic tape files for access and retrieval through standard request forms submitted to the data processing department
- Setting up computer files to support a hospital chart tracking system.

In conversion of a manually maintained MPI, preparation for information conversion can be initiated long before actual conversion takes place. This can be accomplished by assigning the MRD staff the responsibility of auditing all the cards in the system. The information on the MPI cards could also be formatted for data entry.

This format could then be used for input into the computer system by a number of methods.

- The format could be used with punchcards and scheduled for batch input prior to new system start-up.
- The format could also be used with alternative forms of data entry such as key-to-tape or key-to-disk. The resulting MPI data tape or disk could then be entered into the system prior to new system start-up.
- The format could be typed on forms via special font typewriters. The forms could then be read into the system by optical scanning devices prior to new system start-up.
- The format might be used in conjunction with the latest interactive technology in which data entry could be performed directly on line via CRT terminals. In this case, data entry could be completed prior to new system start-up or it could be integrated into new system operations.

In cases where the volume of MPI cards is judged to be too large to convert, a low-cost way of handling the conversion is similar to converting a serial number file to a unit number file. New numbers are assigned and entered, and old numbers pulled forward as patients are readmitted. In this case, only new-patient data is entered in the automated MPI file when the computer system is launched. The MRD then accesses two systems. The manual index card data is entered into the computer system gradually as patients reenter the facility. Eventually, entry of readmissions and new admissions will allow the manual system to be phased out. We will see how these alternatives fit into the conversion issue in the discussions that follow. Readers should bear in mind the question: "What would the determining factors be in selecting a particular method of conversion?"

Procedures Conversion

Procedural conversion includes both programming and manual procedure updates or a combination of the two. For example, changing operations from a manual discharge analysis procedure to an on-line discharge analysis method will require both programming and manual procedure changes. Here again, user participation enhances the process. Medical record practitioners can direct and develop revised procedures. The people who perform the daily tasks that achieve the job objectives are often highly motivated and best qualified to help prepare conversion procedures. They can also provide effective review of documented descriptions of the new tasks. As indicated elsewhere in this text, user involvement helps make the new system "our system" instead of just "the system."

Equipment Conversion

There are two types of equipment conversion. One kind entails the exchange of old equipment for new equipment of the same type; the other entails replacement of one type of equipment with a new type. In computers in place today, equipment conversion often requires extensive equipment reprogramming. Newer computers are being designed with a strong emphasis on equipment conversion without extensive reprogramming. In some situations new computers and peripheral devices for new systems are "plug-to-plug" compatible with the old hardware. There are two reasons for this. The physical structure of the new hardware may be constructed to have little or no effect on the logical working of the software programs so those programs may not need to be rewritten. Data communications equipment can be made to work if standard interfaces have been used. Also, new computing equipment can be made to emulate, or behave like, old machines. This may be system inefficient but may serve short-term needs. These two techniques have greatly improved the conversion process. Along with the considerations just discussed the appropriate strategies of conversion must also be selected.

Strategies for System Conversion

A *direct conversion* is the implementation of all parts of a new system all at one time. It includes implementation of a new system where no other system previously existed or where another system is discontinued. The direct conversion approach is meaningful when: (1) the old system is judged absolutely without current or future value, (2) the new system is either very small or simple and consequently not complex, and (3) the design of the new system is drastically different from the old system and comparisons between systems would be meaningless. The primary advantage to this approach is that it is relatively inexpensive. The primary disadvantage is that it entails a high risk of failure and, possibly, subsequent expense. When direct conversion is to be utilized, systems testing activity takes on even greater importance. MRAs might employ direct conversion when implementing a computerized word processing system for a transcription service.

In *parallel conversion* the old and the new system operate simultaneously for some period of time. The outputs from each system are compared and differences reconciled. The advantage to this approach is that it provides a high degree of protection from any failures that occur in the new system. The obvious disadvantages are the costs associated with duplicating the facilities and personnel to maintain dual systems. Because of the difficulties experienced in the past by organizations implementing new systems, this approach to conversion is very popular. When the conversion process of a system includes parallel operations, the project manager should plan for periodic reviews with operating personnel and users concerning the performance of the new system. A reasonable date for acceptance of the new system and discontinuance of the old system must be designated. When hospitals implement in-house discharge abstracting systems, they may continue to parallel participation with a commercial discharge abstract service until they are sure the new system is working satisfactorily.

Modular conversion, sometimes termed the "pilot approach," refers to the implementation of a system *into the organization* on a piecemeal basis. The advantages to this approach are that (1) the risk of a system's failure is localized, (2) the problems identified in the system can be corrected before further implementation is attempted, and (3) other operating personnel can be trained in a "live" environment before the system is implemented at their location. One disadvantage to this approach is that the conversion period for the organization can be extremely lengthy. More important, this approach is not always feasible for a particular system or organization.

The *phased-in approach* is similar to the modular approach. However, this approach differs in that the system itself is segmented, *not the organization*. For example, the new data collection activities are implemented and an interface mechanism with the old system is developed. This interface allows the old system to operate with the new input data. Later, the new data base access, storage, and retrieval activities are implemented. Once again, an interface mechanism with the old system is installed until the entire system is implemented. Each time a new segment is added, an interface with the old system must be developed. The advantages to this approach are that the rate of change in a given organization can be minimized and data processing resources can be acquired gradually over an extended period of time. The disadvantages to this approach include the costs incurred to develop temporary interfaces with old systems, limited applicability, and a demoralizing atmosphere in the organization of "never completing a system." When the end users do not see anything happening, it is important for the user-managers to require user status reports.

Health information managers need to consider seriously the methods available for information conversion. Choice of direct conversion, phased-in conversion, or one of the other methods will depend on the following:

- Volume of data to be converted
- Cost of one-time batch conversion versus integrating a phase in conversion program
- Anticipated activity of the new system
- Planned date of conversion
- Intricacy of the new system

Figure 6–3 compares these methods. The reader can see how important the user input is in determining the most appropriate method of conversion.

MRAs should map out each alternative's effect on departmental operations. Once this is done, the list of effects can be pooled with the systems designer and adequate and appropriate conversion can take place. This thorough approach to coordinating conversion activities cannot be stressed too much. Excellent computer systems analysis and design products have been lost in cost overruns caused by inadequate conversion methods. Securing coordinated changeover among designers, users, and operations personnel will go a long way to avoid such disasters and will help potentiate effective computer applications designs.

Our considerations of conversion methods will conclude with an excerpt illustrating how a phased-in conversion method was used to develop an in-house patient data system. Conversion of a manual record tracking

Figure 6–3 A Graphic Representation of the Basic Approaches to Systems Conversion

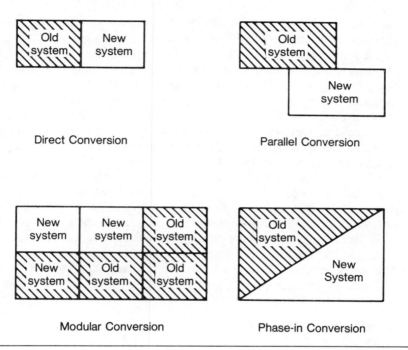

Direct Conversion

Parallel Conversion

Modular Conversion

Phase-in Conversion

system (MRTS) to a computerized system is described. Readers may notice that many of the principles outlined in the discussion of development phase elements were employed in the project.

Considerable thought was given to selecting the method of conversion from the manual charge-out system to the automated system that would have the least amount of impact on daily operations. The obvious option was to start charging out records on the MRTS when the system was ready for use. However, this would have required the file clerks to check record locations on the MRTS and in the card file for an indeterminable length of time.

The most desirable alternative was to prepare the data base for the MRTS before this system was operational so that the current locations of the out-of-file records would be available on the first day of system use. This approach would also require working with a parallel charge-out system, but the time period for system implementation was finite.

The decision was made to keypunch data on the 12,000 charge-out cards and store this information on a computer file that would eventually become the MRTS data base. Two keypunch units were rented and placed in Patient Files. Several file clerks were trained to use the machine and performed the keypunching task on a rotating basis. The cards were keypunched in terminal digit sequence. Location changes during this conversion period were keypunched instead of adding new cards to the charge-out file. Records that were returned to the file were listed and cards were keypunched from the list. At the end of each day, the keypunched cards were batched and sent to the Data Processing Department.

A printout listing of all out-of-file medical records that had been keypunched to date was available to Patient Files personnel by the beginning of each workday. The list contained the patient's name, clinic number, location and date sent. After the keypunching effort commenced, each location change was keypunched and the new location recorded on the printout. The transition period between the time the keypunching effort started and the MRTS became operational was six months.[4]

Program Development and Testing

The second major element of the development phase is software or program development and testing. During this period the computer programs are written, system procedures are developed, and a formalized system test plan is established. No hospital information system or portion thereof should be put into operation without complete testing of the system. The testing should be carefully planned to cover all aspects of the new system. Testing should be as realistic as possible and include the use of operations personnel.

Good systems design will include testing of the following elements:

- Objectives
- Output reports and screen displays
- Input forms and procedures
- Error and correction procedures
- Throughput times and system loading
- Computer times

- Cost estimates
- System documentation[5]

As program development takes place, it should follow the criteria or specifications laid down in the analysis phase.

The system test plan should parallel the analysis phase report in terms of major component sections so that the items listed above can be covered in the test plan. Health information managers should anticipate making a determination of how they can participate in carrying out testing procedures for the major sections.

Investing in User Testing

Users are a very important facet of system testing. The operations personnel who will carry out the on-line MPI operation, on-line statistical analysis operation, or on-line chart request and dissemination activities must be people who understand and validate operations level testing of new systems. A careful scheduling of user testing should be included in an overall system test plan. If it is missing from the projected plan developed by the project managers, user managers should make the additions and schedule the appropriate testing activities in their departmental areas. A model system test plan is shown in Table 6–2.

System testing is a major activity in computer systems development. The recognition of testing as a major component has acquired significant weight for the following reasons:

- The trend toward a higher degree of integration of systems within an organization requires each system implemented to perform successfully right from the start, not only for its own purposes but so as not to degrade other existing systems.
- All levels of users within the same organization depend increasingly upon computer-generated data for making decisions and solving problems. The organization's performance is thus directly tied to the system's performance.
- Increased usage and familiarity with computer-based systems have resulted in higher expectations by the users of the system.
- The inflationary trend and cost of other development activities can be halted with improved testing procedures.
- Investment in systems maintenance resources can be reduced with improved testing procedures carried out before the system is installed.[6]

Table 6–2 Subsystem/System Test Plan

1. Name	A name or number which identifies the test. Example: DISAB (on-line discharge abstracting).
2. Purpose	Why: Test objectives of the test, including identification of the computer program components involved in the test. Example: To test data entry procedures for the DISAB system. Program Components to include data entry, data entry edit check features, file updates, and file change/data entry correction.
3. Location	Where the test is to be performed. Example: Test to be performed in the medical record department.
4. Schedule	When the test is to be performed. Example: Test to be performed on May 16, 1983 at 10:00 AM.
5. Responsibilities	Who: The individuals involved in the test and their specific duties. Example: Testing will be performed by medical record department data entry clerk and reviewed by medical record department.
6. General Procedures	What: A general overview of the test inputs, events, and anticipated results. Example: Testing will include data entry of a sampling of discharge abstracts for May 14 and 15. Items will be entered and the system will be monitored during data entry. Problems will be numbered and charted during the test. Computer files will be checked as follow-up.

Data Collection and Evaluation

1. Data To Be Input	What: a detailed description or sample of live or simulated input data to be used in the test.
2. Data To Be Output	What: a detailed description or sample of the data to be obtained as test results.
3. Method of Data Entry	How the data is to be entered, e.g., screens, source documents, punched cards, etc.
4. Method of Data Evaluation	How the test results are to be analyzed.

Effective procedures development is a final component in program development and testing. This means that procedures for changes, conversion, and operations are made available during the program development and testing portion of the development phase. Departmental managers are ill advised to proceed with personnel training for new systems until they are certain that effective written procedures explaining the new systems operations are available for the staff.

Effective communications between system developers and users has been stressed throughout this text. Medical record professionals will recognize the sometimes tedious role of documentation. Systems documentation is no less significant, however, than patient record documentation. In testing computerized solutions, which may include automated patient data or may be a forerunner of automated patient information, documentation of results is an important component. The sample test report in Table 6–3 provides follow-up documentation of results for the model system test plan previously shown. There will always be a need to incorporate the unique application elements into such a model. Here again, user interface is a focal point.

Personnel Training

Personnel training is the third major element in the development phase of the computer systems life cycle. The personnel training component in the development phase should be viewed by health information managers in two perspectives. Initially, the importance of developing positive attitudes cannot be overstressed. Second, the selection of effective training methods must be carefully determined and agreed upon prior to initiating personnel training. Let us examine how to foster positive attitudes among staff.

Table 6–3 System Test Report

Scope
1. NAME: The name shown on the corresponding test plan.
2. PURPOSE: The purpose stated on the corresponding test plan.
3. REFERENCES: Identification of the corresponding test plan and other pertinent documents, such as previous test results.

Description of Results
1. TEST METHODS: How the test was performed.
2. OBJECTIVES MET: Identification of specific test accomplishments.
3. PROBLEM AREAS: Discussion of problems encountered.
4. RECOMMENDATIONS: Specific actions to be taken, e.g., accept test results, perform additional tests, revise coding.

Source: Marvin Gore and John Stubbe, *Elements of Systems Analysis* (Dubuque, Iowa: Wm. C. Brown, 1975).

The involvement of all levels of staff in the evolution of computer-based systems is critical for fostering positive attitudes among personnel. Management must assure employees that implementation of the new system will change and enhance job tasks but not eliminate the job or the employee. When personnel reductions are anticipated as a result of the computerized process, it is rare that effective computer implementation cannot be accomplished through normal attrition.

Since a positive attitude toward learning is the keystone to the learning process, early attention to this issue will lessen the need to incorporate attitude readjustment sessions or counseling at this stage. At the same time, it is useful to build into the beginning step of the personnel training module, through survey, individual, or group meetings, exactly what the attitude about computer impact is in the department. By determining the level of fear and concern, as well as antitechnology attitudes, managers can build in additional components in the more formal training sessions.

Choosing Effective Training Methods

Another component of personnel training is the selection of effective training methods. Essentially the most effective method of training personnel is hands-on use with the equipment. If documentation of the procedures is clearly written, actual operation of equipment such as CRT terminals by those who must learn to use them is the most effective method of learning. In addition to the hands-on training of operations personnel for an individual application in the department, it is critical that the users of the information generated from that system be incorporated into a formalized training program. This may mean working with the medical staff and departmental managers on new and more effective ways to use patient data.

One of the most significant opportunities provided by computer technology is its capacity to display more and better information in formats that are superior to those available by manual development. This capacity benefits the users. Therefore, a careful explanation of how the information will be made more available in ways that are more useful to the user is a critical part of personnel training. If users have been involved in the determination of their requirements in the analysis phase of systems analysis then they will expect that the system can do more extensive manipulation of the data than did the previous system. Nevertheless, to spend the time to demonstrate and explain how that will be accomplished and made available to them is a necessary portion of personnel training so as to justify the systems design itself. Because this level of personnel is often an on-again, off-again user (medical staff, clinic director, et al.) or is only a partial system user, it is more difficult to prepare the program training process to cover this category of learners adequately. Health information managers may participate in such training by offering to provide demonstrations to medical staff or other patient information users so as to facilitate the use of the new system and to generate questions and reactions to its expected performance. One method that is very useful in personnel training is the pre- and posttest process. By using the pretest, trainers can determine the knowledge level and sometimes, attitude level, regarding computer entry in user areas. By using the posttest, one can view changes in attitudes and knowledge levels. These tests can be used to determine whether additional training is warranted or whether the type of training should shift. If such indications do result from a tabulation of the test results, then some alternative method of explaining the potential of the computer system would be better included in the personnel training process.

As always, team direction of this phase of the project is most desirable. In personnel training the team approach is again an ideal model. In a personnel training program in which system design personnel and user-manager personnel participate equally, users can explore the system to the degree of specificity that they desire. The system is then presented as a viable process that includes both technical and user components that can be identified and traced throughout the system.

The Development Phase Report

The development phase report should parallel the analysis and design phase reports. This report is another milestone in the overall systems analysis process. Table 6–4 outlines a sample development phase report.[7]

The major purposes of the development phase report are to (1) determine if performance specifications identified in the analysis phase are actually completed in the development process, (2) determine if the existence of satisfactory system test reports such as the models provided in this chapter are available to management, (3) determine how the personnel training process has been accomplished and to what effect, and (4) document a formal description of a conversion plan.

The conclusion of the development phase is the last milestone management can use to verify the adequacy of the overall development of the system project. It is the final checklist before actual implementation. It offers feedback so that whatever has been inadequately prepared, documented, or designed can be readdressed. Development phase reports should be provided to users as are other reports so that any questions they have can be directed to the project managers prior to implementation. As the

Table 6–4 Outline of Development Phase Report

I. System scope
   A. System title
   B. Problem statement and purpose
   C. Constraints
   D. Specific objectives
   E. Method of evaluation
II. Conclusions and recommendations
   A. Conclusions
   B. Recommendations
III. System specification
   A. External system specification
   B. Internal system specification
IV. Plans and cost schedules
   A. Progress plans
      1. Detailed milestones of the development phase
      2. Major milestones of all phases
      3. Changeover plan
      4. Operational plan
   B. Cost schedules
      1. Project cost of the development phase
      2. Project cost for major milestones
      3. Operation phase, that is, recurring costs
V. Appendixes

---

costs of systems design and development increase, significant attention to detail, careful completion of analysis tasks, and coordinated communication of systems analysis in all its phases become more and more important. Health information managers should participate in the development of any computer application that has impact on patient records or patient record support operations. The phases discussed so far may apply to implementation of a hospital computer system, an application program, or a department-specific computer system with direct interaction with the MRD. In each case components of the phases discussed should be included.

REQUEST FOR PROPOSAL ALTERNATIVES

In addition to the analysis and design phase activities that take place when an institution is developing an in-house system or purchasing an application program, a facility may also determine that a formal request for proposal (RFP) should be undertaken. Recalling that the design phase report is the documented blueprint for system development, the reader can easily ascertain that the same blueprint or some document that closely resembles it constitutes an excellent foundation for an RFP document.

An RFP is a formal document detailing the functional requirements of an intended computerized application so as to provide a vendor with a detailed profile. The profile is necessary for the vendor to develop a proposal specific to the needs of a particular health care facility. A brief review of the components of a computer system will enhance the discussion of the RFP.

All computer information systems have four basic functions: input—organization, verification, and control of data at the source; storage—retention of data in organized files for ready access; process—sorting, computation, and manipulation of the data according to operational instructions; and output—generation of reports containing the results, computation, and manipulation. To carry out these functions as identified and discussed in Chapter 2, computer systems have three functional components: hardware—the machinery of the system; software—the sets of programs that instruct the hardware to carry out the operation; and service arrangements—the manner by which the hardware and software are acquired and maintained. The functional capabilities of computer systems vary from small systems dedicated only to data processing for one department to large, comprehensive hospitalwide or clinicwide systems that perform a multitude of tasks. Familiarity with these capabilities and a firm background and knowledge of one's own current system are prerequisites to effective RFP development.

There are two approaches to preparation of an RFP. In the first, the RFP includes as great a degree of specification as is possible. This is in order to establish the fact that the organization knows in advance exactly what it expects to get from the vendor. The other approach calls for great flexibility and generality in the specifications so that the vendors are able to propose either a variety of approaches in meeting the functional requirements defined in the RFP or a prototype of a system used before. There certainly are arguments in favor of both approaches. But, we urge that health information managers develop an RFP with enough specificity to evaluate vendors' proposals and estimate costs for a particular computer application realistically. This suggestion applies whether the application is an abstracting package, chart locator system, discharge analysis package that may be available on an individual basis for their departments, or any other application.

The detail or investigation carried out prior to developing the formal RFP to design, implement, and test a system is exactly as is carried out through the analysis phase for systems analysis. The RFP could also be used for the development of another system or an existing basic design.

The investigation of the present system and the documentation of current volume and activity are all necessary prerequisites to developing this document. Clear delineation of objectives is the major concern just as it is in the analysis phase, in problem definition. For health information managers an example of such objectives might be to reduce the time required for admissions processing, to decrease the turnaround time for chart location, or to maintain or reduce the number of days in chart incomplete status for discharge records.

Management Responsibilities in Preparing an RFP

The purpose of the RFP is to provide a manager with an opportunity to gain the information needed to make a decision about a vendor's ability to supply a particular computer application. The form is a systematic and comprehensive document. The following list is the outline for an RFP document that is recommended by the American Hospital Association (AHA).[8]

1. Hospital profile
2. Functional system requirements
3. Technical system requirements
4. Training requirement
5. Implementation requirement
6. Financial consideration
7. Vendor profile
8. Conditions of bidding
9. Evaluation criteria
10. Decision timetable

Table 6–5 is developed from the AHA book on Preparation for an RFP for a Hospital Computer System and illustrates some of the features and issues just listed in greater detail. Note that evaluation criteria are also listed.[9]

Screening Vendors through Request for Information

In an AHA document on developing request for information (RFI) for a hospital computer system, the authors discuss the value of a request for information as an optional step in the process of selecting any computer system. An RFI is used to ascertain preliminarily the marketplace for vendor supply of a computerized process. It is a screening mechanism and requires a minimal set of information from vendors. The information includes:

Table 6–5 Profile of Request for Proposal Document

Hospital Profile
Physical configuration
Inpatient statistics
Clinic Statistics
Emergency services
Ancillary department services
Medical records files and numbering system

Functional system requirements (departmental)

Administration
Key factors performance statistics
Ongoing projects status reports
Other

Admissions/registration
R-ADT
Census
Patient tracking
Management reports
Other

Central service
Order and inventory control
Demand forecasting
Management reports
Other

Emergency services
Registration
History and charting
Appointment scheduling
Order processing
Obtaining test results
Management reports
Other

Finance
General ledger
Accounts receivable
Billing
Insurance reporting
Revenue analysis
Budgeting
Other

Medical Records
Patient number assignment
Medical record locator
Compiling and analyzing record
Record storage and retrieval
Tumor registry
Utilization review
Management reports
Other

Nursing Service
Order entry and processing
Diet ordering
Drug ordering
Charting
Patient care plan
Management reports
Staffing and scheduling
Other

Pharmacy
Order entry and processing
Drug history
Drug interactions
Charge capture
Drug information
Inventory control
Management reports
Other

Radiology
Order entry and processing
Result reporting
Room scheduling
File maintenance
Other

Other departments

Technical Requirements
Hardware

Implementation Requirements
Institutional data specifications

Table 6–5 continued

CPU description
Main memory capacity
Other storage devices and capacities
Expected hardware lifetime
Response time
Description of terminals
Description of printer(s)
Data access time
Data transfer rates
Character set
Transmission lines
Modems
Hardware expandability
Reliability and maintenance
Environmental need of equipment
Interface ability
Warranties and/or guarantees

Software
Operating system
Data security
Data integrity
Reliability
Response times
Maintenance and downtime
Documentations specifications
System design
File and data base specifications
Data recovery systems
Modularity and expandability
Interface ability
Method of data entry
Warranties or guarantees

Application software
Compiler description
Programming languages
Interfacing techniques
Statistical reporting ability
Maintenance and down time
Warranties or guarantees

Training Requirements
Orientation programs
Users manuals

Functional requirements
Technical considerations

Cabling preparations
Computer mainframe installation
Terminal and printer installation
Screen and output refinements
User education and training
System testing
System conversion
User acceptance testing

Financial Considerations
Main computer system
Communication system
Software
Implementation and benefit
   realization
Set-up and site preparation
Operations

Vendor Profile
Corporate history
Corporate future plans
Corporate financial status
Client list
Previous implementation and training
   method

Conditions of Bidding
Withdrawal of bids
Acceptance or rejections of bids
Time of project completion
Nondiscrimination clause
Performance and default clause
Collusive bidding clause
Taxes
Nonappropriation clause
Site visits
Consultants
Implementation team
Closing date
Evaluation of proposal

Evaluation criteria
Costs
Benefits, both quantitative and
   qualitative
Site visits
Selection of vendor(s)

Table 6–5 continued

Vendor factors, including support and profile	Assistance by vendor in obtaining Certificate of need (CON)
	Contract negotiations
Decision timetable	Vendor selection
Selection of finalists	System(s) implementation

- A summary of the purchasing institution's objectives
- Name and address of the vendor
- Summary of the vendor's corporate profile and history of involvement with similar or like computer applications
- List of vendor clients with references who are willing to be contacted for validation of their experiences with the vendor
- List of mandatory requirements that the vendor must meet if the vendor is considered for a formal proposal
- Summary description of the applications offered by the vendor to determine whether the system can be supplied
- Summary of the architecture available by the vendor[10]

For many health information managers this extensive and formal kind of screening process is not practical. A modified version can be employed in the use of individual applications. In selection of a computerized word processing system, for instance, a limited version of this list can be sent out to vendors to determine their potential for providing a proposal. Such a step would be very helpful and cost effective for the MRA. When such a process is used as a screening device, it provides information that streamlines the process of selecting vendors. It allows managers to rule out unqualified or inappropriate vendors. In a federal facility that must abide by open bidding policies and procedures, an RFI cannot be used effectively. Open bidding requires organizations to extend the RFI opportunity to all interested candidates. It may also rule out some vendor proposal activity if the information provided in the RFI process is not sufficient for the vendor to understand the application that is being sought. Once solicited, proposals must be evaluated by the requesting organization. The following list outlines the process involved in evaluating proposals. Both the quality and content of the proposal is evaluated.

Steps in Evaluating Proposals for Contract Services
1. Review the proposal to see that all RFP criteria are met.
2. Check previous experience: talk to other client hospitals.

3. Review experience and training of specific personnel to be assigned to project including work with similar projects.
4. Check to see that well-established principles and procedures of systems analysis and design are employed. This should compare with systems documentation discussed in this text.
5. Carefully examine cost estimates to see that they are well prepared, complete, and comprehensive.
6. Use a neutral consultant to assist in technical evaluation of proposals.
7. Carefully review the contract approach in the proposal to be sure that the specifications listed in the RFP document are incorporated into the contract.

Remember that many computer applications will require in-house maintenance by the hospital data processing department. Evaluation procedures need to incorporate this fact. It is often wise for hospitals to enlist the help of a neutral consultant to assist in proposal evaluation; many will employ a facilities management consulting firm to perform these tasks. Review teams should have an opportunity to examine the review process in advance and discuss factors that need to be addressed in the evaluation. A preview of this kind facilitates the formal process itself.

An MRD adaptation illustrating this profile in the selection of a word processing system is featured in Exhibit 6–3. It is important to tailor such guidelines to the specific problem. Notice the specific focus of the information in this exhibit.

Software Package Review by MRAs

When MRAs are reviewing software packages, a comparable evaluation process should take place. Notice how the key points in the following list compare with the items we have just listed.[12]

Evaluation of Packaged Systems
1. Does the system meet the hospital's specific needs? If not, will the supplier modify the system to meet the hospital's specific needs? Is the hospital willing and able to adjust to the general system requirements?
2. Who else has used the package? (Careful checking is essential.)
3. Can the system be implemented on the hospital's existing computer? What hardware modifications will be required?
4. How will the system be maintained? How difficult will it be to make changes?
5. How good is system documentation?

Exhibit 6–3 Computerized Word Processing Selection Preparation[11]

Hospital Profile

Charity Medical Center is a 404-bed, nonprofit, acute care institution operated by the Roman Catholic Order of the Sisters of Charity. Located in Tacoma's central Apple Hill area, the facility's philosophy, organization, and physical structure have evolved, adapted, and been renovated to best fit the needs of the community it was established to serve in 1861.

This growth and modification of Charity's operations are readily demonstrable by a brief examination and comparison of selected annual statistics of recent years. Figures chosen to illustrate this shift are those of three consecutive annual periods, as well as the two five-year spans preceding 1978.

	1968	1973	1977	1978	1979	1980	8-Year Variance, %
Admissions	13,387	13,488	13,126	13,104	13,671	13,061	−1
Discharges	13,879	13,547	13,201	13,177	13,674	13,050	−1
Census days	94,410	95,733	102,198	103,128	104,978	102,673	+8
Outpatient visits	——	1,547	6,897	8,756	8,189	12,363	+800*
Emergency room	7,984	9,215	11,688	14,811	14,217	14,832	+186

*Variance since 1973.

The most prominent feature of this table is the immense escalation in outpatient and emergency room encounters in contrast to the relative stability of inpatient treatment. This reflects a nationwide trend, and is directly contributable to the creation of a host of new departments that are essentially ambulatory in nature. These encompass a(n): arthritis clinic, sports clinic, family practice clinic, short stay surgery unit, and others. Other developments that have imposed direct and indirect intensification in reporting requirements are

- Recent completion of a $10+ million construction project that houses the largest and most advanced coronary and medical-surgical intensive care units in this region, as well as an enlarged emergency care division
- Proliferation of diagnostic and therapeutic services, especially those directed toward outpatients. These include clinical physiology, ophthalmology clinic; gastrointestinal laboratory; intravenous therapy; physical, occupation, respiratory, and recreational therapy; and others
- Broadening of residency curricula and research endeavors
- A hospice
- Comprehensive rehabilitation medicine program

Increases in documentation requirements is also influenced by

- JCAH and other accrediting body standards
- Current conscientiousness toward malpractice and its prevention
- Greater administrative information needs
- Quality assurance endeavors
- Federal and state legislative demands

Exhibit 6–3 continued

- Addition of other providers to the health care team (nurse clinicians)
- Greater emphasis on cost effectiveness and containment

Functional System Requirements
The central dictation service of the MRD has as its primary mission the production of a vast array of patient care-oriented reports innate to the inpatient stay, such as: history and physical examinations, consultations, operative procedures, discharge summaries, and the like. In addition the unit has absorbed the task of supporting much of the dictation requirements of the aforementioned new departments and services. This escalation is supported by the following numerical data:

	1975	1976	1977	1978	1979	1980	6-Year Increase, %
Number of reports	25,403	23,719	31,621	37,129	39,074	41,761	39
Number of pages	33,796	32,312	43,610	57,490	60,077	61,918	45
Yearly variance	——	−7%*	+25%	+10%	+5%	+7%	

*Decrease attributed to 10-week nurses' strike, when hospital activity held to virtual standstill, excepting emergency cases.

This tremendous magnification in report outputs has many implications for a corresponding increase in personnel and equipment commodities to keep pace with this drastic increase in services rendered. Hence, it becomes imperative to investigate possible methods of resolving this quandary to ensure maintenance of established production standards as well as cost effectiveness.

Financial Considerations: Cost/Benefit Analysis of First Operational Year[1]

	Typewriter-Based	Word Processing
Personnel		
Payroll salaries	$202,124	$108,836
Equipment		
Supertypewriters		
Cost of typewriters	$ 13,930	
Annual depreciation	$ 2,786	
Maintenance contract	840	
Cost per year	$ 3,626	
A-N-K system 30		
Yearly rental fee		$ 21,600
Sergeant dictating system		
Endless loop recorders	$ 18,000	$ 6,300
Transcribing stations	2,450	875
Annual depreciation	$ 4,090	$ 1,435
Maintenance contract	3,000	1,150
Cost per year	$ 7,090	$ 2,685
Equipment subtotal	$ 10,716	$ 24,285

Exhibit 6–3 continued

Supplies		
Paper, miscellaneous	$ 4,900	$ 4,200
Total Costs:		
Cost per year	$217,750	$137,321
Cost per report produced	$4.87	$3.06

Table Footnotes:
1. Yearly price increases will occur proportionately with inflationary factors.
2. Personnel staffing and cost factors derived from following formulas:

$$\text{Full-time equivalents (FTEs)} = \frac{\text{Lines typed/year} \div \text{work days/year}}{\text{Lines typed/day per FTE}}$$

   Total salary cost = Number of full time equivalents × average salary
   Figures Used:
   - Projected lines typed count for 1981 for
     Charity Medical Center — 2,150,448
   - Working days/transcriptionist/year — 252
   - Average salary ($6.50/hour with 15 percent benefits) — $15,548
   - Average lines typed per day:
     Typewriter-based (University Hospital) — 650
     Word processing system (Charity Hospital) — 1,200
3. Single pitch machine with sound hood at $995 each.
4. Estimated need for 14 machines derived from comparison of productivity with German Hospital, University Hospital, Eleanor Hynes Medical Center, and Seton Hospital, and arriving at anticipated need.
5. Five-year straight line depreciation for office equipment.
6. Supertypewriter quote of $60 per unit.
7. Monthly fee of $1,800 derived from following configuration's purchase price, which includes maintenance fees.

1 CPU disk with archive disk drive	$17,000
5 CRT workstations at $4,500 each	$22,500
2 120-character/second printers at $6,000 each	$12,000
Sort, system security, and mathematics support package	$11,000

8. Based on current Sergeant dictating costs of $900 per endless loop recorder and $175 per transcribing station. The dictating stations were not opted for purchase, as it would be more economically feasible to use hospital's existing Centrex phone system for dictating purposes. Cost for this would be same for either option.
9. Maintenance rates per year at $150 per recorder and $20 per transcribing station.
10. Based on costs of aforementioned facilities (footnote 4) and Charity Medical Center.
11. Projected report total of 44,684 for 1981, Charity Medical Center.

Determinations of number of endless loop recorders based on provided formula that estimates the amount of time taken to clear the dictation tank of quota of words. From this, the following equipment allotments were made:

Supertypewriter-based system	20 loop recorders	14 transcribing stations
A-N-K word processing system	7 loop recorders	5 transcribing stations

6. What will be the costs to
   a. Purchase the package?
   b. Modify it, if necessary?
   c. Implement it?
   d. Maintain it?
7. Does the MRD currently use a package from this vendor?
8. If so, what use has been made of the package from this vendor?
9. Will the MRA have appropriate access to system designers—either in-house staff assigned to work with the package or the vendor representative?
10. Does this package have potential for future MRD computer development integration?
11. Is this package being reviewed in accordance with an overall computer master plan for computer systems development within the organization?
12. What are the particular quality control features available to support effective data security procedures?

Of course, a key question is, Does the proposed packaged system meet the hospital's specific requirements? If not, is the hospital willing and able to adjust to the general requirements of the packaged system? By their very nature, packaged systems must be generalized so that they can be applied in a variety of hospital settings, but they must afford each hospital the opportunity to tailor the package's characteristics to its unique needs within reasonable limits.

Many factors are listed because the decisions are complex and costly. Computer technology must be selected wisely if effective improvements in health care and health information systems are to be accomplished. Accurate health data should be viewed as a precious commodity that is needed in order to effectively plan, assess, and carry out health services. The selection of computer technology should enhance and promote effective health data retrieval and use.

Equipment Leasing—A Method of Choice

When computer systems are selected from vendors, computer hardware is often leased. Health facilities and hospitals considering an investment in computer equipment would be wise to consider leasing as well as rental or outright purchase. Even though hardware costs are decreasing, innovations require a flexibility in hardware use that is often limited with purchased equipment. A current review of the lease-versus-purchase issue follows.

Alexander Grant and Company, a national certified public accounting firm, suggests that for many businesses, computer leasing may be the best alternative to financing computers in terms of both cost and benefits. According to a booklet prepared by the firm's management advisory services, the computer industry historically has been a rental industry.[13] As technological changes made computers more flexible for varied types of programming and expansion, however, computer purchases became more practical. But not for all hospitals. The major disadvantage to outright purchase is that it ties up working capital that a hospital with low cash reserves can ill afford.

A computer lease, on the other hand, allows a company virtually unrestricted use of a computer for a specified term, under an agreement with the lessor. Leasing a computer through a finance agreement rather than renting it from the manufacturer can mean lower total costs if the value of the computer is included. Although monthly lease payments may be equal to or higher than rental payments, the net cost of the leasing, minus the value of the computer, is often lower than the rental costs.

Along with investment tax credit and accelerated depreciation, the total after-tax costs can be reduced significantly with a lease agreement. Several types of leasing arrangements exist, including:

- Finance, equity, or installment sales leases. Under this type of lease, the lessee builds equity in the equipment during the lease term, usually 10 years or less, and takes full ownership after the last payment.
- Leveraged leases. In this type of lease, the lessee makes a downpayment at the beginning of the lease term and builds equity in the equipment, taking full ownership after the last payment. The lease term is usually up to 15 years.
- True operating or risk leases. The lessor retains ownership of the equipment during and after the lease term and no equity interest is built up during the lease term.
- Master leases. This type includes allowing the lessee to acquire additional equipment at future dates with predetermined terms.
- Walk-away leases. In this kind of lease, the lessee is allowed to terminate the lease, with special conditions, at a date earlier than the end of the lease term.

With a long-term lease commitment, a health facility's needs may change and the equipment may be technologically obsolete by the time the lease expires. Therefore, some facilities may be justified in first acquiring a higher cost short-term rental agreement, and considering the long-term lease later, after day-to-day experience with the computer has further defined their needs.

Contract Types

Part of the decision regarding what type of contract to execute is dependent upon the results of the needs assessment developed during the analysis process and the conclusions reached when the RFP was constructed. The basic decision as to whether the equipment is to be rented, leased, or purchased will have a long-term effect on the procuring institution. All of the benefits and drawbacks of each type of contract must be weighed and considered with respect to the objectives established in the needs assessment. The possibility of time-sharing a computer system should not be overlooked. This could be a relatively inexpensive means of satisfying hardware needs. The following sections describe the various types of contracts and reference a typical model contract.[14]

Rental Contracts

The simplest arrangement, representing the lowest risk but usually the highest cost, is rental. The function of a rental contract is to provide equipment to the organization for a short time, from 30 days to one year. In this type of contract the facility renting the equipment has made very little commitment for any significant duration of time and has no intention of owning the equipment. In many cases, a 30-day cancellation clause is all that is required. On such contracts accrual of rental payments against the purchase price is generally not significant. Rental contracts are most beneficial in satisfying short-term needs such as demonstration systems to establish feasibility.

Flexibility should be the primary motivation for entering into a rental contract. If management of a facility is unsure of the long-term need for specific equipment, this type of contract could prove to be beneficial. The drawback of a rental contract is usually found in the higher cost of rental payments. Since there is little risk to the facility renting the equipment, the premium paid in rental payments covers the risk to the vendor providing the equipment.

Rental contracts are usually available from the manufacturer and, in many cases, from leasing firms that specialize in computer equipment leasing. If a third party leasing firm is involved, the lessee should ensure that the equipment provided qualifies for a manufacturer's maintenance agreement and that all warranties applicable to the equipment are provided to the lessee.

In a lease contract some vendors impose a penalty for early termination. If early termination is a possibility, the total of the lease payments plus the penalty payment may equal or exceed the amount paid for the same period on a rental contract basis. This should be considered prior to making a decision to lease.

In many cases rental contracts are only provided by the original manufacturer who can afford to take the equipment back into inventory prior to placing it in another installation. If total flexibility is required and no long-term commitment is desired, a rental contract may prove to be the most desirable type of arrangement. However, it may be better to consider an option-to-purchase contract than a pure rental contract, since greater flexibility is afforded.

Lease Contract

A lease contract is not unlike a rental contract except that in a lease contract the commitment is for a longer period. Lease rates are lower than rental rates since the risk of termination is considerably lessened. Lease contracts typically run for periods of one year to three years. The commitment over the lease period is firm, and a certain amount of flexibility is subsequently lost. Regardless, the savings in lease payments may justify a lease over a rental.

Under a lease contract, the facility that obtains the equipment does not intend to own it in the future but is primarily interested in lower payments during the specified period of time that the equipment will be useful. It should be noted that purchase accruals are usually available on lease contracts; if purchase is even a remote possibility, the accrual should be made part of the contract.

Lease with Option to Purchase

Under this type of contract, the facility may plan eventually to own the equipment. This lease is sometimes called a full-payout lease in that the majority of the cost of the equipment will be paid during the lease period and title will pass at the end of that time if the purchase option is exercised.

The purchase option must be voluntary and the lessee must be free to exercise it at any point in time during the contract period. Leases with option to purchase typically run from three to seven years. The length of time selected is usually dependent upon the institution's desire for lower lease payments over a longer period versus making higher lease payments for a shorter period of time.

In a contract of this type, a definite commitment is required to make the lease payments and to exercise the purchase option upon completion of the lease period. Like the lease contract, this type is not flexible. Leases with option to purchase are usually available from the manufacturer, who in many cases will assign the contract to a leasing firm or financial institution.

It should also be noted that leases with option to purchase may carry the same early termination penalties as ordinary leases. Manufacturers' maintenance agreements and warranties applicable to the equipment must also be considered.

Purchase Contract

The last major contract type is a purchase contract. The purpose of purchase is to acquire title as soon as the specified contract terms and conditions are successfully met. If funds are available and if it is determined that both the equipment's useful life and the needs of the facility justify the purchase, this can be a beneficial choice. However, care must be exercised in any purchase arrangement because, although this type of contract normally has the least overall cost for an extended period of time, the risk is the highest. This is a worthwhile choice for data communication equipment because of its long useful life on a variety of tasks.

Flexibility of management must be considered in a purchase decision since equipment purchase generally represents a long-term commitment. This decision requires long-range planning and must consider the equipment's potential obsolescence, its ability to handle required workloads, and many other factors that could potentially make purchase unattractive. If a decision to purchase is made, a purchase contract rather than a simple purchase order should be considered.

The terms and conditions to be fulfilled by the seller prior to receiving payment must be completely defined as in any contract. The equipment purchased should be new equipment if possible and should qualify for a manufacturer's maintenance agreement in addition to a written warranty statement from the manufacturer. Training, installation, conversion, and many of the items normally considered in a lease must also be considered in a purchase agreement.

It may be preferable to enter into a purchase contract that actually is just a rental agreement for a brief initial period; after the facility is fully satisfied with the product, the purchase option is exercised.

Model Contract

An institution should be aware that vendors usually each have a standard contract; keep in mind that such a contract is always designed to favor the vendor and limit the vendor's responsibilities. For this reason it is suggested that the institution draft its own contract, one providing the protection desired and identifying all responsibilities of both parties.

Statement of Work

This section of the contract need only make reference to either the technical requirements in the RFP or the technical description in the vendor's proposal. By incorporating these requirements "by reference," (see the following sample Reference Clause) the size of the contract itself can be kept to a minimum. This will bind the vendor to those requirements just as if they were a part of the contract itself.

> ### Sample reference clause
> The technical description of the hardware capabilities detailed in the vendor's proposal is incorporated herein by reference with the same effect as if it had been reproduced in its entirety.

If reference is made to the technical requirements in the RFP, make sure that any changes resulting from the negotiations are duly noted in the contract. Caution: One word change can cost thousands of dollars in certain cases.

Contract Clauses

Certain clauses must be considered in the creation of the contract document. These fall into three basic categories: (1) required, (2) elective (but highly recommended), and (3) discretionary. Clauses in the "required" category are necessary to protect the institution from costly or troublesome situations. Table 6–6 lists these clauses and the recommended categorization for each.[15]

Contracts

The final document produced in a procurement cycle is the contract. It is the binding document that obligates the vendor as well as the procuring agency to conform to all of the agreed upon terms, conditions, and requirements. Prior documents such as the specifications, work statement, or verbal agreements are contractually binding only if they are included or referenced in the final signed contract.

The initial draft of the contract should be prepared in conjunction with the RFP document. In order for the required and desired clauses to be included in the final contract, it is suggested that the institution include a

Table 6–6 Contract Clauses

Clause	Required	Elective (Highly Recommended)	Discretionary
A. *Cost section*			
1. Contract terms	X		
2. Payment	X		
3. Price protection/price changes		X	
4. Termination or cancellation	X		
5. Nonfunding termination	X		
6. Purchase option (lease only)	X		
7. Taxes	X		
8. Overtime usage (leased hardware)	X		
9. Maintenance	X		
10. Insurance	X		
11. Supplies		X	
12. Hold back		X	
B. *Delivery and installation*			
1. Delivery	X		
2. Site preparation	X		
3. Installation	X		
4. Relocation		X	
5. Freight	X		
6. Acceptance	X		
C. *Terms and conditions*			
1. Assignable rights		X	
2. Patent protection	X		
3. Standards of performance	X		
4. Equipment modification	X		
5. Component cancellation/substitution	X		
6. Attachments	X		
7. Upgrading of hardware/software			X
8. Manufacturer interface	X		
9. Education and training	X		
10. Key personnel		X	
11. Latent defects/warranty	X		
D. *Miscellaneous clauses*			
1. Conversion		X	
2. Notices	X		
3. Progress reports			X
4. Equal employment opportunity	X		
5. Examination of records	X		
6. Clean Air Act	X		
7. Disputes	X		

copy of the contract in the RFP package. This does not mean that the initial contract cannot be modified, particularly during negotiations, but rather it specifies the major terms and conditions that a final contract will include.

Each organization should use legal counsel to guarantee that the wording of each clause precisely reflects its requirements before the document is included in the bid package and to ensure that all wording of the clauses meets the legal constraints.

THE IMPLEMENTATION PHASE

Implementation (Operation Phase) Overview

Following final review of development phase activities, implementation takes place. Decisions have been made for conversion methods and changeover activities to commence. The implementation phase can generally be divided in two: system changeover activities directed to the implementation task itself and postinstallation audit activities directed to monitoring and evaluating the effectiveness of the implemented system. During these activities the reins are transferred to the hands of the user-managers. At this point the new computer system becomes the new medical record MPI or the MRD chart locator system. Along with this shift, the new computer system moves into focus as a functional operation of the department. The potential effects of this phenomenon will be examined.

System Changeover

Three factors are important in understanding system changeover activities.

1. Changeover activities must be scheduled for integration with ongoing departmental activities.
2. Staff must be instructed in appropriate manual back-up activities in case of computer failure.
3. Working with vendors to carry out changeover activities requires a management consumer role.

Changeover activities will address system performance, generally, and staff preparation. What exactly is involved?

Scheduling Changeover Activities

Changeover activities are the activation of the conversion plan set forth in the development phase. That plan is usually a general strategy that specifies the overall process of implementation. However, there is distinct need for department managers to operate from a more detailed structure. Not only is a specific "how to" procedure important to start a new system smoothly into operation, but employees need to understand how the changeover activities will affect other routine operations.

Department Manager Tasks

Prior to any scheduling activities, departmental managers will want to verify that key tasks have been completed. This checkpoint should be coordinated with the systems team. The listed tasks are necessary for each computer application. Plan time to review each of the following:

1. Check to see that final training procedures are available so that consistent methods can be employed. This is particularly important in training for computer systems, as employee attitudes are significantly affected by training methods.
2. Be sure that final user procedures are capable of directing personnel to carry out the new functions. Make sure all final system features are included.
3. Final system requirements should be documented in the development phase report and specified so that each department has a clear understanding of the system requirements in each area.
4. Be sure that results of system testing supplement procedures by providing samples of expected products for each application. This enables department managers to reference individual operations with data processing staff as the new process commences.
5. Verification should be made to indicate that appropriate manual procedures have been established. An example is checking to see that the medical record deficiency checklists used as source documents in a chart deficiency system correspond to the screen displays planned for data entry.
6. Identify a contact person in the data processing department for troubleshooting communication gaps.

Once these items have been checked, the general functional operations in the department should be considered. A brief list of those that will be affected by the new system operation should be compiled. For instance, all staff may need to use a new chart request form that is compatible with a new computerized chart locator system. In another example, telephone changes may be made to support a new on-line patient registration and MPI system in which the MRD verifies and assigns patient numbers to admitting, ER, and outpatient departments. Such changes may interrupt work flow in adjacent areas. Compiling a brief "impact" list for communication to department personnel can help facilitate the expected change.

Another tool in system changeover is a changeover action log prepared for use with the new system. Such a log will likely be provided for each application by the systems team and should be checked by the department manager to be sure it meets the department's needs. All such logs need to be kept in duplicate as a control for user and data processing staff in handling system problems. Figure 6–4 is a blank sheet from a system changeover action log.

Once these items have been reviewed and questions resolved, departmental managers need to inform all staff and specify who will start the new process. Assuming that staff training has been completed, more people may understand the new system tasks better than they understand how the new process will affect their job. Consequently, the next changeover activity should be a briefing session with the MRD personnel. The agenda might be as follows:

1. Report general status on new system.
2. Identify specific planned start dates.
3. Introduce employees who will be operating the new system.
4. Review each functional operation in the department to point out how the new system may affect its operator.
5. Name a departmental "troubleshooter" for questions once the system starts.
6. Ask for and document questions or any last minute observations that may affect system operations.
7. Explain that the new procedure will be monitored by a system performance log.
8. Review system back-up procedures.

Back-up Provisions Are Important

Staff need to understand how to perform the computer system operations accurately. When computers go "down" or are brought down for scheduled maintenance, routine activities must still go on. All written procedures should include a back-up section to enable staff to handle the changeover "crises" process with a positive attitude. It is the user-manager's responsibility to see that there are appropriate manual back-up

Figure 6–4 Changeover Action Log

			Changeover Action Log		

Action Number	Action Description	Person Responsible	Date Referred to Data Processing	Date Resolved	Scheduled Follow-up Date

procedures in each situation. If these are maintained with the new procedures, the employees who are trained on the new computer system will have an appropriate resource. A calm instruction to switch to the manual back-up when the computer is brought down will inform personnel that the tasks must and will be performed in an established routine regardless of computer status. It also indicates that department personnel are responsible for performing the tasks and the computer is only a tool to assist in the achievement of objectives related to that responsibility.

Changeover activities concurrent with starting the new system are then mapped out. These activities should be closely coordinated with the system team and will often provide the foundation data for carrying out postinstallation audit.

Concurrent changeover activities are as follows.

1. Identify and verify exact system start time and date.
2. Start new operation.
3. When changing from one computer system to another, check new outputs with old outputs as much as possible.
4. Verify output content.
5. Check inputs and outputs to see that they are working as planned.
6. Follow up immediately on all errors.
7. Correct and document the errors. Correct manual or computer processes as necessary.
8. Maintain a log to record actions, responsibilities, assignment, and completion dates.
9. Solve all problems promptly. Work with user groups and the systems team to do so.
10. Defer any refinements or changes in the system until changeover has been completed.[16]

These procedures will go a long way to ease the changeover process. Because changeover is an emotional activity, its occurrence usually heightens tensions and amplifies mistakes. Careful attention to the activities just listed is a major management task.

Vendors Require Management Consumer Role

When vendor systems are being installed, most of the changeover activities also apply. The troubleshooter will be a vendor representative rather than a hospital departmental staff member. In complex hospital computer system installations, a team may handle problems. It is important for user-managers to take particular pains with the changeover activity log. In cases where organizations have selected vendor systems and leased computer hardware, tracking the systems functioning is necessary in order to determine system acceptability.

Postinstallation Audit

Postinstallation audit is the second major element in the operations phase. Postinstallation audit is also referred to as the "evaluation phase" in systems development: "The evaluation phase is primarily concerned with the development of criteria and methodologies for the appraisal of the significance of information systems."[17] As such, the activities included in this phase of systems analysis are multifaceted. The audit is performed to accomplish the following general objectives:

1. Determine whether the system's goals and objectives have been achieved.
2. Determine whether departmental procedures, operating activities, and organizational activities have been improved.
3. Determine whether revised scheduling and resource and file usage have been improved appropriately.
4. Determine whether user performance requirements have been met while simultaneously reducing or containing former costs.
5. Determine whether known or unaccounted for limitations of the system require systems development.
6. Determine whether projected economic benefits in personnel savings have been achieved.
7. Determine if improved information flow directed to patient care services has been accomplished.

In postinstallation audit, the operational system is evaluated for effectiveness. It is concerned with immediate and long-range activities. Medical record managers will need to be conversant with a number of issues to effectively participate in these activities. We will consider four major factors:

1. Relating performance objectives to the computer system performance standards is a fundamental task for managers.
2. Postinstallation evaluation planning is a team function that incorporates interdepartmental tasks.
3. Technical audit methods should be defined and described so that application users can determine if matters of data security and system control procedures are likely to meet their needs.
4. Integrating computer system performance into work standards is an ongoing management responsibility.

Measuring Performance Objectives

Relating performance objectives to standards of performance begins with review of the performance requirements prepared during the analysis phase of systems analysis. Since performance requirements are directed to the tasks that comprise functional operations within departments, individual department managers are responsible to see that the appropriate tasks are carried out. The initial step in postinstallation evaluation is a review of the performance level of the computer system. It should be compared against the original specifications. A simple list of objectives with associated criteria can be used for this comparison. Notice the objectives in Exhibit 6–4. These illustrate how specific objectives direct evaluators to measure system performance. Performance objectives should be listed on an evaluation document with space allowed for comments. Managers should also indicate any additional items to be evaluated on the new computer system. These items would include any recent features or additions to the original performance definition.

Computer terminal operators should be asked to clock or tally appropriate activities to determine how well the new system meets objectives. An example would be clocking patient registration time when a new computerized patient registration system is used. Determining registration time over a two-week period can provide a postinstallation performance measure that can be compared to the time used for manual patient registration.

Exhibit 6–4 Goals and Objectives for the Proposed System

GOAL: To employ an efficient discharge abstracting system to provide timely and relevant internal and external reports and statistics to all required users.

OBJECTIVES
1. Abstract all charts of discharged inpatients within four days of discharge.
2. Be able to compile daily listings of discharge diagnoses, operations and ICD-9CM codes for each discharged patient for the business office for billing purposes.
3. Have discharge statistics compiled by the second Tuesday of the following month for hospital distribution.
4. Be able to produce the following reports on demand:
   a. Monthly disease, operation, and physician indexes available upon completion of the monthly abstracts.
   b. Listing of the top 20 diagnoses or surgical procedures for variable time periods upon request.
   c. Lists of charts with complications, normal tissue removal, nosocomial infections, transfusions, and deaths for monthly medical and surgical staff meetings.
   d. Listing of all patients admitted for treatment of neoplastic disease at the end of each month for the tumor registry.
5. Maintain permanent annual diagnoses, operation, physician, and death indexes.
6. Be able to retrieve charts for studies and MCEs and other special requests within 48 hours.

MRAs should employ comparable performance measurement to all computer and manual operations. Results of these activities should in turn be incorporated into updating periodic management plans and projects.

Postinstallation evaluation planning prepares personnel to carry out a review of tasks that have interdepartmental scope. When integrated systems are installed, for example, information flow between departments in hospitals and other organizations is altered. Part of systems design and development is preparation of a formal postinstallation plan that all team members and application users apply in initial system evaluation. If the postinstallation review plan is carefully detailed, each user-manager will have a clear picture of what is to be counted, monitored, and so on to provide consistent feedback to the systems team.

The sample in Exhibit 6–5 illustrates how documents can be used to assess system performance. Department managers will participate in supplying the answers to such surveys. They will also measure the unique performance requirements for operations in their own departments. Together these activities provide a base for system assessment. Whereas these review documents are user-oriented, other methods rely on data processing staff for completion. Another component in evaluation incorporates a three-part evaluation that supplements the measures already described. This three-part evaluation should remain the primary responsibility of the system design team, which may be augmented by additional members for system evaluation tasks.

In the comparison specified in Table 6–7 the three-part evaluation is featured. Which method is selected depends on the focus of the evaluation. If the evaluation of hardware usage is deemed appropriate, a computer performance evaluation will be chosen. When verification of processing procedures and file contents is to be evaluated, a computer audit may be selected. Data security review would also use a computer audit. Computer feasibility study relates the hardware configuration to the known data processing requirements. Only when the existing requirements dictate a hardware change will computer feasibility be evaluated. Hospital computer systems may require hardware upgrading so that new applications can be added to the system. Recall from Chapter 2 that all computers are limited by the processing capability of the central processing unit. When additional applications are developed, the central processing unit may require upgrading so that existing response standards can be maintained.

Technical Audit Methods Provide System Controls

Technical audit methods are major components in evaluation activities. One of the responsibilities of the system team is to recommend evaluation

Exhibit 6–5 Application Controls Checklist
—Implementation—

5.1 Is there adequate segregation of duties and independence from users?

		Yes	No	In Accordance with Org. Std.	Different from Org. Std.	Specify
a. Are user department and data processing responsibilities segregated?	1.3					
b. Are duties of data processing functions separated?	1.4					
c. Are duties in data processing groups periodically rotated and all personnel required to take vacations?	1.5					

5.2 Do the procedures ensure proper control over the development, maintenance and sufficient documentation of this application?

	Yes	No	In Accordance with Org. Std.	Different from Org. Std.	Specify
a. Was the application approved properly?					
b. Was user department review adequate?					
c. Were internal controls reviewed?					
d. Were proper testing procedures performed?					
e. Were preimplementation reviews conducted?					
f. Is documentation complete?					
g. Were modifications, if any, made by programmers not responsible for original program writing?					

Exhibit 6–5 continued

5.3 Do the procedures ensure effective control over computer operations and access to data files and programs for this application?

	Yes	No	In Accordance with Org. Std.	Different from Org. Std.	Specify
a. Are runs made according to schedule?					
b. Are unscheduled runs properly authorized?					
c. Is access to computer controlled in accordance with company policy?					
d. Is access to data files restricted?					
e. Is access to operational programs controlled?					
f. Is computer rental to outsiders adequately controlled?					

5.4 Do the backup and reconstruction procedures provide for reasonable continuity of processing this application in the event of a disaster?

	Yes	No	In Accordance with Org. Std.	Different from Org. Std.	Specify
a. Have programs been tested on stand-by facilities?					
b. Are duplicate copies of programs stored remotely?					
c. Are backup copies of data files maintained? (Refer 4.10 of file)					

Source: Reprinted from *Business Data Systems: A Practical Guide* by Jack D. Harpool, Ed. (courtesy of Ernst & Ernst, Akron, Ohio) with permission of Wm. C. Brown Co. Publishers, © 1978.

Table 6–7 Comparison of Traditional Evaluation Approaches

Computer Performance Evaluation	Computer Audit	Computer Feasibility Study
Continuous sampling	Regular, periodic sampling	Occurs only when hardware change is contemplated
Uses evaluation hardware or software; little staff time required	Internal auditor or "outside" auditor examines records, procedures—limited interviews with staff	Entire study conducted by staff; requires substantial staff time
Provides numeric measurement of hardware utilization	Provides judgmental opinions on accuracy of records and processing procedures	Provides substantial documentation of all systems and procedures together with alternatives for processing
One-time cost for any hardware or software monitors. Price range is $5,000–$10,000 each.	Repeated cost for each audit. Price range is $10,000–$20,000 for each audit.	Cost for each study is basically the staff time devoted to the study. Cost range is $25,000–$50,000.
Evaluation data collected during daily running time	Requires approximately 4 calendar weeks	Requires approximately 4–6 calendar months for complete study.
Results sometimes not valuable due to lack of utilization standards	Results reflect accountant's viewpoint of recordkeeping and reporting	Results emphasize best hardware to meet needs

criteria to management. It is important for the systems team and the users to coordinate their development of criteria to be used in the evaluation of the system. This must be carried out prior to the start of evaluation tasks, of course. System audits should be incorporated into the evaluation process. MRAs will benefit from a clear picture of the scope of technical systems evaluation. Of particular benefit to MRPs is an understanding of the technical aspects of postinstallation audits on new systems. The outline that follows provides a general summary of appropriate evaluation criteria that would be suitable for all computer applications. However, each computer application will require a unique method to measure these criteria. Again, coordinated effort is needed to define the significant measurement needed for each application. Continued team performance is mandatory.

Many system characteristics must be evaluated when assessing the overall merit of a system. Some of these evaluation criteria are

- Time: processing, elapsed, response, or operations
- Cost: annual, per unit, or maintenance
- Quality: better product or less reworking
- Capacity: average or peak loads
- Efficiency: increased productivity
- Accuracy: fewer errors
- Reliability: fewer breakdowns
- Flexibility: many possible operations
- Acceptance: employee or management
- Controls: fewer operational, accounting control breakdowns, and increased security
- Documentation: written/pictorial descriptions
- Training: how to operate the system[18]

Results of the evaluation activities should be incorporated into documentation maintained in the MRD.

Integrating Computer System Performance into Work Standards

Departmental operations are altered by computer systems developments. Work performance standards routinely maintained by department managers should be revised to include new system activities. Documentation of operational work standards provides a foundation for management evaluation and will provide the basis for determining when and if computer systems require upgrading. Similarly, department managers should identify budget data related to the new system. When possible, budget feedback on new system costs should be monitored along with cost centers for other departmental operations.

In his book *Hospital Information Systems,* Homer Schmidt projects evaluation from three perspectives. He indicates that the computer system's probable effects on the patients must be analyzed. Computer system effects on the behavior of personnel within the organization must be evaluated. And a thorough analysis must be undertaken to determine the economic effect of the system on the hospital. Effective integration of computer systems into work standards will deal with all of these aspects. The responsibility for effective management of patient data and health information will continue to reside with medical record practitioners. The computer is a capable tool to assist in the achievement of the desired results.[19]

Final System Documentation

The results of postinstallation audits should be incorporated into the final system documentation. This final report is subject to review by performance review teams. They do this for the organization generally and the report is required for long-range planning for computer development.

The final documentation package requires the compilation of all the documents that were prepared along the way. It consists of the developmental documentation that was used in problem identification, performance identification, project monitoring documents used to control the project, and the documentation that actually will be used during the operation of the system. A format for the final documentation package is suggested by the following outline.

Final System Documentation

   I. Introduction and table of contents
      A. Name of the system
      B. Purpose and objectives of the system
      C. Who uses the system
      D. Where the system fits into the hospital/department
  II. Explain, in a narrative sequence, how the system operates and include the following:
      A. A systems flowchart of the overall system
      B. Any required documentation sheets that are necessary to explain the system, such as
         1. Cost data
         2. Documented operations flowchart and documentation section
         3. Input/output sheet
         4. Systems requirements
         5. Decision table
         6. Equipment sheet
         7. Personnel sheet
         8. File sheet
 III. Show how the computer programs fit into jobs, how jobs fit into cycles, and how cycles fit into the overall system. Describe, in narrative, what each program, job, and cycle does. Include the following:
      A. A computer listing showing the statements of each program.

B. In-line comment or note statement. For every two or three program coding instructions, put in a comment that explains what these instructions do and how these instructions interact with the rest of the program.

C. One program flowchart for each program and an overall flowchart showing how the various programs interact with the rest of the program. These should be coded to the systems flowchart so that readers can perceive operational relationships.

D. Flowcharts and detailed narrative descriptions of complicated logic or calculations within each program.

E. Pictorial layouts of the files, the outputs, and the inputs.

F. Program run book that contains operating instructions, for example, brief program narrative, computer setup information, tapes, disks, carriage control tapes, special printer forms, or any special restart procedures in case of failure prior to normal program end.

IV. Summarize the implementation plans and the results of the implementation including test results.

V. The appendix should contain all other documentation
   A. Problem identification
   B. Feasibility study report
   C. Summaries of analysis, design, and development reports.
      1. General information on the area under study
      2. Interactions among the areas being studied including copies of user interviews
      3. Understanding the existing system
      4. Definition of the new system's requirements
      5. Detailed new system design
      6. Economic cost comparisons
   D. Final written report including evaluation results
   E. Implementation plan
   F. Any significant notes made by the analyst or programmer

The report should be complete enough to enable the reader to understand the essential characteristics. Such information is critical when future inevitable changes are required.[20]

Questions and Problems for Discussion

1. In what phase would a Gantt chart be an appropriate tool?
2. What types of projects would be more effectively charted by a PERT chart?
3. Contrast the PERT chart with a Gantt chart. Illustrate your answer with examples.

4. Identify the major management tasks for the MRP in the development phase of the systems life cycle.

5. Do the management tasks increase or decrease as the system moves into the implementation phase? Explain your answer.

6. What steps should be taken in developing an effective conversion plan for implementing a computerized discharge abstracting system?

7. Contrast the various conversion methods discussed in the chapter. Select an MRD functional operation for computerization. Select a conversion method that would be appropriate for that application. Include an explanation of the factors involved in your decision.

8. Is an RFI appropriate when investigating various vendors of health information systems? Why or why not?

9. Explain the purpose of an RFP. What role would you expect to play in developing such a document for an automated hospital information system?

10. Equipment may be obtained via rental, leasing, or purchase. Give the advantages of each method and list specific situations that might govern choice.

11. Prepare a plan for personnel training for a developing computer system application of your choice. Present your plan in a formal presentation to an administrative committee or a medical staff advisory group.

12. Why is it important for the user-managers to direct and participate in system testing activities?

13. What is the purpose of postinstallation audit? What specific role should the MRA play in this activity? Why?

14. Explain the significance of final system documentation.

NOTES

1. Holly Shepperd Clark, "Master Tracking System: Internal Medical Record Control," *Topics in Health Record Management* (December 1981).

2. John Dearden and F. Warren McFerlan, *Management Information Systems* (Homewood, Ill.: Richard D. Irwin, 1966).

3. Charles Murray and Jean Wallace, "The Development and Use of a Computerized Cancer Data System," *Topics in Health Record Management* (December 1981).

4. Elizabeth Capozzoli, "An Automated Approach to Medical Record Tracking," *Topics in Health Record Management* (December 1981).

5. Charles Austin, *Information Systems for Hospital Administration,* (Ann Arbor, Mich.: Health Services Press, 1979).

6. John Burch, Jr., Felix R. Strater, and Gary Grudnitski, *Information Systems: Theory and Practice* (New York: John Wiley & Sons, 1979).

7. Marvin Gore and John Stubbe, *Elements of Systems Analysis* (Dubuque, Iowa: Wm. C. Brown, 1975).

8. American Hospital Association, *Hospital Computer Systems Planning: Preparation of Request for Proposal* (Chicago: AHA, 1980).

9. Ibid.

10. Ibid.

11. Austin, *Information Systems for Hospital Administration.*

12. "Product News," *Computers in Hospitals* (March/April 1981).

13. Ibid.

14. National Clearinghouse for Criminal Justice Information Systems, *System Development Guidelines on ICAP Manual, Second Edition* (Sacramento, Calif.: Search Group Inc., 1981), p. 55.

15. Ibid., p. 60.

16. Gore and Stubbe, *Elements of Systems Analysis.*

17. Mervat Abdelhak, "Health Information Specialist," *Journal of AMRA* (April 1980).

18. Ardra F. Fitzgerald, Jerry Fitzgerald, and Warren D. Stallings, Jr., *Fundamentals of Systems Analysis* (New York: John Wiley & Sons, 1981).

19. Homer H. Schmitz, *Hospital Information Systems* (Rockville, Md.: Aspen Systems Corp., 1979).

20. Fitzgerald, Fitzgerald, and Stallings, *Fundamentals of Systems Analysis.*

Chapter 7

How To Implement Computer Technology in Medical Record Operations

Objectives

1. Define and describe medical information structure through the evolution of an individual patient's treatment and its relationship to providers and other users throughout the facility.
2. Explain the significance of the registration-admission, discharge, transfer (R-ADT) function in the development of integrated information in automated hospital information systems (AHIS).
3. Define and describe three basic principles of patient data computerization derived from the systems analysis foundation.
4. Develop descriptive profiles for each functional operation in the MRD in preparation for automation.
5. Define and describe strategic planning for automation in the MRD.
6. Present the R-ADT function and demonstrate extended use of the function in designing additional MRD applications.
7. Describe appropriate roles for MRPs in applying patient data computerization principles and strategic planning techniques.
8. Present an illustration of a strategic plan for computerization in an MRD.

The purpose of this chapter is to define and describe medical information structure and relate it to the R-ADT function. The first two parts of the chapter present an explanation of the R-ADT function as a strong foundation for building computer applications within the hospital and the MRD. Patient data computerization principles will be stated as preparation for computerization in an MRD. Readers will be shown how these principles can be used to prepare documentation of existing functional operations. The documentation can serve as a resource to facilitate computer applications once a decision to automate has been made.

Specific techniques that can be used in application development, such as descriptive profiles, file analysis, and data access issues are featured. These techniques are derived from the systems analysis process but have been condensed in this chapter to provide a streamlined management approach. The systems analysis process continues to correspond to other problem-solving techniques. All analysis tools featured in this text should be viewed as a resource bank from which practitioners can select as needed. Experience with design and development of computer applications will enable managers to select efficiently. The tools and techniques should enable users to meet the organization's or department's objectives most effectively.

The third part of this chapter is directed to consideration of strategic and tactical planning methods designed to enable practitioners to prepare a master automation plan for an MRD. This section explains how strategic planning is accomplished. Four functional operations alternatives for initial computer application development within an MRD are presented. Together with a brief rationale, these alternatives illustrate how different operations might be selected to introduce computers into the MRD. Included is a master plan for MRD automation. At the conclusion of the chapter, readers should be able to explain fundamental principles of patient data computing, prepare descriptive profiles for functional operations demonstrating these principles, and develop a written master plan for MRD automation.

MEDICAL INFORMATION STRUCTURE IN HEALTH FACILITIES

The introduction of computers into hospitals and health care facilities provides a rich opportunity to identify, coordinate, and precisely define the unique medical information structure of the organization. To understand the significance of medical information structure and its role in the establishment of computer applications in medicine, let us define terms and three basic objectives.

Information can be defined as "knowledge or intelligence drawn from facts or data"[1] and "the result of modeling, formatting, organizing, or converting data in a way that increases the level of knowledge for its recipient."[2]

Structure refers to framework, foundation, or the physical base of an entity. Information structure can be described as a framework that incorporates and supports data. The data represent events and elements that take place in a given operation for communication to a recipient. Traditional medical records, for instance, have provided the framework for representing and communicating patient data. The purpose of the information structure in the patient record, of course, is to represent an individual person who is actively receiving diagnostic or therapeutic treatment. Further, it is to yield facts about that treatment to authorized recipients. This purpose transcends verbal, paper, and computerized patient record systems. Medical information structure refers to the framework that incorporates the data that represent treatment modalities, diagnostic and therapeutic action, and the results of services provided to an individual patient during a medical or health care encounter.

Three major objectives are to be achieved by structuring medical information:

1. It should increase the quantity and quality of information available to providers and expand their decision-making base.
2. It should clearly identify and describe medical care or service units that have been provided.
3. It should direct users to choose from available options in order that appropriate action can be taken during a health care encounter or, later, in planning new services based on cumulative or aggregate data.

A thorough documentation of the patient's reason for seeking health services, the plan for health services for that patient, and the progress and results of the services provided can effectively increase the quantity and quality of information available to individual health care providers. This first objective was recognized early in the historical evolution of hospital accreditation standards and associated medical record standards. The objective increased quantity, quality, and timeliness of information and is a fundamental necessity in medicine. It will continue as a primary objective of computerized medical information structure. Indeed, it is a major reason for the origin and development of medical information systems in this country.

When we trace the movement of patients and their records through an entire episode of care, a complete record reflects the flow of information and corresponding services through the hospital, ambulatory care center, or other health facility. An organized program of recordkeeping reflects the points of origin of all events, elements, and feedback that take place in the health services provided. When the information in the record is matched against individual service area counts, such as laboratory count of blood tests given, the second major objective can be accomplished. The medical care can be clearly defined, described, and counted.

Direct and indirect patient information, test results, service analysis, quality review, and statistics used for planning purposes are all examples of health or medical information that illustrate the third major objective. They are used to assist information users when making decisions related to their respective tasks. This objective focuses on the dynamic use of medical information, which is a major goal of medical record professionals.

The three objectives are activated when the various parts of medical information structure are identified and the inclusion and purposes of individual information elements in a sequential process are verified. This includes defining and describing the initiation, evolution, and ongoing use of patient data and the corresponding events and services that occur simultaneously in ancillary service areas. Similar information, for instance, is used in radiology, nuclear medicine, clinical laboratory and in institutional support areas such as housekeeping, accounting, and business office. The challenge in designing medical information systems that deal effectively with patient care, patient record functions, and corresponding requirements in management of health services is to provide for integration of information among these areas. This integration must efficiently support and promote the basic objectives of medical information structure.

Medical record professionals have a unique opportunity to contribute to the identification and verification of medical information structure through their education in the use of patient data and its applicability to resources management and policy formulation. The preceding chapters explained how the systems analysis process provides effective methods for designing computer systems. The question at this point is: How are the areas for automated information management selected? What comes first?

R-ADT: The Most Efficient Beginning

Using the definition of medical information structure, it can be seen that the first or fundamental unit on the framework is the unit of data that identifies a patient. Included in the identification are enough data elements to characterize an individual uniquely. This identification and characterization takes place when the patient initiates a health care encounter, by choice or necessity. This initial set of facts, captured on admission, is the opportunity for information providers and users to establish this individual patient's unit of data as a foundation or core upon which all related data elements will be added or built. The process is known as R-ADT. It is the most direct and efficient method to commence integration and automation of patient data. Data that are registration-specific are entered into the computer as in the case of registration for outpatients or reservation for inpatients. This collection of a basic set of data elements is the essential minimum data upon which all future data entries will be added. It can be verified and updated but need not be recollected. Basic registration data capture always includes identification data and usually includes financial and social identifiers. It provides the essential information "bus" or movable set of common data elements from which other extensive modular and integrated subsystems can be connected. It is the foundation of communication and message-switching features in hospital computer systems. Notice the communication of data elements in the data-switching flowchart in Figure 7-1. Analysis of this figure will reveal that most items are collected during the R-ADT process. These same data elements are in continuous use thereafter. When captured at the point of patient entry, the message-switching communication capability has been established. The R-ADT role in developing medical information systems clearly demonstrates that data capture at the patient entry point will continue to occur.

The patient record is the core of computerized applications in health record management. The record is the source document for all health services rendered. It is also the source document for clinical, analytical, and financial analysis and serves as the communication source among providers. Medical information systems incorporate patient data in three ways. (1) They may computerize the content of the record and provide access to providers via screens, hard copy, and plots. (2) They may computerize selected data from the record to use for management analysis and research; patient demographic information and diagnoses are examples. (3) They may computerize operations that support and extend the availability of patient records to providers. Once the foundation of data is established by means of the R-ADT process, additional functions for computerization become evident. Each data element captured during the R-ADT process is a source for the functional operations carried out in the MRD. R-ADT data singly or cumulatively can provide the foundation for computerizing functional operations in the MRD. The listings that follow demonstrate that the R-ADT function itself can be refined further to capture partial R-ADT data sets or complete R-ADT data sets. The lists also describe the linkage R-ADT has to various functional operations.

Reservation/Registration-Admission

Data elements such as patient name, number, unique demographic data, admission date, and admission diagnosis can be linked to the following tasks:

- Create a computerized MPI.
- Provide computerized admission lists and census reports.
- Initiate a PSRO admission screening and review program.
- Provide an automatic notification to the word processing department of patient admission to facilitate admission dictation.
- Notify outpatient appointment scheduling when a new patient is registered so that clinic appointments can be coordinated with already captured patient identification.

Figure 7–1 Data Switching Flow Chart

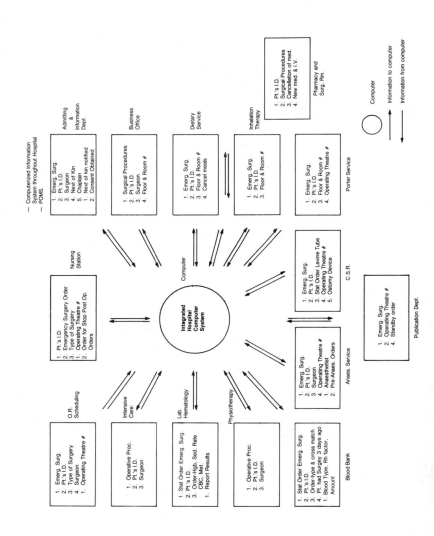

- Provide notification to medical record file room via a printer that creates an automatic log of patient admission chart requests and a simultaneous charge-out slip to use for pulling earlier admissions.
- Generate notification to special registries such as cancer surveillance that track individual admissions.
- Establish patient identification as a part of a chart locator system that can be updated as the patient is transferred or discharged.
- Generate medical staff or physician in-house case list for use as a source for a computerized chart deficiency system.

The notification methods could be as simple as an admission list distributed manually, an automatic printout notice on printers located in individual departments, or on-demand displays available to providers.

Discharge-Transfer

The data elements collected at this point might include discharge data, attending physician, discharge disposition, hospital services utilized, complication or referral data flag code, and other information that could be used in extending computing capability to do the following:

- Initiate a computerized discharge abstracting system that utilizes data elements captured during the hospitalization to provide a partially completed discharge abstract for MRD use.
- Notify business office via printout of discharge list and diagnostic information for third party billing. If such a notification system is coordinated with computerized coding, the third party reimbursement cycle could be improved.
- Initiate or update a computerized chart locator system that tracks the location of the patient record by providing updated charge-out slips to medical record files or an automated log to a record control clerk in the MRD.
- Initiate screening systems for use in quality assurance, medical audits, and risk management programs.
- Establish notification systems for special registries that may track patient hospitalization through an authorized copy of the discharge abstract. Such networking could be provided through tape-to-tape transfer. Examples of such programs would include cancer registries and donor transplant registries.
- Print discharge order and associated nursing care plan to be used for accompanying the patient in direct interfacility transfer. This feature could also be an extension of a computerized order entry program.

- Notify outpatient appointment systems where clinic visit follow-up is planned. Such notification could include automatic medical record files charge-out preparation.
- Where an institution is developing a computerized medical record, patient data stored during the hospitalization could be added to the header information already captured in the R-ADT process.

The information structure for R-ADT has been described by certain Veterans Administration medical centers as one method of choice upon which to build computerized systems for their hospitals.[3] The Public Health Service emphasized the R-ADT function as the first step in developing the Public Health Service Automated Medical Information System (PHAMIS). See the illustrated role and foundation of R-ADT in the PHAMIS features illustrated in Figure 7-2.

Postdischarge Medical Record Use

The reader may recognize the growing significance of data base systems as an expansion of R-ADT. Data base concepts are often key elements in developing computer applications in medical records. Data base management utilizes R-ADT data and additional data as designed by the user-manager. The functions listed below demonstrate additional use of data that can be collected throughout the course of hospitalization or a treatment series in an alternative setting. Notice how data elements reoccur.

- Coordination of MRD-compiled statistics can be accomplished by using data already captured to produce hospital activity reports. Such reports are included in summary form from discharge abstracts and may be provided by census-based data on a monthly basis. If the data are collected during the patient stay, programs can be written to enable users to access the patient data base and create statistical reports according to need. The Command System developed by Shared Medical Systems is based on this concept.
- Mandatory reporting for vital statistics and other epidemiological needs can be transmitted with computer assistance.
- Retrieval of medical information through disease and operation indexes can be accomplished more efficiently when coded cases are stored on magnetic tape accessed on a request basis. Allowing the computer to scan coded data files often is more efficient than using MRD personnel to manually review computer index printouts to identify cases.
- Record request systems may be coordinated with outpatient appointment scheduling systems, chart deficiency systems, or stand-alone

Figure 7–2 Seattle Public Health Hospital

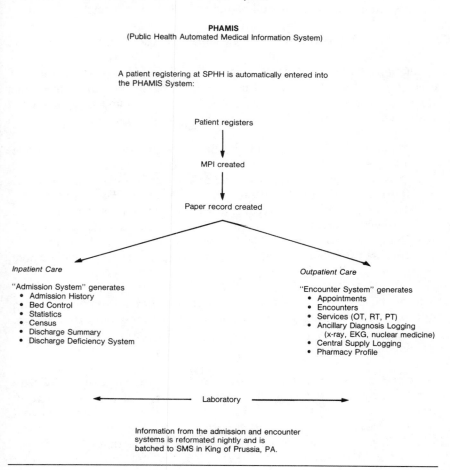

PHAMIS
(Public Health Automated Medical Information System)

A patient registering at SPHH is automatically entered into the PHAMIS System:

Patient registers

↓

MPI created

↓

Paper record created

Inpatient Care

"Admission System" generates
- Admission History
- Bed Control
- Statistics
- Census
- Discharge Summary
- Discharge Deficiency System

Outpatient Care

"Encounter System" generates
- Appointments
- Encounters
- Services (OT, RT, PT)
- Ancillary Diagnosis Logging
  (x-ray, EKG, nuclear medicine)
- Central Supply Logging
- Pharmacy Profile

← Laboratory →

Information from the admission and encounter systems is reformated nightly and is batched to SMS in King of Prussia, PA.

applications. Manual record retrieval can be accomplished with excellent control if requesting parties can utilize a CRT display to enter the request and the CRT also displays the data requested. The request is clocked, logged into a monitor report for medical records, and automatically printed in a duplicate charge-out that can serve as a routing slip for the patient record and a charge-out slip for the files. In systems with automated clinical record capability, inquiry, or printing of limited segments of the patient record at the appointment site might be used. If programmed accordingly, such a system would enable MRPs to determine record activity by type and user and integrate such information into the overall planning for MRD services.

PRINCIPLES OF PATIENT DATA COMPUTERIZATION

Emerging from the R-ADT concept are three principles that directly affect the MRA. They affect the planning, organizing, and controlling processes that must be relied on in order to provide a flow of patient data that is accurate, complete, timely, and accessible.

1. Planning for future use of the data requires that the practitioner address the patient record as soon as it enters the system.
2. Each functional operation of the MRD that is initiated through the R-ADT foundation necessitates the formation of a descriptive profile. The descriptive profile includes, but is not limited to, data entry and exit points and identification of all data elements and potential users.
3. Computerization of R-ADT and medical record functions must be considered in the context of computerization planning for the whole institution.

Each of these will now be examined in more detail in conjunction with additional application examples that help illustrate the principles in action.

Entering the System

The first principle requires that the record professional have a familiarity with the details of data capture, retention, use, and distribution in all applicable departments. For R-ADT, the departments or departmental units involved in initial data capture are outpatient, emergency, reservations and, of course, admissions. In some institutions that have a *decentralized outpatient registration,* the initial data capture function is carried out at the point of care such as laboratory or radiology. The registration process has a significant impact on the future use of the medical record and its accurate and timely development. For it signifies the opening of a new case, or *patient encounter,* and must be accompanied by the development of a new or additional entry on an MPI listing. Other significant results of data capture at registration include determination of methods to notify all necessary departments of the patient's impending relationship with the department. Notification usually goes to dietary, nursing units, housekeeping, business office, and the laboratory. However, the information processing is an individualized decision and varies from institution to institution.

Another look at Figure 7-1 will show how extensive the notification process can be. The MRP's identification and documentation of patient data used in manual R-ADT systems will be drawn from the entry points just listed. This process should reveal exact initiation and use of the MPI. It will probably also reveal several manual patient index files throughout the institution. Analysis of these will in turn provide practitioners with the descriptions of common data elements required by all users that will define the base data necessary to establish a common MPI. Enough data will also be found to establish a registration data dictionary. A data dictionary is a listing of all the terms used in an application and includes the meanings of the terms. A registration data dictionary would then be a listing of all the terms associated with R-ADT, including those of the MPI. In this connotation, it is a more powerful informational unit than the MPI. Consider the improved efficiency that is gained when one unique patient number identifier is available through a common MPI to satisfy patient identification and verification needs for all the ancillary services within the hospital. Using the analysis tools explained in Chapter 3, practitioners can develop a rationale for creating and maintaining an on-line MPI. The key to developing such an application resides in understanding and communicating the potential of effective R-ADT development.

Creating Descriptive Profiles

The second principle requires more extensive discussion and illustration. A descriptive profile is a written document that describes a functional operation through a detailed explanation of the operation's characteristics. It includes the objective of the operation, the specific sequences including entry/exit points, required outputs, input techniques, data element descriptions, volume characteristics, and user identification. In concept such profiles are comparable to the products of a full systems analysis process. In recognition, however, that many applications will require a more concise problem identification, the use of a modified product of this type will suffice.

Descriptive profiles can be used by practitioners to prepare for automation. As will be seen in the methodology featured in the following pages, descriptive profiles offer a precise accounting of an individual operation. They can and should be prepared by MRPs or under their supervision. We recommend that all functional operations be analyzed in a profile methodology to acquire a precise understanding of the information flow and user needs. Key functional operations should be done first and others methodically prepared in accordance with strategic and long range planning for the department. This process should be considered a management function similar to budgeting, operational quality assurance monitoring, and other administrative activities. Certainly components of the descriptive profile are already performed in other tasks. Objectives are an example of this. The character of automation, however, requires a precise description of the operation and its associated information so that the detail and logical sequence can be built in to the application from the beginning.

Along with the characteristics listed in the definition, a profile offers additional potential for MRPs. A thoroughly documented profile can serve two or more simultaneous purposes. For instance, when a method is set up to track and account for sources of incomplete record data, simultaneous tracking of detailed complete record data, as well as performance patterns of users can be accomplished. The type of deficiency as well as the person responsible for its correction can be identified. One can view this process as clinical (the person(s) or service(s) responsible for the omission or error) and analytical (the classification of the error or omission into a category or type). It is analogous to clinical care provided to patients and keeping track of the specific types of care such as is performed through service analysis in hospitals. One use of these data is a specific and personal record. Another use is the general tracking that places single events in a larger context. In this example, both the individual provider or department performance and collective reporting of general deficiency areas, such as a rate or percentage of completed discharge summaries per discharge, can also be determined.

The computer enhances a user's ability to view data in different ways. In another example current features available in computerized word processing systems may provide for two products. Not only are outputs for the primary functional operation of a transcription unit produced in medical record reports but analytical tracking can automatically be performed on volume, operator, and report categories. Features such as these can be recognized by the practitioner and should be included in the profile in order to plan and develop an automated system specific to the desired needs and to expand management details needed to develop job standards and procedures. MRD personnel who have routinely performed this dual task in the manual system must continue to understand and carry it out with the assistance of computers.

Users of the information are also included in the profile. The R-ADT process envelops a vast array of data users: patients, physicians, nurses, third party payers, and various hospital departments. Identification of the users at this point can pinpoint duplication of reporting, assist in streamlining formats for output documents, and provide critical review of proposed data design early on while a new method is still in the planning stages.

How are comprehensive descriptive profiles developed? A look is in order at a methodology that the reader will recognize as drawn from the systems analysis process explained in earlier chapters.

Descriptive Profile Methodology

The following items listed in Exhibit 7-1 should be completed by the user-manager for each functional operation under consideration for computerization. The purpose is to prepare information about the operation that ensures a clear description of the user's perceived and defined requirements in a manner that the systems analyst can accurately interpret.

When practitioners are participating in more detailed systems analysis as explained in Chapters 4–6, the products listed will be developed in the analysis phase with some variation in sequence. As MRPs become more familiar with the methods used for communicating information and procedure needs to computer system designers, adaptations may be used. Careful attention to detail helps develop a strong communication foundation so that future shortcuts are mutually acceptable and streamlined methods become more efficient. The next two examples are adaptations.

Exhibits 7-2–7-4 illustrate a descriptive profile used in planning for a computerized discharge abstracting system. This partial profile includes objectives, input/output description, and information source document analysis. By mapping out the information in this way, the MRA was able to communicate efficiently the functional operation needs to the programmers working with the system.

In another method for developing profiles for MRD applications, the hospital computer systems department provided a detailed breakdown by program description and associated information. Exhibit 7-5 shows this. Both contribute to the process of computerizing the operation.

The descriptive profile model may be used to perform information analysis necessary in order to computerize medical record operations. As discussed in Chapter 4, individual system teams may select from a variety of analysis tools. The samples in Exhibits 7-6 and 7-7 are forms that also perform this task. Notice these examples are used to analyze input documents and reports. Practitioners will be inserting such items into the overall descriptive profile development.

Manual File Analysis

No discussion of medical record computer applications is complete without a particular focus on the process of file analysis. Of course, file analysis is performed in setting up computer system files. Many of the

Exhibit 7–1   Data Needed To Complete a Descriptive Profile of a
Functional Operation

Steps in Completing a Descriptive Profile for a Functional Operation	Example Describing a Master Patient Index Operation
1. Prepare a statement of the operational objective	"To establish a master patient index which can be used to retrieve and/or confirm correct name and number for all patients admitted to the facility."
2. Identify correct operation sequence	Prepare flowchart of existing operation.
3. Define required outputs	Prepare exact models of current cards and desired screen displays, microfiche format and/or reports.
4. Describe input process	Include written account of information source and procedure used. For instance, patient interview by admit clerk with subsequent summary sheet review by MPI clerk is used to prepare MPI cards. (This information should be traceable on the flowchart in Step 2.)
5. Prepare user data element definition and characteristics (include both input and output data).	There may be several tools* used at this step. Both input and output forms need to be examined to be sure all the data elements are identified. Once that is done, each data element should be defined and described in the format that follows:

Format for Preparing Master User Data Element
Definition and Characteristic Descriptions

Name	Definition	Alpha-Numeric	Used for Data Input Documents (List)	Used for Output (List)	Total Number of Characters	Format Method	Edits

	This format is a consolidation of other tools. It would provide a comprehensive summary. Generally, it is more efficient to organize *all* data elements to help ensure that the systems analyst accurately interprets user needs. In individual situations, users may be asked to prepare additional forms that describe data element use in alternative ways. Some examples of these will be featured at the conclusion of this model.
6. Document proposed operation sequence	Prepare a flowchart of the proposed computerized operation.

Exhibit 7–1 continued

7. Document user definition and role	Practitioners may use several tools to accomplish this. A forms distribution chart could illustrate how admission/census reports are used within a hospital. For an on-line MPI project, an input/output chart could be used to demonstrate the exact data elements required from the MPI by all hospital departments. This information is needed to develop an integrated hospitalwide MPI that could be used by all departments. It may be that simple incorporation of an account number would enable both billing and medical records to use the same index. Addition of a one- or two-character code might allow other ancillary departments to use a common index. Such a code might serve as a "flag" to the respective departments. Other files may print to the patient file by use of the registration or unit number.

*Chapter 3 describes appropriate tools.

items included will parallel the information gathered in the descriptive profile model. File analysis itself occurs in the next technical step. It is initiated in the analysis phase and completed in the design phase of systems analysis. The focus of this text is on the process of investigating, designing, and developing computer applications in medical record functional operations. Automation of the entire patient record is beyond the scope of this discussion. However, we are providing a detailed technical discussion of file analysis at this point for two reasons:

1. To illustrate one detailed technical method used in the process.
2. To encourage readers to imagine the analysis as it might apply in our future efforts to automate the patient record; these efforts will require the MRA to aid in extracting and abstracting information from these automated records not only for clinical and health services research but also for policy formulation.

Another example will be helpful in viewing the file analysis process. Consider that the MPI makes up one file in a computer system. Because it is the header information in a patient record, it provides an excellent beginning. It, in fact, is the embryo that will first be taken through the process.

Exhibit 7–2 Application Objectives

---

1. Review the format for discharge abstract on a CRT terminal to allow for additional information.
2. Number the fields of input to show where the area of information is located so that the computer programmer can design a program.
3. Provide a source document for the discharge abstract listing items found in the CRT format and to what fields they are assigned.
4. Designate on the discharge abstract what departments are to be responsible for information to be inputed via CRT terminal.
5. Review operation, physicians, diagnosis, and discharge analysis indexes, and provide a source document for each one. Provide the necessary instructions for computing statistics when needed.

Exhibit 7–3 Preliminary Discharge Abstract

DATE: 08-13-79   HOSPITAL
                              DISCHARGE ABSTRACT

PAT NO 60-64-01   NAME   ZIP 98841
AGE 56   SEX M   RACE W   BIRTH DATE 08-26-22   FC 10-00
ADMIT 07-17-79   TIME 14:05   ADM TYPE ER   ER ADM N
DISCH 08-11-79   LOS 025   DISP—EXP NO +48 HRS - SRV 00—
ATTD MD Y054 YARRINGTON, CT G104 GORTNER, D
CONSULTS
TRANSFUSIONS: WB— PC— SCU DAYS 04   CCU DAYS 00

PRNCPL DX (FIELD NO. 70)*
ADDTNL DX (FIELD NOS. 71–76)*

DAT 07-18-79 SURGEON YARRINGTON, CT ASST BALLARD, ROBERT
PRMRY OP 31365 LARYNGECTOMY, WITH RADICAL NECK DISSECTION
OTHER OPS

DATE—   SURGEON   ASST
PRMRY OP
OTHER OPS

DATE—   SURGEON   ASST
PRMRY OP
OTHER OPS

COMPILED BY: (R)   (I)

*See Exhibit 7-4, Source Document Analysis, Col. 1.

Exhibit 7–4 Source Document Analysis

ABSTRACT FIELD NO.	ITEM	SOURCE	EXPLANATION
33	Service	Medical Records	Determined by the following code

Code 02 Pediatrics
10 Medicine, general
14 Communicable diseases
18 Dermatology
22 Cardiology
32 Neurology
33 Respiratory
38 Psychiatry
40 Surgery, general
46 Thoracic surgery
48 Ophthalmology

49 Otorhinolaryngology
56 Neurosurgery
58 Orthopedics
61 Traumatic
62 Urology
70 Gynecology
75 Abortion
76 Obstetrics, not delivered
77 Obstetrics, delivered
80 Newborn

ABSTRACT FIELD NO.	ITEM	SOURCE	EXPLANATION
40	Attending MD	Admission form	
41	Attending MD	Admission form	
50	Consultations	Nursing staff	Medical records staff will fill in consultations not previously verified by nursing staff
51-54	Additional consultations		same as above
60	Transfusions	Central service	Whole blood-packed cells (number of units)
61	SCU-CCU days	Census system	Total number of days
70*	Principal diagnosis	Medical records	To be filled in manually using HICDA code number and description until dictionary completed. Test field to see if diagnosis appears before input via terminal
71-76*	Additional diagnosis	Medical records	Same procedure as above
80	Date of operation	Surgery	
81	Surgeon	Surgery	
82	Assist. Surgeon	Surgery	
83	Primary operation	Surgery	CRV numbering used
84-87	Other operations	Surgery	
90	Date of operation	Surgery	
91	Surgeon	Surgery	
92	Assist. Surgeon	Surgery	
93-97	Additional operations	Surgery	Operation procedures performed on a different day than original procedure

*See Exhibit 7-3.

Exhibit 7–5 Incomplete Charts Status Report

SUBJECT: Incomplete Charts Status Report by Primary Responsible Physicians and/or Surgeon

A. TITLE: Incomplete Charts Status Report by Primary Responsible Physician and/or Surgeon

B. PROGRAM NO: 74

C. FREQUENCY: WEEKLY

D. PURPOSE: To supply the Medical Records Department with a list of incomplete charts by department (Medical Service) and attending physician. Note: 1 copy of this report will go to the Medical Director by Department.

E. FORMAT:

1. Patient Name and Number	The patient's last name, first name, and middle initial as recorded in the Patient Master File, with its unique 6-digit patient identifier and check digit.
2. Primary Responsible Physician/Surgeon	The primary responsible physician as entered by Medical Records in the Incomplete Charts System. The code will print if present.
3. Disch/Op/Trans Date	The discharge or expiration as entered by Admitting or the date of surgery or transfer as entered by Medical Records in the Incomplete Charts System.
4. Days Deficient	Number of days since the date of discharge, surgery, or transfer that the record has been incomplete.
5. Deficiency Reason	The 30-character description of the deficiency. All deficiencies are listed for a record. See Appendix II for list of deficiencies and descriptions.

    6. Totals are produced for each primary responsible physician and/or surgeon and dept.

F. PROCESSING

    1. All discharged patients with incomplete records are listed in order of oldest discharge date first within primary responsible physician and/or surgeon. Primary responsible physicians and/or surgeons are listed in alphabetical order within the departments. The patients with the same discharge date are listed in terminal digit patient number order.

Cancels: Prepared By: Approved By:

Exhibit 7–6 Input Document Analysis Form

Form no.	Name:
Prepared by (department):	
Purpose:	
How is data obtained?	
Distribution - copies to: (List departments)	Volume: _____forms per _____ (time period)
	Schedule for completing form:
Problems	Other comments:

Source: Reprinted from INFORMATION SYSTEMS FOR HOSPITAL ADMINISTRA-TION by Charles J. Austin by permission of Health Administration Press, © 1979.

File Analysis

A comprehensive file analysis includes an evaluation of contents, format, physical storage, utilization, and economic factors. The following items may be included in specific file analyses.

Content
- Data elements—definitions, ranges of value, units
- Amount of information—average number of characters per record; maximum number of characters per record

Exhibit 7–7 Report Analysis Form—Discharge Analysis Summary

Report no.   Title:

Generated by (department)

Purpose:

General content: Source of data contained
                                 in the report:

Distributed to: Schedule:

                                 –Daily   _____
                                 –Weekly   _____
                                 –Monthly   _____
                                 –Quarterly   _____
                                 –Annual   _____
                                 –Other (specify)

                                 _____

Volume (average number   Processing:
of pages)   –Manual   _____
                                 –Computer   _____

Problems: Other comments:

Source: Reprinted from INFORMATION SYSTEMS FOR HOSPITAL ADMINISTRA-
TION by Charles J. Austin by permission of Health Administration Press, © 1979.

- Number of records (how many names in the MPI)
- Duplication of data elements with other files. (Similar data is maintained in other files.)
- Identification of data elements that serve as retrieval keys or linkages with other files and records (patient name or number)

Formats
- File structure and organizational characteristics—in the case of automated records. (Record organization: How is the information to be stored and retrieved?)
- Storage media—in the case of manual records

Physical storage
- Space occupied by a file
- Physical layout; in the case of manual records this would include file drawers, shelf storage, etc.

Utilization
- Sources of updating information
- Users
- Frequency and number of accesses for purpose of retrieval or update
- File growth rate as in new patients admitted
- Historical changes in access frequency and volume such as occur when overall facility objectives change. A hospital that begins an outpatient cancer therapy program is an example
- Reliability of file data
- Access times by day, week, etc.
- Purging characteristics
- Access controls for data security
- Efficiency measures such as relevance ratio; scanning ratio (How relevant is the data? Perhaps only a portion needs to be accessible and the balance can be stored on historical tape.)

Economics
- Cost of personnel and equipment associated with file maintenance including access and quality control
- Cost of space
- Incremental access costs
- Cost of file reorganization or periodic purging
- Cost of access control
- Cost of file analysis
- Data conversion costs
- File consolidation costs

When looking at manual files, the easiest procedure for estimating volume of information is simply to calculate as follows:

1. volume =
   Number of records in file 1 × average number of characters of information in a file 1 record
   plus
   Number of records in file 2 × average number of characters of information in a file 2 record
   plus
   .
   .
   .
   Number of records in file n × average number of characters of information in a file n record
2. The resulting figure must usually be reduced by some fraction representing the anticipated reduction in file sizes resulting from a purging of the files.
3. When analyzing automated files, a similar computation may be carried out. The resulting figure following (1) and (2) may have to be increased by a fraction that represents "overhead"—information stored for the purpose of linking records to one another or for the purpose of indexing the computerized records.

Estimate Space Requirements

Space requirement estimates should always incorporate anticipated growth rates. Generally the following statements apply:

1. One may assume that space requirements will increase at a rate at least equal to that experienced up to the time of the file analysis.
2. Space requirements should be calculated over an interval of time equal to the value period of the cost/benefit analysis.
3. In the case of manual files, an ample growth rate can be extrapolated based upon current file volumes, past growth rates, and (for example) current admission rate and population trends.
4. In the case of automated files, because of (a) the likelihood that additional data elements may be added at a later time, and (b) the effect of file "overhead," a compound growth rate should be used. The resulting figures could then be adjusted downward to reflect purging and reorganization. The computation could be shown as

   Volume in n years =
   Current volume × (1 + overhead factor) × (1 + growth rate)^n, e.g.,
   Current volume   = 600

Overhead   =   5 percent
Growth rate   =   5 percent
Estimated volume in 3 years desired
$$= 600 \times (1 + .05) \times (1 + .05)^3$$
$$771 = 600 \times 1.05 \times 1.22$$

Physical space requirements for manual files also need to be estimated.

Estimate File Accuracy

It is always appropriate to reestimate the accuracy of files as a part of a system redesign or improvement effort. This should be done carefully. Generally the procedures involve doing the following:

1. Establish criteria for determining that a particular record is accurate. For example, What reference will be used? How will missing data be treated? Will all data elements be considered or only selected elements?
2. Decide how many records will be examined. This usually depends upon staff and time available. In an MPI, all entries should be checked prior to conversion to an automated system.
3. Establish a procedure for selecting records to be examined. If the records are numbered sequentially, for example, one may want to select records randomly, using a random number table.
4. Calculate the estimated accuracy when the record examination is completed. For example:

$$\text{Percent accuracy} = \frac{\text{Number of records found accurate}}{\text{Number of records examined}} \times 100.$$

Once the file analysis process has been completed, one can determine how the requirements listed in the process could be used to plan an on-line MPI. Medical record practitioners will need to determine the file size and access information as noted in the file analysis content and utilization sections.

File Size and Access

Preparing for a new computer system or application entails determining the size of the file. This can be ascertained by the following:

• Determine what items are absolutely necessary for the desired output.

- Count the characters needed for those items constituting a record (name, number, birthdate, etc.).
- Determine the number of incoming records by historical experience (admission rate).
- Determine the quantity of the backlog data you desire to be entered.
- Estimate any suspected increase due to automation.
- Count the length of items that will be used as keys to the file.

After the preceding data are collected, an analysis of the file length can be estimated as in Table 7-1. The figures would take into account the MPI file only. All computer space must be examined in detail including: permanent files, working files, temporary scratch files, output files, application programs systems utilities, and operating system overhead. Some of this allocated space should be explained by the data processing staff.

In addition to figuring current levels of need, managers should try to envision the future and allow for computer expansion corresponding with the projected increase in departmental activity. The concepts reviewed in this section are essential components of descriptive profiles. Practitioners should expect to work closely with technical resource people in file analysis and design tasks. Similar teamwork is necesary in determining appro-

---

Table 7–1 Analysis of File Length for an MPI Record of 158 Characters

16,000	Admissions
320	Increase due to automation (preadmission reservations, for example)
+ 1,330	Backlog entered
17,650	Total records, increase, and backlog
× 158	Character record
2,788,700	Bytes of MPI file storage
17,650	MPI name, increase, and backlog
× 7	Character key
123,550	Bytes for key
2,788,700	Bytes of MPI storage
+ 123,550	Bytes for key
2,912,250	Total bytes for MPI file and Key

priate data access needs for individual medical record applications. The next section describes data access in detail.

Data Access

Speed of data access, relevant to software development, is an indicator of systems performance that is directly related to systems design. The type of access method used will reflect the speed of response to on-line inquiries. Three different classical computer access methods will be discussed: sequential, index-sequential, and direct addressing. These have useful manual analogies and are thus useful to analysis of manual files.

Sequential access implies "one right after the other" or "in sequence." This access method can be demonstrated in a manual setting: an MPI has 40,000 cards filed alphabetically in its drawers. Given that the card desired is *Smith, James R.,* instead of going to the "S" drawer, start at the beginning of the "A" drawer and read every card until you find *Smith, James R.* Using this type of access method is also very time consuming for on-line applications and should be avoided. Sequential access is appropriate where a monthly or weekly report is generated, or where an entire pass through the file is necessitated.

Index-sequential files search an index rather than a file. It is similar to the practice in a library, where a person uses the card file to locate a book's catalogue number, which shows the general location where a book will be found. Once the shelf is located, a sequential search is initiated.

On larger files, however, it is beneficial to utilize several indexes that index the indexes. This procedure consumes some storage space but saves time in processing on-line requests, as it points to the general location where the desired information will be found.

Direct addressing schemes provide the fastest means of addressing available. With this procedure, there is no need for file searches or index preparations; the inquiry itself contains the machine address. Almost as fast as direct address inquiry, and used quite often, it uses an algorithm to convert the key to the machine address. Recall that an algorithm is a stored set of rules that leads and assures development of a desired output. Algorithms often contain a number of branch points where the next step taken depends on the outcome of the preceding step. This technique is used where the input item is not compatible with the machine address initially or, as an alternative entry procedure when the address is not known.

The effectiveness of computerized applications is dependent on response time to the users. If people are to be satisfied when using computer terminals, the speed of response must be fast enough to allow them to carry out daily tasks at an effective pace. Effective planning for the necessary response is accomplished at the level of analysis we have just examined. For effective use of computerized MPI, the response to the terminal operator should not exceed five seconds from one display to the next. One and one-half to three seconds is optimal.

The foregoing discussion focused on some of the elements and tasks involved in developing effective descriptive profiles. As with formal systems analysis, these should be viewed as documentation methods for preparing for computerization of individual functional operations in MRDs and related areas. The more skilled user-managers can become in defining, describing, and documenting individual operations, the more effectively automation can take place. Let us turn now to discussion of the third principle in Patient Data Computerization. The third principle leads us to place individual projects in the context of organizational planning.

Fitting Subsystem Plans into Master Plan

The third principle is directed to effective integration of department specific and facility master plans for patient data computerization. The MRA's commitment to the institution's master plan for computerization is activated in determining how the MRD functional operations can be implemented so as not to conflict with the master plan. There is tacit understanding in the third principle, that the MRA is an active member of the institution's management team and is involved in working with others—administration and medical staff—in the development of cohesive computer objectives. This team approach to planning is crucial when one considers the implication for cost benefits, or without such planning, deficits. Compatibility of hardware, development of software, decisions regarding computer language, selection of an in-house versus a time-shared system, determination of computer personnel categories, and roles and relationships of system plans to users are all addressed by an institutional committee or team. The conclusions reached by the group are critical to the individual objectives and plans of each department. R-ADT, for instance, affects nearly every department in the organization. A user committee is imperative in the planning as well as monitoring and postinstallation audit.

This chapter has thus far identified information benefits from applying information management technology to the R-ADT foundation. We have included many possible MRD operations. A number of these functions may be performed in other areas of the institution. Quality assurance, for example, is a separate department in some hospitals. If the information relationship is clearly understood, computer applications can be developed to facilitate information exchange for all departments that work with the same patient data. The challenge to practitioners is to recognize the potentials and prepare for effective computer development within their department. The following section will address strategic and tactical planning for individual computer applications in MRDs.

PLANNING FOR INDIVIDUAL COMPUTER APPLICATIONS IN MRDs

"Planning is the process of making decisions in the present to bring about an outcome in the future. It is the determining of proper goals and the means to achieve them. It involves the statement of assumptions, the development of premises, and the review of alternate courses of actions."[4] Planning can be for short range, that is, one year or less. It can be intermediate range, that is, more than one but less than five years. Or it can be long range, that is, five years and over. Health record practitioners can participate in all of these plans. Traditionally, short-range and intermediate-range planning are the processes most common to middle managers. MRAs usually take part in long-range planning only when the total organization undertakes a long-range planning program. However, the introduction of computer technology is changing the schedule of planning activities.

Strategic Planning as a Process

Rapid technological development and increased use of information are affecting the use of planning. Planning is more important than ever before, and long-range planning is imperative at the midmanagement level. It can be carried out at this organizational level and coordinated with broader organizational goals.

> It is true that practically every basic management decision is a long-range decision—ten years is a rather short time span these days. . . . Management has no choice but to anticipate the future, to attempt to mold it, and to balance short-range and long-range goals. It is not given to mortals to do well any of these things. But lacking divine guidance, management must make sure that these difficult responsibilities are not overlooked or neglected but taken care of as well as is humanly possible. "Short range" and "long range" are not determined by any given time span. A decision is not short range because it takes only a few months to carry it out. What matters is the time span over which it is effective. . . .

The idea behind long-range planning is that "What should our business be?" can and should be worked on and decided by itself, independent of the thinking on "What is our business?" and "What will it be?" There is some sense to this. It is necessary in strategic planning to start with all three questions. What is the business? What will it be? What should it be? These are, and should be, separate conceptual approaches. With respect to "What should the business be?" the first assumption must be that it will be different.

Long-range planning should prevent managers from uncritically extending present trends into the future, from assuming that today's products, services, markets, and technologies will be the products, services, markets, and technologies of tomorrow, and above all, from dedicating their resources and energies to the defense of yesterday.

Planning what is our business, planning what will it be, and planning what should it be have to be integrated. What is short range and what is long range is then decided by the time span and futurity of the decision. Everything that is planned becomes immediate work and commitment. The skill we need is not long-range planning. It is strategic decision making, or perhaps strategic planning.[5]

Strategic planning is the continuous process of making present entrepreneurial (risk-taking) decisions systematically and with the greatest knowledge of their futurity; organizing systematically the efforts needed to carry out these decisions; and measuring the results of these decisions against the expectations through organized, systematic feedback.[6] In considering automation for an MRD or health record service, strategic and tactical planning should be integrated with one of the time plans listed previously. To illustrate, the following list of questions relates to a health care setting.

- What should the health care facility be?
- What services will be offered by the health care facility in the year 1990, the year 2000?
- What is unique about what the facility does?
- Should this uniqueness be exploited?
- What services does the health care facility want to provide?
- What parts of what the facility is doing now should it continue to do?
- Who is the competition?

- Are more or fewer services required?
- What is the environmental influence under which the facility operates?
- What is happening technologically, socially, and politically that could have impact?
- How and to what degree will this affect the facility?
- What is the health information department (HID)?
- What should the HID be?
- What services will be offered by the HID in 1990, in 2000?

To answer these questions the strategic planning process consists of four distinct parts. First, strategic dimensions of the environment that affect the organization are examined. An example of this is the regulatory requirements on patient information handling in hospitals.

Second, a resource audit of the organization is performed that lists strengths and weaknesses. For example, an urban HMO may provide an excellent outpatient cancer therapy program that is viewed favorably by other facilities and the medical community at large.

Third, strategic alternatives are identified that include consideration of specific characteristics of the organization such as size, services provided, adding or subtracting services, etc. For example, an MRD in a large university hospital that provides support for an active outpatient clinic will have different characteristics than an MRD in a 150-bed community hospital.

Fourth, a strategic choice is made. This is the last step in the strategic planning process and usually the most difficult. This is influenced by organization attitudes, judgment, inter- and intradepartmental tradeoffs, and the political climate of the organization. A director of an MRD in a 200-bed community hospital may select a discharge abstracting package from Shared Medical Systems in order to participate in a hospitalwide coordinated approach to purchasing computer services.

Strategic choices involve risk-taking. "In selecting a strategic choice, willingness to assume risk is paramount and timing is critical. . . . The potential gains are worth the potential chances for failure, or nothing ventured nothing gained viewpoints are followed. In contrast, unwillingness to assume risk leads to a protected, conservative, strategic choice. Commonly some compromise is followed—a combination of the most opportunities with the least threats."[7]

How does this translate into a specific example for the health record manager? It deals with the question: Given an opportunity to participate in the computer systems planning in general and computer systems planning for the MRD specifically, how will the practitioner proceed? What will be the first operation automated? How is that decision made? What strategies justify the choice? If managers view the answers to these questions as terminal outcomes of a thorough strategic planning process that focuses on the future, they will be better prepared to answer the questions. For the health record manager, the course of action identified in this chapter might be as follows:

1. Examine the patient data computerization principles and determine how they relate to the functional operations in a particular MRD.
2. Methodically prepare descriptive profiles for the major functional operations in the department in anticipation of preparation for automation.
3. Participate in organizationwide planning for automation as a representative of the MRD and patient information processing in general.
4. Engage in strategic planning activities specifically for the MRD that results in a plan for implementing computer technology over the long range period.

Initiating Computer Technology in the MRD

Let us briefly examine how these activities might result in alternative starting points for initiating computer technology in an MRD. The following pages outline and illustrate how a master plan for automation is developed. Given particular circumstances of individual departments, the first functional operation to be computerized could be any one of the following.

The MPI is a natural spinoff of the R-ADT foundation presented and explained in this chapter. A manager may select it first because it is the key to most other functions in the department and because it is a step in developing an in-house hospital information system. (Recall how the R-ADT foundation can be extended to many functions in the MRD.)

Computerized word processing may be the first functional operation selected in order to promote the advantage of increased service to the medical staff and perhaps other hospital personnel. It may offer a very positive computer experience to people who are unsure of the benefits of automation. This in turn may translate into future support for additional applications.

A *combined computerized record control and outpatient appointment scheduling system* may be chosen because it provides improved medical record access to support staff who are responsible for providing the patient record to clinical users. Like word processing, this operation has the potential to "sell" other users on the merits of automation.

On-line discharge abstracting is a service offering an improved discharge abstracting operation by providing more timely and more specifically tailored reports for clinical and administrative use. In some situations, this will also reduce costs. Further, if the hospital is already operating a computerized R-ADT or ADT system, perhaps an in-house discharge abstract system could be developed from existing resources.

Selection of one of these examples constitutes a strategic choice. Let us move this choice to the next step.

Identifying Constraints

Remember the constraints discussed as a component of present system documentation in the analysis phase of the systems life cycle? They were explained in Chapter 4. Here, again, the manager takes another look at constraints. They are considered in a systematic fashion that identifies and recognizes the total organizational climate. The following factors show an alternative method to identify constraints.

1. The general organizational setting is examined. This includes a determination of the level and particular emphasis of care; the goal of an institution may be acute care in specialized diagnostic categories. Physical location, population characteristics, and general community health resources are also examined.
2. The legal and accrediting agency mandates that govern the organization's operations and services are examined. State licensing, federal certification for Medicare and Medicaid, and JCAH are examples.
3. The characteristics of the clients are studied. General patterns of mortality and morbidity, length of inpatient stay, frequency of outpatient visits, emergency services required, and readmission rate are considered in this factor.
4. The employees or practitioners are considered. Included in this factor are licensure laws for health care practitioners and physicians that are incorporated into the operational activities of the organization. Also included are personnel practices mandated in the standards and guidelines of the accrediting and licensing agencies and contractual agreements of collective bargaining processes.[8]

With these planning factors in mind, department managers prepare a list of functional operations they wish to computerize, beginning with the application they consider the most appropriate for their setting. Each item on the list should include a premise statement or justification that briefly explains the rationale for placing it in a particular position. This is illustrated as follows:

Case Discussion

A. Background

EKH Memorial Hospital is a 1,000-bed teaching and research hospital located in the inner city. Each year approximately 50,000 inpatients and 225,000 outpatients are taken care of within the 20 different clinical departments and the 20 different ancillary service units. The permanent staff of the hospital is approximately 3,000 including 300 physicians. Administration is in the early talking stages of an in-house computer with maximum storage space. The MRD uses PAS abstracting.

[In determining what data or functions to computerize initially, the first step is evaluation of the present manual system to see where money, manpower, and time are being spent. It is wasteful to design and operate a system that will be more elaborate than is needed to satisfy the requests of the users of the system. It is also expensive, and sometimes impossible, to retrieve desired information which has not been properly stored. Therefore, a knowledge of the needs for processed information and of the various uses to which the information will be put is as necessary as the knowledge of the various methods of storing and retrieving information. Charting of the existing needs and information flows in all aspects of the hospital support system is required. By first noting the accuracy, timeliness and efficiency of the present system, one can better evaluate the potential impact of the computerized system once it is installed.]

In order to list and prioritize the problems properly, some immediate objectives of the computerization of the medical record department should be formulated. General objectives are:

B. Objectives

1. Improve the quality, quantity, utility, and speed of medical data communication while at the same time containing costs.
2. Establish a data base for administrative and business functions.
3. Establish scheduling and booking files and communicate such information for patients, personnel and medical care services.

4. Provide data necessary for projection of health care needs and planning for hospital and medical services, including the community.

[At any time data should be able to be extracted according to many criteria and grouped with other data or classified to satisfy the needs of any user. A system must be selected, tested, re-evaluated and revised as necessary. In hospital MRDs, factors to be considered are (1) the number of records stored and the rate of growth of the file; (2) the number and types of requests for information; (3) the number and types of reports to be issued; (4) the extent of cooperation and correlation with other departments; (5) the amount of space available for equipment; (6) the number of personnel needed to operate the system; and (7) the cost of maintaining the system.[9] Our master plan illustration continues as follows:]

C. Prioritized functional operations and their projected performance objectives are listed below. Note the time estimates included.

Objectives
1. MPI—projected performance objectives
   a. Convert manual card system to on-line in six months (time estimate).
   b. Eliminate issuing of duplicate numbers.
   c. Have edit checks in system to stop function if duplicate number issued.
   d. Limit patient waiting time to maximum of three minutes in order to check for previous admissions.
   e. Introduce the same patient number for both medical and financial purposes.
2. ADT—projected performance objectives
   a. Notify ancillary departments of all new admissions, transfers, and discharges immediately.
   b. Train ER staff to use terminal and issue patient numbers as well as admissions and MRD staff.
   c. Produce accurate and complete census reports.
   d. Provide accurate and on-line bed control.
   e. Provide statistical data for administrative reports and studies.
   f. Provide data base for eventual computerization of entire medical record.

3. Word Processing—projected performance objectives
   a. Convert transcription unit from IBM Selectric typewriters to word processing in six months (time estimate).
   b. Improve operator production by 30–50 percent.
   c. Eliminate all outside transcription services within six months.
   d. Reduce document turnaround time to 24 hours for histories and physicals and operation reports and 46 hours for discharge summaries.
   e. Improve document quality.
   f. Standardize formatting of documents.
   g. Provide services for other departments thereby increasing work load at no additional cost within six months.
   h. Facilitate more efficient management reporting.
4. Discharge Abstract—projected performance objectives
   a. Implement on-line abstracting system within one year (time estimate).
   b. Reduce abstracting errors to maximum of 1 percent.
   c. Produce information for PSRO on a monthly basis without additional abstracting.
   d. Increase abstracting productivity by a minimum of 20 percent.
   e. Have available data for research and studies on a timely basis.
   f. Furnish billing with diagnoses within 48 hours of discharge.
   g. Produce management reports for facilitating closer control over abstracting processes.
   h. Eliminate redundant collection of information.
   i. Make up an abstract history file to accommodate requests for research and administrative purposes.
   j. Produce indexes and reports specific to needs.
5. Chart Deficiency—projected performance objectives
   a. Provide accurate and complete listings by physicians of chart deficiencies on twice weekly basis.
   b. Provide service chiefs with weekly listings of incomplete charts per physician.
   c. Decrease the number of incomplete charts and retain an acceptable level.
   d. Comply with regulations governing completion of charts.

6. Chart Locator—projected performance objectives
    a. Implement system within two years (time estimate).
    b. Add bar codes to all records.
    c. Decrease number of misfiles and lost charts.
    d. Facilitate immediate retrieval of medical record in emergency situations.
    e. Locate all records requested for a project or study within a three-day period.
    f. Give responsibility for accurate order to requesting party.
    g. Automatically order charts according to time frame requested.
    h. Be able to track records while they are "out of file."

The next section is included so that any key functions or special feature ideas can be included as part of the plan. These features represent some of the desired components of the operation after it is computerized.

D. Systems described with some key functions identified.
    1. MPI
        a. Entire MPI to be entered into system.
        b. New admits entered directly into system with 3- × 5-inch card typed for backup for one week and then microfiched.
        c. Microfiche all cards and then destroy cards.
        d. Index to include name, hospital number, birth date, zip code, social security number, admission and discharge dates.
        e. Edit checks built into system to prevent duplication of numbers.
    2. ADT
        a. Terminals in admissions, MRD, and ER.
        b. Terminals in nursing stations, dietary, laboratory, radiology, and billing.
        c. Computer printout to housekeeping daily with updates as necessary.
        d. Daily printout to chaplain.
        e. Preadmission information entered into system that has been mailed in prior to admission from physician's office or patient.
        f. Produces census reports.
        g. Statistics available for administrative purposes and studies.

3. Word Processing
   a. Four work stations and two printers in MRD transcription unit.
   b. Train two operators per shift on two stand-alone units and familiarize others in first three months (conversion estimate).
   c. Install other two units and complete personnel training in second three-month period.
   d. Operations and procedures transcribed within 24 hours of dictation.
   e. Histories and physicals transcribed by next morning after dictation.
   f. Discharge summaries transcribed within 48 hours of dictation.

4. Discharge Abstracting, Deficiency, and Physicians' Incomplete
   a. Operator access by special code for security.
   b. Demographic data in computer from ADT system.
   c. Days of care in special units, admit and discharge dates, physicians generated from nursing stations.
   d. Deficiency lists abstracted directly to CRT at time of analysis of charts.
   e. Physicians' incomplete list generated weekly for service chiefs.
   f. Coding personnel code on face sheet (for backup) and enter data directly to CRT within 48 hours of discharge. (Primary and secondary diagnoses indicated.)
   g. All codes to billing within 48 hours of discharge.
   h. Program designates information to tape to be sent monthly to PSRO.
   i. Program to have built-in edit checks for increased accuracy of coding.

5. Chart Locator
   a. Use bar codes and light wands for record tracking.
   b. CRTs for entry and inquiry.
   c. Printer for order reporting.
   d. Implement order entry function designating immediate' response required, response in 24 hours or scheduled appointments requiring three-day response time.
   e. Printer prints slips sent to requesting party with chart or location if not in file room.

E. Evaluation
1. Monitor each system on a continuous basis until evidence that operation is functioning smoothly and accurately.
2. Monitor on a periodic basis thereafter.
3. Inverview users to gain reactions to new system and suggestions for improvement.
F. Benefits
1. Improved patient care.
2. Establishment of data base.
3. Basis for computerization of entire medical record.
4. Builds esteem of employees, patients and community in forward approach.[10]

The foregoing plan provides one organized method for mapping out a master plan for automation of the MRD. Like systems analysis, it is a critical tool for MRPs today.

Tactical Planning

One of the questions addressed in strategic planning is, What should the business or department be? In support, tactical planning asks, "How will we get there?" Tactical planning can be defined as the identification of the major tasks required to achieve strategic objectives. Supportive to strategic planning, it must, in fact honor the directions given by strategic planners.[11] For health record managers, tactical planning must begin with a firm adherence to the third principle of patient data computerization—coordination with an organization's master plan. This coordination takes into account any plans for automation that will properly integrate functional operations of the MRD either individually or as portions of other applications developments. We emphasize the importance of coordinated planning so that future information integration can be accomplished. Four other tactical plans can be suggested:

1. Given a master departmental plan and appropriately developed descriptive profiles, practitioners may elect to purchase a software package to meet their needs. As discussed elsewhere in the text, these can include the use of computer resources off-site, time-sharing systems, and using in-house hardware. The decision must be made on the basis of how well the package fits the objectives of the application.
2. Assuming the same planning activities have taken place, practitioners may elect to purchase their own microcomputer for installation in

the MRD. As today's technology allows information transfer from large computers to the microcomputers, information integration is now possible in ways that were not considered practical prior to 1980. A look at microcomputers, their potential, and their limitations is featured in Chapter 9.

3. Health record managers may choose to participate in in-house development in conjunction with an in-house automated hospital information system.
4. A combination of the three previously described tactical choices can be selected. The variety of such combinations are described throughout this text.

The next ten years will bring amazing changes to patient record systems and to the managerial activities of health record practitioners and other health care professionals. It is certain that computer literacy will increase and applications will multiply. The infinite variety of choice demands that professionals grow and change with the opportunities. Patient information systems will continue to play a key role in the ongoing delivery of health care.

Vendor Resources in Patient Data Computerization

Many hospital and other health facility systems will make use of existing packages rather than develop their own in-house applications. Today's market affords many choices for such an approach. Practitioners will be participating in a selection process for single packages as well as for larger, integrated subsystems and total information systems. Some current examples of the variety of software packages currently available from vendors are listed in Exhibit 7-8.

All of the applications listed are available from at least one of the major sources of hospital software—computer manufacturers, systems companies, and software companies. This list will be quickly outdated, however, and should be considered an example rather than a reference source. In addition to the packages listed, commercial systems provide a variety of integrated packages. It is not uncommon today to find a hospital employing packages from more than one vendor to satisfy its particular needs. Medical record practitioners and other health care professionals must keep abreast of the general picture and state of the art in computer science. Table 7-2 illustrates current efforts in systems developments according to major vendors.

The reader can see by the data in Table 7-3 that the same vendors that identified specific packages featured in Table 7-2 consider themselves to

Exhibit 7–8 Applications Packages for Hospitals

Application Packages for Hospitals*	Medical Record Department Interface (Data or Functions)
1. Batch census, admitting, and billing -----------------------→	Statistics
2. Accounts receivable	
3. Payroll/personnel	
4. General ledger	
5. Accounts payable	
6. Budgeting	
7. Demand forecasting	
8. On-line admitting ---------------------------------------→	Pt. record
9. On-line outpatient/ER registration and billing ----------→	# Assign
10. On-line maintenance of the medical record ------------→	Pt. record
11. Medical records statistics ------------------------------→	Statistics
12. PSRO reporting ---→	Statistics
13. Programmed learning (computer-aided instruction)	
14. Work management for clinical laboratories	
15. Scheduling ---→	Record charge
16. Generation of radiology reports ------------------------→	Pt. record (Demograph)
17. Clinical order-entry -------------------------------------→	Chart analysis
18. Equipment inventory recordkeeping	for completeness
19. Inventory management for pharmacies	
20. Maintenance of medication profiles ---------------------→	Pt. record
21. Drug interaction reporting ------------------------------→	Pt. record
22. Simulation and modeling	
23. Menu planning and printing	
24. Patient diet planning and food management ------------→	Pt. record
25. EKG interpretation --→	Pt. record
26. Data collection for ICU monitoring systems ------------→	Pt. record
27. Laboratory result reporting -----------------------------→	Pt. record
28. Laboratory specimen logging ---------------------------→	Pt. record
29. Quality control for clinical laboratories	Assign #
30. Medicare cost reporting	
31. Timekeeping	
32. Preventive maintenance records	
33. Physician billing	

*Thirty-six percent of the packages have interface or impact with the MRD.

be offering hospital information systems. Many of the vendor names are familiar to MRPs in their experiences with specific medical record applications.

Table 7–2   Major Vendor Systems Developments

Department/Application	Number of Vendors Now Offer	Plan to Offer
Nursing		
Order set entry	5	1
Auto-Kardex (charting)	3	2
Patient care level tracking	3	2
Staffing requirements	5	2
Nursing notes*	5	0
Care plan*	1	6
Medication schedules*	6	2
Medication monitoring*	3	2
Pharmacy		
Patient medical profile*	7	1
Drug precaution/interaction	2	4
Unit dose care replenishment	5	2
Pharmacy inventory	3	4
Lab		
Specimen collection lists	6	1
Specimen collection labels	8	0
Results reporting*	7	0
Cumulative results*	6	1
Interface to lab computer	6	1
Radiology		
Scheduling	4	2
Report normals only*	3	3
Report all results*	2	5
Radiology index	4	2
Medical Records*		
Medical record index	5	2
Medical record abstract	5	2
Chart locator and delinquency tracking	3	3
ER/Outpatient		
Registration*	8	0
Order entry*	8	0
Demand billing	7	0
Patient scheduling	6	2
Miscellaneous		
Preadmitting*	8	0
Computer assigned patient ID*	8	0
Time clocking	5	0
Surgery scheduling	2	3
Diet list preparation*	7	1
Utilization review*	7	0
Cash receipts entry	6	0
Doctors' registry*	6	1

Source: The table is based upon the responses of eight vendors to a questionnaire listing the enhancement applications and asking whether the vendor now offered or planned to offer the application. The eight vendors are

| HBO | McAuto | Burroughs | Pentamation |
| SMS | NCR | DATX | HDC |

*Patient record impact.

Techniques for Evaluating Vendor's Packages

How can vendor packages be evaluated? It is here again that management responsibility is critical. As discussed in Chapter 6, highly specific evaluation methods can be used to select an appropriate package. Evaluation criteria for vendor-supplied services, hardware, software, and any other commodity offered can be equated with the criteria established to evaluate RFPs. The process of evaluating a software package to perform statistical analysis, for instance, utilizes the same guidelines, as the evaluation process used for examining a solicited proposal to implement an on-line discharge abstracting system. A validation process that guarantees that a minimum level of basic RFP requirements have been met sets the stage for evaluation. Evaluation methodology must be fair and unbiased. If it is not, vendors will not spend the time and money necessary to submit a proposal. Or, if they have submitted a proposal but feel they have not been fairly evaluated, they will protest the selection. If upheld, a protest can lead to considerable delay and embarrassment for the procuring organization. In any event, it will require considerable time, effort, and dollars to resolve the issues. Therefore, the selection methodology must be completely fair and incontestable.

As noted in the discussions in Chapter 6, weighted evaluation criteria should be stipulated in the RFP to give the vendors an opportunity to assess which areas are considered to be most important when responding with their proposals. The actual evaluation technique used in evaluating the proposals, although more detailed, must be in concordance with the overall weighted evaluation criteria stipulated in the RFP.

The evaluation technique should differentiate between mandatory requirements and other optional system requirements. The evaluation technique uses mandatory requirements as a determination of responsiveness; the other system requirements are used as the basis for actual evaluation. The technique must preestablish relative values for the optional system requirements, and these values should be established in weighted scores supported by meaningful rationale and fully described in value matrices as discussed in Chapter 6. When all of these steps have been completed, the system that attains the highest score in meeting the requirements for the full system life cycle is selected. The following is a list of questions for evaluating packaged systems:

1. Does the packaged system meet the hospital's specific needs? If not, will the supplier modify the system to meet the hospital's specific needs? Is the hospital willing and able to adjust to the general system requirements?

Table 7–3 Present Vendors of Level 1 HIS

VENDOR	Number Installed or in Process, APRIL 1980	Number of Beds, Least–Most	Order Entry Terminal
HBO & Company (HBO)	237	100–629	CRT
Shared Medical Systems Corporation (SMS)	100	154–1,725	CRT
McDonnell Douglas Automation Co. (McAuto)	53	103–650	CRT
IBM Corporation (IBM)	50 (est)		CRT
NCR Corporation (NCR)	25	284–641	BCWR or OCR[1]
Burroughs Corporation	20	191–737	CRT
DATX Corporation (DATX)	10	176–534	CARD READER
Pentamation Enterprises	9	280–800	CRT and OCR
Tymshare Medical Systems	5	80–436	CRT
Compucare, Inc.	4	250–440	CRT
Space Age Computer Systems, Inc.	4	349–815	CRT
Hospital Data Center of Virginia (HDC)	4	256–800	BCWR[2]
Spectra Medical Systems, Inc.	2	140–600	CRT
Technicon Medical Information Systems Corporation	3	446–719	CRT
Interpretive Data Systems, Inc. (IDS)	1	250–250	CRT
National Data Communications, Inc. (NADACOM)			CRT
Genitron, Inc.			BCWR
Information Resource Electronics Corp. (IREC)			CRT (Intelligent)

1. bar code wand reader (BCWR); optical character reader (OCR).
2. HDC also utilized CRT's for inquiry purposes.

2. Who else has used the package? (Careful checking is essential.) Can the system be implemented on the hospital's existing computer? What hardware modifications will be required?
3. How will the system be maintained? How difficult will it be to make changes?
4. How good is system documentation?
5. What will be the costs to:
   a. Purchase the package?
   b. Modify it, if necessary?
   c. Implement it?
   d. Maintain it?
6. Does the department currently use a package from this vendor?
7. If so, what use has been made of the package from this vendor?
8. Will the MRA have appropriate access to system designers—either in-house staff assigned to work with the package or the vendor representative?

9. Does this package have potential for future medical record departmental computer development integration?
10. Is this package being reviewed in accordance with an overall computer master plan for computer systems development within the organization?
11. What are the particular quality control features available to support effective data security procedures?

Management Information Systems for Medical Records

Consider using computers to store JCAH standards in a form that can be monitored quarterly by the MRP. This allows planning for the data needed to demonstrate compliance with the standards. These data can then be captured and stored as it is collected so that current status is always accessible. Status reports on individual standards would assist in developing management action plans that effectively integrate these requirements.

Computers can also be used to store work standards and track productivity. We are most aware of the productivity monitored in word processing systems. Attention to similar analytical tasks with other departmental functions is warranted if best use is to be made of human resources and talents.

Coordinating Systems Analysis and Planning beyond the MRD

Thus far in this chapter the focus has been on coordinating the systems analysis foundation into a comprehensive planning process for developing a master plan of automation for medical record departments. It is useful, also, to view a model of strategies for extended users such as clinical providers and other allied health midmanagement personnel to use in the development of more comprehensive systems that are to incorporate computerized patient records. The following strategies are intended as a coordination of appropriate organizational, technical, and management steps that could be used to guide such efforts. Of particular interest is inclusion of patient record computerization in these strategies.

1. *Employ a rigorous systems analysis process with a strong user team emphasis.* This requires team identification and analysis of patient information problems in the existing medical record operation.
2. *Extend the systems analysis process to include a clear definition of the proposed design that is a computerized model or segment of the patient record.* Include its location and impact within an overall information flow in the organization. Flowcharting the evolution of the patient record from its inception will provide a ready and accurate resource to users.

3. *Apply flowchart overlays to determine where patient record data elements interface and merge with subsystems.* There are a variety of users of patient records, and the record evolves in a certain series of steps. Users enter data into the record at specific intervals for particular and specialized use. By applying flowchart overlays of each process, common data entry and exit points can be identified in conjunction with common data elements.

4. *Require user validation of all selected data elements, as well as of flowcharts and other tools.* One validation method is to create a basic dictionary or list of data elements to be contained in the patient record, define each data element, describe its use, and require user concurrence and verification of the dictionary.

5. *Examine user needs.* Create a written document that includes a list of desired data elements compiled by the anticipated users.

6. *Analyze existing user operations for matching and extending systems development.* This step examines the communication requirements of all users in the light of existing computer developments of a given organization. By looking at the entire picture, attention can be directed to a unified patient record system rather than limited territorial pieces. There must be an understanding among clinical, analytical, and administrative users on the use of patient record information. This must extend to include an awareness of individual modules within overall computer applications in the organization.

7. *Build in systems communications techniques at all levels.* This step includes a notification to all users of the process that is underway and seeks users' questions and concerns for feedback to the design team.

8. *Employ automation first to provide improved access to the existing system before selling users on a computerized patient record.* In keeping with the basic principle of automation, the manual system, including the manual patient record, should be fully evaluated. Maximum efficiencies should be provided for the manual system along with detailed informational analysis before any computer applications are selected. Documented computer developments in medicine have clearly demonstrated that computerization of inadequately analyzed operations produces ineffective and costly systems. This is disastrous to the ultimate goal of developing effective computerized patient records. One of the ways that computer potential can be identified, improved, and introduced to all levels of users is to employ it in improving the access and processing from the manual record. For example, computerized word processing systems can be used to expedite clinical summaries. Also, computerized clinic

charge-out and record locator systems can reduce the retrieval time of the medical record through better record management.

9. *Prepare users to work with an interim record for some time.* An interim record can be defined as products of computer applications generated through computer printout and pulled together to comprise a paper form of the patient's medical record. For a number of years medical and health settings will continue to work with partially computerized patient data so that the record may include typewritten entries, handwritten entries, and computer printout forms. The printout forms will increase as the computer stored records are developed. Monitoring and organizing the interim record document will provide a continuing control as well as experience for MRAs that will enable them to translate the necessary processing concepts and requirements to computerize patient records fully.

10. *Employ known strategies for measuring user criteria before and after systems implementation.* This one recommends that those developing computerized medical records work with known methodologies for master planning in which the computerized patient record will be developed so that evaluation criteria are established as a required portion of the proposed model definition.

11. *Share the potential of improved formats and better data with users as soon as the developments are available.* When users' information exchange can be enhanced by providing better formats in existing computer applications, positive attitudes about computers are reinforced in the users. Time spent to improve the computer's role as a positive tool in daily operations and to initiate the imagination for building future data formats in alternate displays will be returned when input is required to develop model computerized medical records.

12. *Plan index networks to help users locate other patient record segments.* Function keys on terminals can be used to link inpatient and outpatient information. They can also be used to link other medical record locations. By planning for specific linkage, capability for maintaining a unified patient medical record resource is identified.

Questions and Problems for Discussion

1. Should a knowledge of medical information structure be a prerequisite to planning for patient data computerization in MRDs?
2. How does the R-ADT foundation provide the most efficient beginning to developing facility patient data systems in hospitals and clinics?
3. Prepare a report for an administrative medical computing task force that outlines potential extensions from a computerized R-ADT system. Include a rationale for your choices.

4. What are the principles of patient data computerization? Explain how they apply in three situations likely to occur in MRD operations.
5. Prepare a sample descriptive profile for a functional operation in an MRD. Present the results for review in a discussion group.
6. Calculate the storage needs for an MPI for a 200-bed community hospital that provides an active outpatient cancer therapy program. Allow a code on the index to refer users to both inpatient and outpatient records. The present file contains 10,000 patient names. The outpatient visits are 26,000 per year. The readmission rate is 40 percent for hospital admissions.
7. Why is planning so important in developing patient data systems? What is the relationship to computerized patient data systems and master plan for information systems that is developed for a hospital?
8. Explain the need and purpose of strategic planning in medical computing. What effect can this process have on information systems development?
9. Should the medical record department master plan for automation be influenced by the hospital master plan for systems development? Why?
10. Develop a plan for implementing patient data computerization into a medical record department. Include computerization of a minimum of six functional operations. Include project team composition, sequence of operations addressed, and a rationale for your selection. Project a long-range completion date for your plan.
11. Prepare a letter to your administrator requesting information on the master planning process for your hospital. Include a general promotion of the need for this process and a description of the health record manager's participation.
12. What are some things to look for in evaluating vendor packages for hospital use?
13. Define and describe specific ways that the principles of management information systems can be implemented in MRDs.

NOTES

1. Kathleen A. Waters and Gretchen Frederick Murphy, *Medical Records in Health Information* (Rockville, Md.: Aspen Systems Corp., 1979).
2. John Burch, Jr.; Felix R. Strater; and Gary Grudnitski, *Information Systems: Theory and Practice* (New York: John Wiley & Sons, 1979), p. 4.
3. Gretchen Murphy and Arden Forrey, "Systems Analysis Documentation for R-ADT," unpublished report, 1979.
4. Joan Liebler, *Managing Health Records: Administrative Principles* (Rockville, Md.: Aspen Systems Corp., 1979), p. 99.
5. Peter F. Drucker, *Management, Tasks, Responsibilities, Practices* (New York: Harper & Row, 1974), pp. 121–122.
6. Ibid., p. 125.
7. George R. Terry, *Principles of Management* (Homewood, Ill.: Richard Irwin, Inc., 1977), p. 183.
8. Liebler, *Managing Health Records*, p. 102.
9. Mary Potts, "Action Plan for Computerized Functional Operations" unpublished paper, November 1980.
10. Darla Betz, "Prioritized Functional Operations in a MRD" unpublished paper, April 1981.
11. Terry, *Principles of Management*, p. 188.

Chapter 8

How To Safeguard Privacy through Data Security

Objectives

1. Recognize the unique character of a patient's records as inseparable from the patient.
2. Describe the ownership of a patient record.
3. Describe the components of a data security program.
4. Describe the role of the MRA as a data security officer.
5. Describe the features included for security in a software program.
6. Describe procedural and practical measures for assuring data security.
7. List three methods to authenticate terminal users and those who enter data.
8. Explain the use of an authorization table.
9. Describe surveillance procedures in a computer system.
10. Describe the methods available to provide security in a data bank.

The core of patient data computerization is the exchange of information between patient and physician or between patient and any direct health care provider. Another, less visible part of this exchange, but of equal importance, is the documented evidence of the exchange. The evidence of the patient's health care encounter is the record that describes and extends the information exchanged by the two parties involved. Results of laboratory tests, x-rays, and pathology reports and objective and subjective evaluations made by patient care providers collectively demonstrate that the record is a unique extension of the patient. Like a painting, the record has unique, identifiable characteristics. Prior to the recording of the information exchanged in the encounter, there was no exact or even similar source of information. Not even the patient is a source of this exact, documented data. This is true because the record is a result of elicited and analyzed data and specimens. The record achieves its uniqueness and usefulness only as a result of the exchange process between the patient and the physician.

From this point of view the confidential information exchanged during diagnosis or treatment is a fundamental concern for those who have the authority, responsibility, and challenge to maintain privacy for this information in a manual or computerized data system. This concern for the patient's privacy among the medical record professionals who must continue to develop methods to provide privacy for the information under their control is not new. What is new is the increasing use of computers to process and retrieve confidential patient data. Manual methods used to release information from manual records do not apply to access or retrieval from computerized records.

In this chapter information is presented that will prepare readers to understand and apply ethical principles when faced with an opportunity to develop a data security program. The topics covered in the chapter are based on the following premises:

- Primary data retain their characteristic of inseparateness from the patient whenever it is the account of a care encounter.
- Secondary patient data must be carefully monitored for potential privacy leaks.
- The information on a patient's record is considered to be in the ownership of the patient.
- Freedom of information laws tend to open up data previously considered confidential. Privacy laws tend to promote protection of data considered confidential.
- Abuses in privacy of patient data continue to occur with increasing use of the computer.
- Data security requires a comprehensive program of policies, procedures, technology, personnel, and monitoring.
- Release of information policies and procedures for manual record systems does not apply to computerized medical record systems.

The premises will be described in detail in conjunction with the chapter topics.

Patient and Record Are Inseparable

It becomes apparent when we consider the patient and the record that the two are inseparable. There is an obvious physical separateness of the two, but when studying the information recorded about a patient it becomes evident that the recorded information that represents a patient encounter is an extension of that individual patient. That document becomes the primary source of information about the patient besides the patient himself or herself. In fact, before the data were recorded and documented, the only other source of information on the same topic was the patient. And it was limited only by the patient's ability to express the information or supply the specimens. Those who treat patients and who use the information generated from patient data must always keep in mind the inseparateness of the two elements.[1]

Closely related to this concept of uniqueness of the individual record is the longevity of data entries. Because of space needs there have been various attempts in manual record systems to purge old and no longer relevant data. Decisions regarding the usefulness of old data according to relevancy or age itself have been a crazy quilt mix of institutional needs (space and money for storage) and legislative decision on retention (minor children's records should be retained until the child reaches majority, for example). There is no state-by-state uniformity of legislative ruling or management practice in resolving manual record retention or purging questions. Legibility, for instance, is an obvious quality of recording that can be used to determine which records can be destroyed. Nurses' notes, however (for no apparent logical reason), have frequently been targeted for purging. Yet, they contain important observations on a patient's condition. The vagaries of purging used for manual records in the past cannot be relied on for determining retention programs for computerized patient data.

Those faced with designing computer-stored records must determine what can and cannot be purged when computer storage becomes the same problem that manual storage has been for so many years. This determination would be made during the analysis phase of systems analysis. Since computers have the capacity to store an individual's lifetime health record, a predictable purging pattern must be determined. This relates to privacy specifically. Certain diagnoses and problems do not need to be retained forever and for privacy's sake it would be best if they were not. Episodic drug use, venereal disease data, pregnancy data are all important for care but probably not necessary for a lifetime record on all patients for all time.

One other thought on this point should be considered. As computers become more and more the right hand of the medical community, there will be more sophisticated, decisive, and expanded uses for computers. At some point in the not too distant future, traditional laboratory work that we now feel fairly familiar with will be added to or replaced by new and as yet unknown tests and procedures. Who will be familiar with all of them? It is likely that only the specialists whose disciplines utilize these tests will be thoroughly trained in their interpretation. If summarized at the end of a care episode, it is likely there will be no need to retain single reports. Perhaps summarization of the test results as related to particular incidents of care will be a method allowing effective retention and purging policies. The previous scenario is one attempt to predict what might be coming with increased computerization in medicine. Specialty areas will have procedures and tests important only for specific conditions at specific times. A determination of which of these test results must be stored forever will be the result of an increasing amount of tough decision making within the various specialties.

Traditional Medical Ethics and Computers. In the past patients, providers, and users dealt with manually developed data; in the future they will deal with automated data. Current endeavors regarding privacy reflect attempts to develop security measures that apply to computer technology. This attempt to develop adequate protection for the mushrooming technology is taking place at the same time as two other technologically related developments: (1) entry of previously unlinked, manually stored data into computers and (2) evolution and development of automated health data banks. These two activities will accelerate in the years to come—the former in the 1980s and the latter in the 1990s. By 1990 it is anticipated that there will be far less manual patient data to convert. Systems developed during the 1980s will start with automated data. Many systems analysis processes will no longer even address conversion of manual data.

Past practice featured two other important components: the educational background of those responsible for the control of patient records and the traditional method of voluntary monitoring utilized by the health care profession. Those who treat patients and those who work with patients must meet prescribed educational criteria. Many must further demonstrate professional ability through credentialing. A significant part of the medical record professionals' training, education, and credentialing is focused on the privacy of patient information. Future practice should place the same emphasis in education and so forth so that those who work with computerized patient data can offer patients the same strict adherence to the ethical and technical aspects required in releasing confidential information from a manual system. Similarly, accrediting agencies, professional associations, licensing and review organizations, and all groups who traditionally have provided the standards and the means to measure the standards should direct their attention to the development of standards for computerized patient data. Direction should emanate from this group so that patients and users of the data can be assured that those who have the most experience and recognized valid purpose provide the guidelines for this most important development in the provision of modern health care.[2]

Computer Systems Require New Privacy Policies

Currently, and more so in the future, those who design and develop computerized data systems must be involved in the formulation of policies that can be effectively used by those who work with computer systems. The computers will, after all, perform only as programmed. They cannot be expected to function as policy activators if software is not designed, hardware not constructed, and users not educated as to the necessity for specific privacy policies. Perhaps there is a need to be reminded continually that computers, like the typewriter, the renal dialysis machine, the file folder, the dental pick, etc., are tools that only assist in the provision of patient care. Even with its special aspects—memory and logic—the computer will work only as efficiently as it is programmed. And policies regarding its use must be just as clearly defined as they are for the other tools.

Medical record professionals have demonstrated a sensitivity to medical privacy and the confidential character of the data under their control through their educational programs and familiarity with applied medical ethics. Physicians, nurses, administrators, and medical record personnel have worked together for many years on the issue of medical ethics and the application of ethical principles to the policies and procedures safeguarding patient privacy. There was an attitude on the part of all who were involved in patient care—patients, direct and indirect providers of care—that information exchanged to accomplish diagnosis and treatment would be held in confidence by those who document the information.[3] Computers and the reimbursement by third party payers have altered that attitude, and with reason. Today a threat exists to optimal information exchange in the practice of medicine and optimal use of computers to perform to capacity in this aspect of medical practice. These two threats exist because patient privacy has been jeopardized by incidents of release of confidential information to unscrupulous individuals who used it for reasons unrelated to the reason for which it was given. The following are examples of results of that release.

- Disclosure that an employee (patient) was an alcoholic
- Sale of abortion patient names to "pro-life" groups
- Extortion based on knowledge of a patient's treatment for VD
- Questioning of neighbors about another neighbor's (patient's) mental state
- Informing an employer of an employee's (patient's) mental status
- Disallowance of health insurance eligibility on the basis of an unconfirmed diagnosis of "cancer"

The last three were insurance related incidents. Especially noteworthy about these examples is the fact the patients all had willingly signed consents to allow this information to be released.[4] That is, they signed routine consents for release of information upon admission for hospitalization. Consents that are considered to be "informed consents" or those that represent in writing what was explained to the patient orally are not useful for the average citizen, for the most part. We are not a particularly literate society. The lack of clarity in the wording of written consents is widely recognized. In one study it was demonstrated that most patients without a college education were unable to read consent forms with any recognizable understanding. When one notes that only 28 percent of our population has the benefit of a college degree, it is easy to understand the plight of patients who willingly sign consent forms in order to be treated while being totally unaware of the implications of the signed consent.[5]

Publicized breaches of confidential data are only one piece of evidence of the changing attitude about privacy. The other less conspicuous turn of events permeates many phases of individuals' lives. That is an increasing awareness of loss of privacy regarding many aspects of life besides health— financial, social, marital, and so on. It is known that over 7,000 computers are used by the federal government to store data on its citizens. The names, and in most cases, many other facts on over 150 million Americans are stored in government data banks. The government has an average of 18 files per person on as many as 3 billion people.[6] In the private sector, one computer-based advertising firm has 72 million names cross-indexed to driver's license, special purchases, and magazine subscriptions.[7]

A change of the attitudes about privacy held by patients and providers in the recording of their exchange of information can be seen. Some health care providers are reluctant to use computers to process and disseminate complete and accurate health care data on grounds that violations of privacy are more likely. But the improvement in information handling afforded by the computer need not be foregone if health record professionals make preserving patients' privacy a tenet of modern, computerized recordkeeping and information processing.

Privacy and Ownership

"Privacy: the right of an individual to be left alone, to withdraw from the influence of his environment; to be secluded, not annoyed, and not intruded upon by extension of the right to be protected against physical or psychological invasion or against the misuse or abuse of something legally owned by an individual or normally considered by society to be his property."[8] "Privacy is a condition or status. It is the maintenance of a personal life-span within which the individual has a chance to be an individual, to exercise and experience. It is the capacity to control information about oneself or one's experiences. In most general terms, it is the right of the individual to be left alone. . . . The individual is a creature who lives in society. In living in the human community, he/she being at the heart of the mystery which is each human person, is something which cannot be communicated. The center of a person is a mystery and privacy preserves that mystery which is me."[9]

Let us discuss some characteristics of ownership since it is one part of the privacy definition. A tangible object is usually evidence of the owner's possession. Control and guardianship are also characteristics of those who own something. The ability to change, add, or delete some part of the tangible object is also at the direction of the owner. Regarding patient data, what the health care facility owns is only the record itself. The patient, on the other hand, does not have a clearly defined owner relationship with the information recorded on the document the facility owns. Information is intangible, out of the patient's control when written on someone else's document. It is pretty clear that when changes are made in the information, this change is not known to the patient. But, those who have made ownership decisions have always agreed that the information, intangible though it is, is the property of the patient.[10] Recent legislation in many states supports the patient's ownership of the information, for it allows patients to have access to their records for the purpose of examination, correction, or additions. States with laws concerning confidentiality of and patient access to medical records are: Alaska, California, Colorado, Connecticut, Florida, Hawaii, Illinois, Indiana, Louisiana, Maine, Maryland, Massachusetts, Michigan, Minnesota, Mississippi, Montana, Nevada, New York, Ohio, Oklahoma, Oregon, Rhode Island, South Dakota, Tennessee, Texas, Virginia, Washington, and Wisconsin.[11]

The statutes vary in intent and scope and must be examined carefully, like all laws, to be understood. Study of the various statutes reveals that the term "medical record" in some cases refers to specialty records only, such as those for psychiatric facilities, or single reports only, such as histories or emergency reports. Other unique aspects of the statutes are the various options or lack of options regarding examination of the records, copying rights, correction, and so on. Recent legislation seems to reflect that ownership as we have traditionally been taught to consider it, is being supported by those who enact the laws of the land. That is, ownership of the patient's information is ascribed to the patient.

Support from Freedom of Information and the Federal Privacy Act

Other legislation germane to this topic is the Freedom of Information Act and the federal Privacy Act. Both freedom of information and privacy laws, based as they are on sound principles derived from basic rights, tend to allay fears and skepticism about how public officials and employers use their power; about the ways in which information about the individual members of society is used; about the often secret decision-making process by which bureaucrats decide what is the public good.

The more established of the two types of information laws is freedom of information. Nearly all the states now recognize the principle that there should be a statutory basis for public access to records of government procedure. All 50 states do have some form of legal authorization for public attendance at policy-making meetings, and for the most part judges have upheld that right. Informational privacy laws are still developing and have not yet become as widely accepted as the freedom of information measures.

It is generally uncontested that government agencies and private businesses need some information in order to conduct their business and to offer services to members of the public. In order to qualify for those services, individuals must occasionally reveal sensitive personal information. The object of the information privacy laws is to prevent unwarranted uses and disclosure of that information and to let the individual know how it is being used. Those laws recognize that certain information simply does not belong in the public arena. To this extent information can be separated into two clearly defined groups: public information that determines how the government functions and private information on individuals that is used only to accomplish a specific purpose for which the individual has given consent.

Inevitably, there are conflicts. In 1979 in the state of Virginia, a woman discovered that the state privacy act prevented her from seeing her own medical records even though the records would have been required to be disclosed under the state Freedom of Information Act.[12]

Similar conflicts exist in other states. In 1978 an attempt was made to correct inconsistencies in the Connecticut Personal Data Act and Freedom of Information Act (FOIA). A bill to combine the acts was introduced in the legislature but it did not gain much momentum. Privacy proponents argued that FOIA advocates were taking a much too broad approach to release of personal data, whereas, conversely, access advocates accused the other side of excessive secrecy. What should be remembered about both the freedom of information laws and privacy laws is that at their best both are essentially antisecrecy laws. Freedom of information acts assure the public that there will be no secrets regarding information that is stored and collected and used in the conduct of public business. Privacy laws assure that there will be no secrets regarding information stored on private individuals and what constitutes that information regarding content, accuracy, and who may use it.

The federal Privacy Act gives individuals the right to inspect, copy, or contest inaccuracies in their own medical records, provided those records are for care received in a federal medical care facility. Because of exceptions to this law, one must be familiar with the various rules and procedures of the individual federal agencies in order to exercise the rights expressed in the act successfully.[13]

The report of the Privacy Protection Study Commission completed and published in 1977 and titled *Personal Privacy in an Information Society* was the most exhaustive and conclusive document produced on the topic of privacy in this century.[14] Its recommendations have still not been embraced by any organized group. Its recommendations regarding medical records, just one type of information it studied, have not been adopted as a goal by any group whose professional focus is medical records. In fact no group or organization is intent on establishing legislation or education that will curb the abuses of privacy and the increasing use of computers that can proliferate privacy problems at such a speedy rate.

Patient Data Security Programs

Patient data security is protection of computerized data or information from accidental or intentional disclosure to unauthorized persons or from unauthorized modification. Security should not be construed, as some writers and researchers contend, to be a technical matter. Security as it is needed today is also a sociolegal and ethical matter: It cannot be adequately dealt with at the software and hardware level but must also be addressed by those who are empowered to hire personnel and put into action policies that affect the development and use of software and hardware. The techniques for data security are varied and include levels of skills, attitudes, and knowledge. Computer hardware, legal issues, software design, institutional procedures, and personnel all affect data security within any organization or with any computer installation. For the contemporary medical record manager, the obligation to understand the necessity and components of a data security program cannot be overemphasized. The computer expands the scope of data development and retrievability far beyond the limits of manual data systems. It increases the use of the data and makes the information available at more sites and to more users. The MRA whose department is sizable and processes a large volume of patient data will need to consider the data security function as deserving the attention of a full-time professionally credentialed medical record supervisor.

Data Security Program Components

The best data security program

- Includes organizational policy structure.
- Involves professionals practicing in the facility.
- Builds in maximum physical security for hardware use.
- Provides comprehensive software support.

Development of data security policies reflects recognition of the privacy issue. Because policy is a formal declaration of intent and a guide to action within an organization, subsequent development of appropriate procedures is dependent upon formulation of policies. A statement that a data security program is in effect in the organization serves a dual purpose. It notifies the organization's staff and the public that there are policies directed to data security. Data security policies are the foundation upon which managers build department-specific objectives and procedures. The professional medical record manager utilizes organizational policies when formulating measurable objectives for secure patient data systems. Procedures followed by employees in their routine daily work tasks should reflect the data security program.

Procedural and practical protections within the institution must provide for physical security of hardware and enactment of operational procedures related to data entry, systems operations, and data retrieval. Protection of the installation and maintenance of data processing hardware includes policies and procedures that address prevention, detection, and recovery. In dealing with prevention, appropriate location, trained personnel and access procedures must be considered. Numerous data processing industry resources affirm this principle.

Detection includes maintaining and verifying physical access as well as system access. Back-up facilities adequate for providing protection of vital patient information are the cornerstone to recovery when a data breach has occurred. Within central computer centers and at data entry or retrieval points security depends upon the following factors:

- Personal trustworthiness of hardware operations personnel in computer department management
- Technical competency and an understanding of privacy issues
- Overlapping responsibilities for central processor operators
- Restriction of access to the computer center and terminal locations. Sign-in logs and badges indicating authorization for access to specific areas should be used

- Locked terminals controlled through central authority or user data security officer if remote location is employed
- Magnetic strip cards interpreted by input device
- Physical security for the central processor and environs to protect against natural disasters such as fire, water, and so on
- Back-up facilities in the event of electrical power failures and provide duplicate data files where specified[15]

The Role of the Health Record Practitioner as a Data Security Administrator

The health record practitioner whose job responsibility and job title indicate the primary duty of the position is to function in a patient data security role needs to be familiar with security for on-line systems and data base management. The patient data security officer or manager must recognize all the components of an effective data security program and be prepared to establish, implement, and monitor data security measures. For instance, the introduction of interim records raised an important question regarding the computerization of hospital medical records. That is, how are signatures verified and recognized? Individual states will resolve this legal issue according to their own rules and regulations. In Washington State, this question is resolved by the following statute taken from Washington Administrative Code 248-18-440:

> Each entry in a patient's medical record shall be dated and shall be authenticated by the person who gave the order, provided the care or performed the observation, examination, assessment, treatment or other service to which the entry pertains. . . . All entries in a patient's medical record shall be legibly written in ink, typewritten or recorded on a computer terminal which is designed to receive such information. Entries recorded and stored in a computer may be stored on magnetic tapes, discs, or other devices suited to the storage of data. . . .

The answer to this issue then, seems to be found in the term *authentication*. The methodology of authenticating, or identifying, terminal users is the same for those who enter data as for those who access data. (See the next subsection.)

In a hospital or other health care facility the MRA will very likely fulfill the role of security officer alone. In a larger organization such as a PSRO regional data bank, the security position may be shared with a data processing or information manager. When responsibility is to be shared, it must be carefully defined and designated for each individual so that there is no potential for conflict or breach of data privacy. Included in data security measures are software design, identification of terminal users, authorization tables, surveillance procedures, and security features of terminals.

Identifying Software Security Measures

Several considerations are important in the design of software security measures. The security system may be an integral part of an operating system, or it may be a series of tasks that the operating system performs like any other task. It must be designed in such a way that no input job or message can escape the scrutiny of the security programs, however. Five aspects are built into security systems programs:

1. *Identification*. The identification of a terminal user must be established. The terminal being used must also be positively identified. On a batch job not submitted at a terminal, the job card will carry the identification code.
2. *Authorization*. The security software must establish whether the user or terminal is authorized to do what is being attempted.
3. *Alarm*. When security procedures are violated in a manner that appears deliberate or unnecessarily careless, an appropriate authority should be notified immediately. Would be intruders should be made aware that the system has burglar alarms.
4. *Surveillance*. Details of all procedural violations must be logged, and statistics or charts of how the system is being used should be compiled. Both should be perused by the system security officer each morning.
5. *Integrity*. The operating system and the security programs must be designed in such a way that they cannot be compromised by ingenious techniques. As new techniques become apparent for bypassing the security facilities, the programs have to be modified to seal the leak, and this is likely to be a continuing process.[16]

Data security systems software is available in vendor-developed packages, facility-developed system features, and individual applications. In order to accommodate efficient increases in data processing and establish opportunities for direct access devices in telecommunications while maintaining and containing system cost through reduced operator intervention in data security programs, it is necessary to turn to software security systems design. Software design programs must deter improper access attempts. They must try to prevent such attempts from succeeding and must detect accesses to protected data should an attempt be successful. They must do so within a prescribed cost limitation and with appropriate acceptance by users of the system. Features to be included in overall software programs include the following:

- Assurance that authorized persons can access their data easily without being uncomfortably aware of the data security system. The system should be comfortable enough that systems designers and users can work with it and be unconscious of the security provisions.
- Assurance that unauthorized persons cannot access restricted data.
- Immediate notification to a designated computer operator when unauthorized data access attempts are made. This can be a console message that identifies the user and the data set with a comparable message sent to the user to discourage casual attempts to break the system.
- Maintenance of a record of all unsuccessful data access attempts to enable security officers and system managers to track attempts and note patterns of access attempts.
- Maintenance of identification and verification of all system users by category and by specific allowable functional operations.

Managing Data Security

It is generally accepted that control of unique information should be in the hands of a user-manager. As users work with data entry and retrieval they deal with the day-to-day operations of data security programs as related to the hardware in their particular work area. For instance, an admissions clerk deals with the registration of patients. This entails retrieving financial and other identification information from previous records for patients who are being readmitted to the hospital. It is most probable that no clinical data needs to be displayed on admitting department terminals. In contrast, terminals on the nursing stations may be used to display current status of diagnostic and therapeutic work on given patients. In that environment there may be no need to retrieve financial data. There may only be a need to view the diagnostic and therapeutic data, for example lab work, so that physicians and nurses have immediate and total access to clinical facts. Both of these examples illustrate particular applications of data protection in a medical environment. Both illustrate retrieval of selected information only, by means of a computer system. They also illustrate the comprehensive scope of data security and the need for a data security program.

Technical protective measures should be based on the probability of a potential breach of information, the value of what is being protected, and the probable effects of loss. Potential breach is anticipated in designing overall structure of data security programs. Value of information is determined through coordinated efforts of users as just described; and probable loss effects include not only cost for reproducing information but the potential damage to individual patient services that could occur. For instance, when health information is released inappropriately but routinely to employers who pay for medical insurance and the information is subsequently used in screening for job promotion, the potential damage to the consumer/patient may be both economic and social.

Planning and Organizing Specific Data Security Programs

Systems designers can usually provide pretty precise cost factors so that users will know how expensive the data security program will be. It often begins with hardware considerations, since terminals are the external doorway into the computer system. Simple measures such as locks on the terminal, badges, and identification cards are the first and least sophisticated level of data security. More elaborate security—protecting data through display means or requiring user codes for making certain procedural changes—begins to increase the cost to the user. Cost depends upon the volume of information to be handled and the use of constant monitoring as opposed to periodic monitoring or to special audits on actual operations of computer applications as data are both entered and retrieved. Managers are concerned with the cost of protecting data against unauthorized modification and unauthorized disclosure. They are interested in determining the potential cost to the organization when an intruder modifies or acquires the data and determining the value of information with respect to total system replacement.

They are also interested in knowing how responsible the users will be in implementing data security policy established by administration. This includes recognition of the importance of data security by the users; and security controls in terms of job and function for all employees with appropriate sanctions built in. It is in this area that the least has been developed for computerized patient data systems. There is need to determine the responsibility of all the people involved, their level of expertise, their experience, their motivation, and their ability to understand the significance of data security as a critical element in designing information systems for medical care. The following data security measures afford procedural and practical protection that should be incorporated in any system designed to protect data:

1. Establish and use user codes.
2. Monitor all inquiries automatically.

3. Use cryptography for protection of information in particular.
4. Establish user files.
5. Employ passwords.
6. Use fingerprints, voice prints, signatures, handshake, and retina prints to authenticate user identity.
7. Keep a record (an audit trail) of all attempts to breach security, for both ongoing and special runs.
8. Confine routines by blocking out data paths that unauthorized persons may exploit.
9. Check data consistency to make sure information variables are within appropriate ranges.
10. Require users to input checklist of operations to be done. Data bank routine interprets the checklist to the user if there are no security violations.
11. Segregate data and store on different files and give users access rights only to certain paths.
12. Refuse to answer query in a statistical bank when fewer than three individuals are in a category and queries that seem uncertain as to data as they are placed in the data bank and uncertain as to answers.
13. Use the following procedures in a health data bank when obtaining individual patient records:

    a. Substitute individual code for individual name.
    b. Separate identifying information into a new file assigning a unique number to each individual and place this number with the information about the individual in both files. Use a third file and a third party to link the identifying information to the data. A researcher who wishes to add information to the data bank looks up the individual identifier in the file and obtains the unique code for the individual. The researcher sends the code and the data to the third party, who looks up a number corresponding to the code in the link file and sends the number and data to the data bank. The data bank administrator then files the data using the number from the third party as the identifier.[17]

Passwords

Let us examine the role of password protection in software security programs. A password data set can be stored in the system in a keyed direct access file. The keys can be data set name and password combination. The presence of a record within the particular data set name and the password combination as its key can indicate that the password is valid for that data set name. A given password may authorize "read only" access or "read/write" access to a data set. Any number of passwords may exist for a single data set and any number of data sets may have the same password. The password data set must be named and reside on a series volume so it cannot be shared between systems. Then the utility and the protection program within the system can provide the ability to add, modify, delete, or list entries in the password data set. Application data set's labels may then contain flags indicating whether a password is required to access them and whether a specific data set is protected for all accesses or only for update. A flag can be set by utility routines in the system or by the password data set maintained as utility. It is important to link the validation process of users and data sets. An open security routine can be initiated whenever an attempt is made to open, delete, or rename a data set whose label indicates password protection. The security routine can prompt the operator and the user and request a rekeying of the password to reverify or scan the password data set to see that a record with the key of the same data set name and password combination exists. If not, the access attempt can be recorded and aborted. A full data set name can be used for the search. If a search fails, one more try may be allowed. If the second try fails, the access attempt may be aborted. Password functions must be broken down in order of increasing power so that users understand their role in use of the password and how it will affect a restricted access to their information. First, the *read* password allows retrieval of only specified records, such as lab information at a nursing station. The *update* password allows retrieval, update, insertion, and deletion of records in the data set. This would allow a lab summary report to be updated for new information or modified for corrections by authorized personnel in the lab only. A *control* password allows update privileges and controls interval access to the data set. The *master* password allows control privileges and deletion or alteration of entry to the data set catalog itself. The reader can see that the password system can be very structured, very controlled and include a wide variety and levels of user and data protection.

A full understanding of the potential of software security protection is an essential goal for data managers. Software security system protection incorporates many other features. The password feature illustrates how such a design can be structured to protect individual and uniquely identified information generally and specifically. Additional features in software design may be structured through other techniques. For example, specialized audit programs, system monitors, data file back-up, or cryptographics can be used. Although there must be a limit on data security information provided to potential users so that data protection is not adversely affected by such disclosure, basic features need to be explained to all users. Protection must be explained in terms that users and patients or clients of the system can understand. In this way the unique features of software design programs in data security can become major building blocks in establishing accepted and acceptable medical data banks that will be necessary in order to carry out medical care in the future. Such programs are critical to designers and security officers.

Identifying Terminal Users

There are three basic methods by which the identity of an individual may be established:

1. Something *known* by the individual
2. Something *possessed* by the individual
3. Something *about* the individual

The first category includes such things as a password, the combination to a lock, or facts from an individual's personal background. The second category includes artifacts, such as badges, passes, cards with machine-readable information, and keys to locks. The third category includes physiological attributes, such as an individual's appearance, voice, fingerprints, and hand geometry.[18]

An authorization table or algorithm (program) may indicate what transaction types the user is permitted to enter, or it may relate to some other aspect of system use, such as what programs the individual is permitted to use, which data sets the individual is permitted to read, and which individual is permitted to modify. Figure 8–1 is an authorization table.

When the security system uses more elaborate structure than simple stratification or compartmentalization, tables may be used to indicate what an identified user is permitted to do or which users are permitted to read which records. This chapter discusses the structure and use of such tables.

The authorization tables may be short and simple, or they may be so lengthy that careful attention must be paid to their structure to avoid excessive overhead. The need for highly complex authorization schemes should be questioned.

A basic form of authorization table has an entry for each user stating what he is entitled to do. Such a table may give the security code assigned to each user and a field indicating his category of authorization.

The computer's first operation would be to check the security code or password. Rather than using one category of authorization, it may be desirable to structure the authorization table according to the facilities

Figure 8–1 An Authorization Table

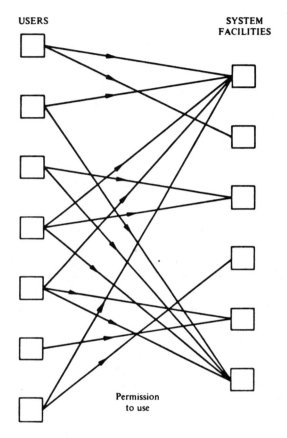

that may be used or the actions that may be taken. The authorization table may list for each items such as the following:

1. Programs that one may use
2. Types of transactions that one might enter
3. Data sets that one may read
4. Data sets that one may modify
5. Categories of data within a data set that one may read
6. Categories of data that one may modify[19]

Notice how authorization tables are used in the following excerpt.

A Multihospital System Incorporates Confidentiality into Practice

In one multihospital computerized cancer data system it was considered critical that there be appropriate patient confidentiality at each hospital.

> Each authorized person has a secret code to access the system. Each can determine through identifying information whether the patient is already in the data base, and to indicate if data are to be added to that patient's file.
>
> On shared patients, hospital lawyers evaluated the legalities of printed shared data and determined that a patient being treated in one institution who goes to another has authorized that those data from the first institution be shared, since this is necessary for the patient's most appropriate care.
>
> Also considered critical was the securing of the confidentiality of each hospital's data. The system was designed so that each hospital can look at its own information and the anonymous pooled information from all hospitals on the system. However, on shared cases where Hospital A has information to add on a patient already entered by Hospital B, Hospital A can request and get only enough information to properly identify the patient in the data base but then must first "register" the case for its hospital before being allowed to proceed. In other words, unless a hospital registers a case as its own, it cannot input information and is prohibited from accessing the information already entered.[20]

Surveillance Methods in Data Security Programs

On a computer system surveillance procedures can be of two types: real time and non–real time. A well-protected system will utilize both. Real-time checks detect violations of security procedures such as the following:

- Users type in an invalid identification (personnel number or social security number).
- Users type in a security number that does not conform to the would be user's identification.
- Users try to employ a terminal for which they are not authorized.
- A terminal is not identifiable as one that should be connected to the system.
- Users request a program they are not authorized to use.

- Users send a transaction type they are not authorized to send.
- Users try to read data they are not authorized to read.
- Users attempt to modify data they are not authorized to modify.
- Users take some action at a time when it is not permitted.
- Users attempt to gain access to a highly secure region such as the authorization tables.
- Users exceed a specified amount of activity at their terminals.
- Users take more than a given number of incorrect actions at a terminal.

Many of the foregoing violations will occur accidentally. The user mis-keys a number, or a transmission error occurs. Most terminal users make occasional mistakes. The action of the computer, therefore, on receiving the first unauthorized request will be to ask the user to reenter it.

If the computer then receives a correct number, it will allow the user to proceed. If it receives the same unauthorized number, the user may be inadvertently trying to do something that is not supposed to be done, but without malice aforethought. If a new number that is also unauthorized is entered, the user may be trying to break into the system by trial and error. If the terminal is unidentifiable, if the user is attempting to reach the authorization tables, or if anything is astray with the security officer's terminal, this may be serious indication of felonious intent.

When a second violation in succession is detected the computer can take one of the following actions:

- It can caution the user, allowing only one more chance to take a correct action.
- It can lock the terminal keyboard in such a way that it can only be unlocked by an appropriate authority. As it does this, it will notify a person responsible for security.
- It can refuse to accept new input for a period, perhaps one minute, and then start again. This would impede a user trying to obtain pass-words with some high-speed trial-and-error scheme.
- It can "keep the user talking" while preventing him or her from obtaining any sensitive data or from modifying any data or programs in the system. At the same time it will contact a person responsible for security in hope that the miscreant will be caught red-handed.[21]

Figure 8–2 shows a possible sequence on a system that does not lock the terminal. The computer in this case checks the terminal identification, the user's identification and a once-only security number. If the user identification is invalid or the security number is not correct, the computer

Figure 8–2 Possible Sequence on Unlocked Terminal

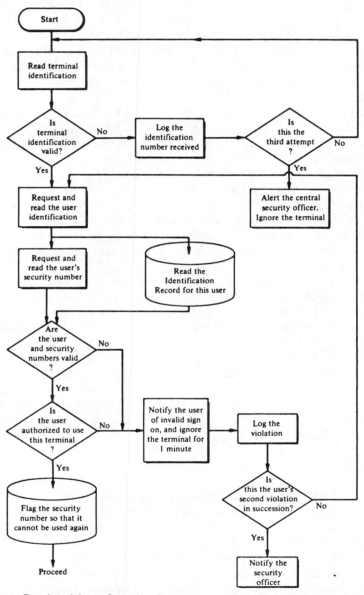

Source: Reprinted from *Security, Accuracy, and Privacy in Computer Systems* by James Martin with permission of Prentice-Hall, Inc., © 1973, p. 181.

notifies the user that the sign-on procedure has not met the requirements. The computer then ignores the terminal for one minute. If the user has committed two violations in succession, a security officer is notified and the activities of the user may be monitored. If the terminal being used is not a valid one, more serious action will be taken. The computer will ignore the terminal and alert the security officer. It will only do this after several transmission retrievals in case the mis-identification was caused by a transmission error.

Figure 8–3 shows another approach, which is more secure and could be used on an in-house commercial system. Here the keyboard is locked if the user types invalid numbers twice in attempts to sign on. A record of the violation is written on the file so that it can be inspected by an office manager or security officer when the terminal is to be unlocked again.

If the user is not authorized to employ the terminal where the sign-on was attempted, the user will be notified of this and told that a security officer has been informed. If one attempts to use data for which one is not authorized, one will again be cautioned that a security officer has been informed, and if one persists in trying to gain access to unauthorized data, the terminal keyboard will be locked.[22]

No system can be regarded as secure unless both the hardware and the systems programs were designed for security. Some hardware-software combinations that were not originally designed for security were subsequently made more secure by means of a series of modifications and additions—like burglarproofing a house.

The hardware additions have included positive identification of terminals and other components, access protection to prevent a program in one user's partition of main memory reading information from another user's partition, and operator logs of such a nature that no operator action can go unrecorded. Table 8–1 lists the features that are included on terminals for security reasons.

Data Banks

Data banks for patient data are technically possible and currently operated in various sites. The sites do not include hospitals or other traditionally recognized health-care-related organizations. Most patient data banks lack uniformity and must be studied as individual entities in order to gain an understanding of their objectives, scope of data retention, access policies, population served, and so on. Perhaps the two most widely recognized data banks demonstrate the diversity that exists in these developing data organizations. PSROs are federally established organizations that receive, process, and retain a predetermined and limited set of data ele-

Figure 8–3 Possible Sequence on Locked Terminal

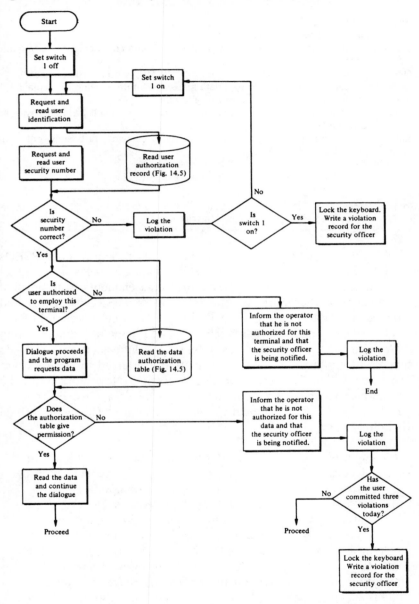

Source: Reprinted from *Security, Accuracy, and Privacy in Computer Systems* by James Martin with permission of Prentice-Hall, Inc., © 1973, p. 182.

Table 8–1 Features Included on Terminals for Security Reasons

1. Physical lock and key
2. Keyboard lockable by the computer
3. A unique terminal identification number that is transmitted on receipt of a command from the computer
4. Operator's identification card reader (for magnetic strip credit card or ID card)
5. A means of automatically checking a data message before printing or displaying it to ensure that it has not reached the wrong terminal
6. A key that enables the operator to prevent the printing or display of items entered (e.g., security code)
7. Automatic prevention of the printing or display of certain input items, either by terminal editing or computer control
8. Automatic erasure of data in control unit buffers
9. Cryptographic enciphering and deciphering

Source: James Martin, *Security, Accuracy, and Privacy in Computer Systems* (Englewood Cliffs, N.J.: Prentice-Hall, 1973), pp. 186–87.

ments that are abstracted or collected from primary patient records. PSROs act to reduce unnecessary expenditures by reducing unnecessary hospital stays or unnecessary services. They utilize data elements from hospital records to achieve their purpose. The data bank of a PSRO is regional and includes selected data elements from the hospitals in one geographic region for a selected set of patient records. Health Care Financing Administration (HCFA) is also in the data bank business. HCFA data banks are statewide in scope and contain the data on patients hospitalized in states where HCFA is operating demonstration data banks. In New York State the HCFA-sponsored data bank contains discharge data on every patient hospitalized in that state. The data bank of the future has been described in the following way:

> While it is readily accepted that the computer is necessary for the collection and arrangement of data on large numbers of patients it is controversial if, and to what extent, the computer should be used for maintenance of individual patient records. Nevertheless, in the dream system the computer stores individual patient records. The physician providing the care interacts directly with the system. Errors in recording data are corrected at or nearly at the time the care is provided because of the interaction with the computer by the physician using English language. The interaction is acceptable to the physician   because the physician's documentation burden has been reduced and information's ready availability helps the physician's decision-making process.

Once computerized, the information would be logical and consistent, since this is a prerequisite for computerization. Although communicating with the physician would be in words, the inherent logic would permit these words to be converted to codes for other functions such as billing and computer-assisted analyses. Such coding would reduce the cost of data transfer to other systems by reducing the electronic "space" required to transmit the information.

The system would be the source of billing information, development of statistics for planning and regulation, and regional quality assurance through links with other systems. Outside users such as PSROs would have authorization for use of certain elements of the computerized medical record and the probability of error would be low since the original information is selected and transmitted electronically.

The link would be through a data broker, a regional data bank under the management of an independent corporation managed by a board of directors representing sources and providers and users of the data. The source information system will be linked directly to the data broker's system with the source maintaining control over what elements are transferable to the regional system.

Control of accessibility to any data element would be a stepwise function beginning at the initiator of the information. The patient would control release of information solely provided by the patient. The same would be true for a physician. Once the data is disidentified, its release would be controlled by the institution providing the data. The data corporation could not release such data without authority of the organization providing it.

The regional data corporation would act as the intermediary between data generators and data users. Where the need for data is defined, a mutually acceptable release would be obtained by the corporation. Certain data generated in each institution would not be moved to the regional data bank but data used for commonly defined multiple purposes would be added to the data bank as the need is defined.

In the dream system, the physician interacting with the computer will assure that the information is reliable and actually reflects the care provided. The individual physician would be able to compare his or her practice with others in the community and against community norms and standards. But, where are they established?

The function of the PSRO, one user of the regional system, would be to provide the analytical services, facilitate the establishment of the norms, criteria, and standards, and to provide accountability to the community it serves on the care provided. Regional quality assurance would become a reality![23]

Data banks provide an opportunity for confidential data to be linked with other types of data such as financial, social, and even other sources of medical data. One area of concern about data banks is the possibility of users retrieving data without the consent or knowledge of the individual represented by the data—patient or client. With proper design and development, however, data and the storage media for data can be secure as we have pointed out in the earlier parts of this chapter. An area that still needs development is the education of patients regarding both the risks and the safeguards in computerized patient data. Users and clients such as patients should be made aware of the protection possibilities as well as the access possibilities. It is the responsibility of professional computer system designers and the social and medical ethicists to see that patients (1) have an understanding of protections necessary, (2) can relate the privacy issue to risk, and (3) recognize that appropriate action (policies, procedures, sanctions) must be taken by any given medical data bank to deal with these issues. Perhaps the most appropriate suggestion for those who are faced with the necessity of gaining the knowledge required to work with a patient data bank is to suggest that one study the statutes, rules, regulations, and other forms of legislation in the state where one works or resides.

Conclusion

The ethical guidelines developed by the Joint Task Group for Medical Computing, Medical Society of the County of Erie New York, is a must for anyone planning to work with or be responsible for data bank security systems.[24] Another set of useful guidelines is the following:

Ten Principles of Privacy
1. Any person should be able to inspect his own file maintained by agencies or firms and have copies made at a reasonable cost.
2. Any person should have the right to supplement the information in his personal data file.
3. A method should be established to allow for the removal of erroneous or irrelevant information and provide that agencies, organizations, and persons to whom material had been previously transferred, be notified of its removal.

4. The disclosure of information in a personal data file to individuals in agencies or organizations other than those who need to examine the file in connection with their duties should be prohibited.
5. A dissemination record should be maintained to identify all persons inspecting such files including their identity and their purpose.
6. Agencies or firms maintaining personal data files should ensure that the information be maintained completely and competently with adequate security safeguards.
7. Agencies or firms collecting data from individuals must apprise them if the request is mandatory or voluntary and what penalty or loss of benefit will result from noncompliance.
8. It should be required that personnel involved in handling personal information act under a code of fair information practices, know the security procedures, and be subject to penalties for any breaches.
9. Any individual wishing to stop receiving mail sent because his name is on a mailing list should have the right to do so.
10. Agencies or organizations should be prohibited from requiring individuals to give their social security number for any purpose not related to their social security account or not mandated by federal statute.[25]

Questions and Problems for Discussion

1. How will patient privacy be protected in the computer environment? What policies can be formulated by health record practitioners to support this need?
2. Given an automated hospital information system that stores the patient record on-line, what information should be available to admitting clerks, unit secretaries, floor nurses, physicians, and medical record practitioners?
3. Discuss possible limits on a data security system in a health information system. What are limits? Who would determine them?
4. What are some of the problems associated with the patients' right to see and make corrections in their medical records?
5. Prepare a report to your administrator outlining the components of a data security program for your hospital. Consider this a fact-gathering process currently underway to investigate the purchase of a vendor-supplied hospital information system that stores the patient on-line during hospitalization.
6. Develop a list of specific safeguards that could be used with CRT terminals in a hospital environment.
7. Prepare a plan to introduce surveillance methods into a medication administration program in a community hospital.
8. How are passwords used in maintaining data security?

NOTES

1. Kathleen A. Waters and Gretchen F. Murphy, "Technical and Legal Aspects of Medical Privacy," Computers and Medical Privacy Monograph, *Journal of Clinical Computing* (September 1980).

2. Elemer Gabrieli, *Journal of Clinical Computing* (June 1980).

3. Susan C. Manning, Susan Grant, and Stanton J. Price, "THRM Forum," *Topics in Health Record Management,* 2, no.1 (September 1981)(published by Aspen Systems Corp., Rockville, Md.).

4. Joann Ellison Rogers, *Mademoiselle* (May 1980).

5. Gary A. Morrow, "How Readable Are Subject Consent Forms?" *JAMA* 244, no. 1 (July 4, 1980).

6. "Computers, Spies and Private Lives," NOVA #811, Public Broadcasting System, 1981.

7. Ibid.

8. Alan Westin, *Computers, Health Records and Citizen Rights,* National Bureau of Standards Monograph 157 (Washington, D.C.: U.S. Government Printing Office, 1976), p. 348.

9. Lawrence E. Burns, "Confidentiality and Medical Record Keeping," Computers and Medical Privacy Monograph, *Journal of Clinical Computing* (September 1980).

10. Gabrieli.

11. *State Health Legislation Report* 9, no. 1 (May 1981).

12. Wallis E. McClain, (ed.), "Access Reports, A Summary of Freedom of Information and Privacy Laws of the 50 States," October 1978.

13. Ibid.

14. *Personal Privacy in an Information Society; The Report of the Privacy Protection Study Commission* (Washington, D.C.: U.S. Government Printing Office, July 1977).

15. Homer Schmitz *Hospital Information Systems* (Rockville, Md.: Aspen Systems Corp., 1979), and Randall W. Jensen and Charles C. Tories, *Software Engineering* (Englewood Cliffs, N.J.: Prentice-Hall, 1979).

16. James Martin, *Security, Accuracy, and Privacy in Computer Systems* (Englewood Cliffs, N.J.: Prentice-Hall, 1973), pp. 186–87.

17. Gabrieli.

18. Guidelines on Evaluation of Techniques for Automated Personal Identification, Federal Information Processing Standards Publication 48, April 1, 1977.

19. Martin, *Security, Accuracy, and Privacy,* p. 159.

20. Charles Murray and Jean Wallace, "The Development and Use of a Computerized Cancer Data System," *Topics in Health Record Management* 2, no. 2 (December 1981).

21. Martin, *Security, Accuracy, and Privacy,* p. 179.

22. Ibid, p. 180.

23. Robert W. Schoenecker, "One View of Information Systems for Regional Quality Assurance (PSRO); Computers in Health Care: Are They Worth It?," in *Proceedings of the 8th Annual Conference of the Society for Computer Medicine, 1978.*

24. Kathleen A. Waters and Gretchen Murphy, *Medical Records in Health Information* (Rockville, Md.: Aspen Systems Corp., 1979), pp. 267, 268.

25. Robert J. Bradley, "Privacy: What Kind and How Tight the Lock," *Data Management* (May 1980).

Chapter 9

How To Relate Current Technology to Future Practice

Objectives

1. Relate the general evolution of computer technology to developments in medical information systems.
2. Identify the reasons for historical failures in hospital computer system developments.
3. Define and describe major trends in developing medical information systems.
4. Illustrate major trend developments in automated hospital information systems (AHIS) and ambulatory medical record systems (AAMRS).
5. Explain the role of data banks in medical information systems.
6. Identify current vendor commitments in major computer system development and marketing.
7. Relate major medical information system trends to organizational planning.
8. Identify an appropriate role for health record practitioners in understanding and participating in technological developments.
9. Explain the relationship of data quality to computer technology in health care settings.
10. Explain the role of microcomputers in planning and developing computer applications in health information systems and MRDs.
11. Define and describe prototyping as a technological innovation with implications for the future.
12. Present recommendations for standards of design and development in emerging medical information systems and patient data computerization.

This chapter explains the role of computer technology in medical information system developments today. It also defines and describes some key information and technological issues that have direct impact on the health record profession now and in the future.

First, we will examine the reasons why many of the initial efforts in hospital computer systems failed. Major trends in medical information systems developments will then be described. These will be illustrated by examples in automated hospital information systems (AHIS) and automated ambulatory medical record systems (AAMRS). The relationship of health care organizations and their departments to the current trends will then be discussed, with a picture of industry patterns and commitments for developing computer systems emerging.

Three significant issues for health record practitioners in the current and future developments of medical and patient data computerization will be raised: (1) the relationship of computers to data quality, (2) current trends in microcomputers as alternatives or components of medical and patient data computerization, and (3) the emerging technology of programmerless systems design. Finally, specific recommendations are made on patient data computerization for the future. These recommendations are made from the perspective of the health record professional as a specialist in patient data collection, retention, retrieval, and use. As such, they represent the primary purpose of the health record profession.

MEDICAL INFORMATION SYSTEMS AND HEALTH CARE TODAY

Overview

The 1970s witnessed great strides in the use of computers in our society. At least three factors contributed to this growth. First, the technology (hardware and software) improved dramatically. Second, the cost of hardware, in particular, decreased significantly. And third, there was a significant investment in research and development in computer applications stemming from public and private sources.[1] Developments in the computer industry have been so rapid and far reaching that any number of different ways exist to accomplish a particular set of objectives. Choosing the right computer system for a medical application for a hospital, clinic, primary care facility, and other setting depends on the particular facility's unique organizational structure and stage of development. It also requires a great deal of knowledge of the capabilities of computers and their configurations.

Computer systems evolved from small stand-alone computer systems in the 1950s to centralized large ones in the 1960s. These in turn began to support a number of high-speed remote batch and interactive time-sharing systems leading to computer networks. Today minicomputers can serve as an in-house system or as a satellite computer to communicate with a larger computer in another location. Computers in distributed processing systems are used to integrate information and functions with other computers to process information in networks. With the advent of the microcomputer, computing capability is available to diverse users for less than $3,000. Today, people enjoy the desktop microcomputers at home and on the job. For health professionals the computer choices and opportunities are increasingly varied and complex.

Should each medical facility worry only about its own application? Should a group of hospitals or health facilities contract for a shared computer system? Should an individual hospital consider participation in a large computer network? As we have seen, analysis for any new computer application is a process, not an isolated action. It should be viewed as a sequence of steps to achieve a well-defined goal. The analysis should attempt to satisfy existing requirements for the operation application as an integral part of other data- and information-handling needs of the institution.

Many applications in medical care today depend on computer technology to process information in diverse areas. What should the contemporary health information practitioner know about medical computing in general and its relationship to patient data computerization? There is clearly a need to acquire a general picture of the role of computer technology in health care if this question is to be addressed effectively. How do individual health record applications fit into the overall picture of computer technology in medicine? In Chapter 7 we examined the relationship between the MRD and its parent organization's overall technological planning. At this point, we will see how the practices of departments and organizations compare to overall trends. Just as the study of history aids in understanding the present, understanding the role of computer technology in health care today begins with a look at the people and events that have gone before.

Influences on the Development of Computerized Medical Data

Medical and health information computing today is the result of the intermingling of people, technological development, electronic achievement, and the demands of the users of information to have faster, more appropriately organized, concise, and useful data for professional decision making. "In clinical medicine, the selection of the most accurate diagnosis or the choice of the most effective therapy are information-dependent decisions."[2] This axiom has long been espoused by E.R. Gabrieli, a pioneer in computerized medical information systems development. Gabrieli and many others identified the need to provide information more efficiently. However, the process required to meet this need was not well understood. Limits on the technology, the expertise of people, and knowledge of the medical arena were major stumbling blocks in early attempts to use computers to perform the tasks required. Nevertheless, a number of major projects were initiated in the 1960s. These increased throughout the next decade. In the 1970s, approximately $29 million was spent on automated hospital information systems and developing clinical information systems by the federal government alone.[3] Progress was slow, however, because many of these early attempts resulted in failures. Nonetheless, it is important to note them for they are recent enough to have implications for today's plans for patient data computerization. In the following discussion of reasons why past efforts to implement hospital computer systems failed, we quote from the writings of Morris Collen.[4]

1. A suboptimal mix of medical and computer specialists was the most common cause of failure. Generally, the project staff consisted of (a) well-motivated physicians who had little experience with computers and (b) computer and systems experts with little experience in medical applications. To this cause, we add what occurred in the hospital setting when there was a suboptimal mix of hospital personnel and computer specialists. Because of the communication difficulties between these disciplines, the computer staff was frequently unable to accurately estimate the needs of the users. The users were not cognizant of the computer specialists' lack of comprehension of the scope and detail of the subsystems that were the targeted applications.

2. Inadequate commitment of capital for long-term investment was the second most frequent cause for failure. Most organizations grossly underestimated the large amounts of money and time involved in implementing an HCS [hospital computer system]. Several projects in the United States were terminated after three to five years because several million dollars had already been spent and the HCS was still far from completion. Unknown is the amount of money spent by individual hospitals across the nation. What is known is that everyone who worked in hospitals in the 1960s seems to have a story that recalls a computer project failure that accounted for a sizable financial loss to at least one hospital with which they were familiar.

3. A suboptimized systems approach was frequent. Several HCS projects failed because they had successfully implemented one or more subsystem components for the administrative, laboratory, bed census, patient scheduling, or pharmacy units and now desired to integrate them all into a HCS. At this point, they discovered serious incompatibilities between the various modules that would require major reprogramming at prohibitive costs to achieve an integrated patient file. These projects usually continued the individual subsystem modules as independent computerized units.

   At the other extreme were some projects that began with the global systems approach to implement a total HCS; the sheer enormity of such an approach has not yet found anyone with the vast resources capable of successfully following this course.

4. Many of the early HCS projects were never implemented because of unacceptable terminals. The first systems required physicians and nurses to use keyboard terminals such as typewriters. It was soon established that physicians would not accept such means for communicating with the computer, and clerical-type personnel had to take over the process of data entry. As a result, current HCS projects all use visual display-type terminals, since physicians will accept touch-wire or light-pen methods of selecting data from a terminal display.

5. An occasional cause of failure was inadequate management organization. Included in this reason for failure was the failure to recognize the necessity to involve all potential users at the outset of and during project development. Another related problem was the tendency to overlook administrative responsibility for the project. No one individual in the facility was designated as the final authority on project planning and decisions. An odd mixture of vendor/institution decisions regarding department user operations was often the disastrous result of the omission by the hospital to name a hospital-employed project manager.

Perhaps the reasons for the failure of hospital computer systems as cited by Collen can be expanded. The 1960s introduced the health care industry to the availability of computers. Computers had proven to be successful in other sectors. Hospitals, in particular, opted to utilize the new technology. The good intentions of all parties were to be offset by the inherent loopholes of any pioneer effort of the scope computers comprise. The allied health professionals, administrators, physicians, computer vendors, and systems analysts just did not share enough common ability of work experience, education, and language to carry out developmental projects successfully. They had no shared record of success and failure. They did not have the ability to estimate time needed for project planning, implementation, and evaluation; nor did they recognize the complexities of the projects they had undertaken. Furthermore, there was no previous systems experience from which to draw. For example, systems were often designed without a recognition that modular growth should be a design principle. The medical and health professions regarded the new area of computer science with reluctance. Today there are many educational programs in medicine and allied health that still do not include computer courses in the required curricula for those who major in these disciplines.

The incredible technological growth that demonstrates the adage "New today—obsolete tomorrow" continues to be a problem. Microelectronics has expanded into new hardware and off-the-shelf software that offers users an opportunity to work with applications using already developed programs. Early system lessons have been learned and measurable progress can be seen as the 1980s move along. A look at current philosophical trends in developing medical information systems will help illustrate the present and future for health professionals concerned with medical and health information computing. How is computer impact translated into trends in medicine and health care today?

Trends in Medical Information Systems in Health Care

Three major philosophical and developmental directions are underway in American medical computing today. The first involves on-line, real-time, communication-oriented systems with partial interdepartmental data integration. In these systems users interact with computers on-line and obtain immediate "real-time" response from a computer file or data base update for each transaction performed. The system also relies on computer-directed exchange of some common information between two or more separate departments in the organization. Such systems make use of common data elements and use data base design to provide access to the same data by several users. There is usually no plan to develop these systems into comprehensive ones incorporating all users and including computer-stored patient medical data over time. These systems presently exist as level I automated hospital information systems (AHIS) in hospitals. They also have been introduced in ambulatory care settings and other health care environments.

Second, some on-line, real-time systems have been designed from the start to capture and process part or all of the patient record. These systems incorporate the features identified in the first illustration and include a computer-stored medical record as a major component in systems design. These systems are considered level II AHIS applications. The CoStar program is an example of this trend in ambulatory care.

Third, comprehensive, on-line systems have been designed to capture patient information and link or combine patient data elements with medical resources to create a higher level of information. Medical decision-making features designed to provide the benefit of medical experiences, research, and new technology as an interactive resource for rendering patient care are examples of major components of these systems. These may be considered level III systems. A more detailed look at these trends is in order.

Level I AHIS

The first trend is illustrated by health care facilities that are constructing or purchasing computer systems that utilize electronic data processing and communications equipment to provide some on-line processing with real-time responses within the facility.[5] In hospitals, this definition equates with the level I automated hospital information systems (AHIS) described by Marian Ball and Stanley Jacobs.[6] These systems are communications oriented and drawn from unique, stand-alone systems such as billing subsystems. They are built to integrate data in a basic set of applications. Communications in this sense means allowing users to enter messages via terminal keyboards and thus send queries, comments, and instructions to other individuals in physically separate departments. Answers and other responses can be sent back, via terminal, to the initiator of the message by those who received the message from the terminal in their area.

The basic set of applications that incorporate integrated data in a level I system are

- An ADT application with bed and census reporting
- An order/requisition entry, communication, and charge collection application
- An inquiry application for today's charges for demand bill purposes[7]

In level I, a hospital uses the computer system to reuse information collected by one department for an additional application. For example, hospitals have often computerized billing operations initially. Many are upgrading these applications so that information collected for billing purposes can be used simultaneously for other needs, say census preparation.

The second trend is characterized by medical information systems that are being constructed primarily for clinical use. These systems often incorporate an extensive medical data base. This type of system may or may not include all parts of the medical record. That is, some of the systems include the collection and processing of the complete medical record and others do not. None of the systems at this level perform the function of creating a higher level of information from the data base elements they collect and store. This direction in medical computing can be illustrated in the following examples.

The *Computer-Stored Ambulatory Record System* (CoStar) is a powerful and flexible information system designed for widespread use in ambulatory care organizations. The most unique feature of CoStar is its ability to automate ambulatory medical record data. The system was developed by the Laboratory of Computer Science at Massachusetts General Hospital, the National Center for Health Service Research, and the Digital Equipment Corporation. As a computerized medical management information system designed for small to medium-sized clinics and group practices, one of its major features is its ability to modify the classical "encounter-oriented" presentation of data and display summaries of medical items organized by problems or classes of data.

CoStar can process patient registration, medical record, and financial data. Patient registration data can be entered on-line by clerical personnel. *Encounter* data can be recorded by the health provider on the appropriate encounter form. Data entry clerks can enter all encounter data. All patient complaints, physical exam, vital signs, diagnoses, medications, immunizations, injectables, therapies, supplies, X-rays, lab data, and procedural data are stored in the system. Condensations of health history and referral data are also stored. The system can generate patient statements and bills for Medicare, Workmen's Compensation, Blue Cross/Blue Shield, and other private insurance carriers. All accounts receivable and general ledger data can be processed by the system.[8]

The *Technicon Management Information System* (TMIS) is another example that can also be considered a level II AHIS. It was designed to assume the substantial clerical burden spanning the originating point of medical directives (the doctors' orders) to the documentation of their completion (medications given or test result values). The system also takes over the amplification necessary to carry out orders, to document patient status, and to communicate patient information. In order to perform these functions, the system provides three basics: (1) a means for capturing information at its source; (2) a means for computer processing and storage of these data, and (3) a means for display and printout of these data.

At El Camino Hospital in California, TMIS is a hospitalwide system that uses a computer to store data and to send it, automatically or upon request, to the people who need to act upon it. The computer adds speed and accuracy to the transmission of information throughout the hospital. It also performs many data processing tasks such as sorting, copying, filing, summarizing, checking for abnormal data, pricing, and a variety of other functions usually done by physicians, nurses, technologists, clerks, and other hospital personnel. A broad range of medical data (such as physicians' orders and test results) and administrative data (responsible party, insurance coverage) are processed by the computer.

Hospital personnel communicate with the computer by means of two devices: the *Video Matrix Terminal* (VMT) and *Multiprinter*. The VMT consists of a television screen, a light pen, and a keyboard. VMTs are located throughout the hospital and are used to send new information to the computer, or to retrieve information such as laboratory results, the time of the last dose of medication, and so forth. Each person who is authorized to use the VMT has a unique identification code. This code limits the capability to send and receive information, based upon the individual's position and the hospital's policies concerning data access control. Use of the VMT is based on the elementary communication technique of pointing. The television screen is used to display lists of items; for example, laboratory tests. A specific item is selected by pointing the light pen at the desired word (or phrase) and pressing a switch on the barrel of the pen. Using the light pen, a physician can select the specific patient, and then "write" a full set of medical orders (for laboratory work, medications, x-rays, diet, activity, vital signs). The computer then stores the orders and sends appropriate documents via printer or screen (laboratory requisitions, pharmacy labels, and x-ray requisitions) to the proper hospital departments.

The keyboard on the VMT is used to supplement the light pen. In departments such as admitting, the keyboard is used extensively to record the patient's name, address, responsible party, and other such information that cannot be written using the light pen technique. Physicians, nurses, and other hospital personnel use the light pen technique extensively and employ the keyboard only occasionally. The keyboard permits complete freedom to write special instructions or unique orders that are not available for light pen selection.

Multiprinters are located throughout the hospital so that each document prints where it is needed. Computer-produced printouts, in addition to those referred to previously, include the following: patient care plans, medications-due lists, laboratory specimen pick-up lists, cumulative test results summaries, radiology reports, discharge summaries, and more than 100 other documents.

In summary, TMIS is a comprehensive data system for patient care and hospital management functions. It uses a large-scale computer and advanced input/output devices to record information at its source and to make this information available throughout the hospital when and where it is needed. Its basic objective is to provide better patient care and more efficient hospital operations through an improved communication system. TMIS substitutes computer processing for manual data processing, thus gaining advantages in terms of speed, accuracy, cost, legibility, and completeness and consistency of data.[9]

This type of system places a major emphasis on the handling and communication of medical, administrative, and fiscal information. Systems in this second category are generally limited to the establishment of an electronic medical record (none are as yet complete) and the preparation of summary lists (such as drug and clinical laboratory profiles). Although useful, this level of processing is essentially limited to making clinical data more concise and readable. It does not, generally speaking, use these data to create higher level information that physicians and other medical professionals can utilize to identify and apply state-of-the-art diagnostic and therapeutic procedures. This leads us to the next direction currently underway.[10]

Medical Information Systems Generate a Higher Level of Knowledge

The third trend is illustrated by the construction of a medical information system that combines the patient medical data base and analytical uses of clinical and service delivery information. Two major components of this level of development are (1) a medical knowledge data base and (2) a rule base or structure by which to apply the medical data base knowledge to clinical data so that higher level information is created for medical decision making.

The methods or rules by which the clinical information systems manipulate clinical data are varied and include various forms of artificial intelligence, the application of Bayes' theorem, elaborate "tree structures," and statistical models. However, most if not all of these systems share a common distinction—the requirement of establishing medical standards or norms often when such standards are not clear or even do not exist.[11]

Two examples are described to amplify our discussion of this direction in medical computing. PROMIS (Problem Oriented Medical Information System) represents an ambitious attempt to develop a health care information system with an extensive medical knowledge data base and guidance system for use both in inpatient and outpatient settings. After more than ten years' effort, Lawrence Weed and his coworkers have developed a system unique in at least three respects.

First, it provides for a complete electronically stored medical record. Second, the record itself is based on the problem-oriented format. Finally, the system has been designed to direct or guide the user through a reasoning process. According to Weed, these capabilities were developed to counteract problems that currently hinder the delivery of high-quality care: namely, dependence on the physician's memory, ineffective organization for massive amounts of medical data, and lack of meaningful feedback about appropriateness of care.

The MYCIN system at Stanford University is another example. It is a program designed to be a consultative aid to physicians in their quest for selecting the most appropriate antimicrobial agent (drug). Using a rule-based approach and an in-depth knowledge base of infectious diseases, in many instances MYCIN can be as effective as infectious disease experts in prescribing antibiotics (and sometimes better). MYCIN is one of several projects using a so-called artificial intelligence approach. Its developers claim that the usefulness of MYCIN is greatest when only limited facts are known about the infection and its causative agent. Two features of this system are worth noting. The first is that since it is rule-based, new rules can be added relatively easily (as opposed to changing "trees"). Thus MYCIN can be taught new information rather easily and quickly. The second feature is that when requested to do so MYCIN will present in a proper sequence the rules used to make a recommendation. This capability allows the system to be a powerful teaching tool as well as building confidence in the user as to its validity.[12]

Data Banks As a Resource

Data banks offer alternative information services and often serve as a component of the level III systems just described. Two types of data banks serve in this capacity. The first is concerned with recorded medical knowledge (books, periodicals, lectures), the second with clinical experience (clinical trials and the experience of colleagues). The computer offers a mechanism by which clinical experience can be collected, stored, stratified, and later made available to interested providers. Thus, through an appropriately constructed data base physicians in a particular area such as coronary artery disease now have the prospect of being able to share an aggregated clinical experience. As more clinical information is gathered over time, a data bank has the potential to become more accurate and useful. This type of clinical information system creates its own medical knowledge base. The following examples illustrate how data banks are employed in this way.

Data banks can be used to support the clinical management of the individual patient. For example, the physician can query a system to ascertain how groups of patients similar to the particular patient of interest have reacted to various treatment protocols (approaches). This added information can be very useful in cases where multiple chronic disease conditions exist. Carefully constructed data banks are also very useful in research and in the assessment of emerging medical technologies (say coronary bypass surgery). The National Center for Health Services Research currently supports three projects dealing with data banks. One of these, the rheumatic disease data bank established by Dr. Fries at Stanford University, contains data on over 10,000 patients with various rheumatic diseases, such as rheumatoid arthritis, lupus, scleroderma, and juvenile rheumatoid arthritis.[13] A data bank established by Dr. Rosati at Duke University contains information on over 5,000 patients with coronary artery disease and myocardial infarction, including all of the history, signs, symptoms, laboratory tests, exercise electrocardiograms, catherization laboratory data, and progress notes obtained in the course of the patients' clinic visits.[14]

This information is analyzed to assist the physician in making the choice between treating the patient medically or treating him surgically. The system picks out all patients in the data bank who are similar to the one currently under treatment and presents the information in tabular form. The physician then is in a position to contrast the outcomes of the patients in the two groups, surgical versus medical, in terms of survival, freedom from pain, and restoration of function.

A somewhat different approach is being taken by Dr. Bleich in Boston.[15] His system will permit a user to search the clinical data base resident in the Beth Israel Hospital Information System to identify similar patient groups and their "experiences." This approach should not be confused with the work being done by Fries, Rosati, and other investigators who are developing disease-specific data banks. The Bleich system only uses objective data (laboratory, radiology, drug orders, etc., plus registration and fiscal data); Fries and others also use subjective data (amount of swelling, limb mobility, etc.). Although more limited in the amount of data collected on a specific disease, the Bleich system will have the advantage of not being disease-specific and thus should prove useful to a broad spectrum of medical specialties.

Integrating Medical Information System Trends into an Organization's Future

In summary, the major trends featured in this chapter typify the technological developments in medical and health information systems today. These are major trends in general systems developments. Investigation would reveal that each trend has been supported by federally funded grant projects, individual health organization investments, and major vendor efforts. How are these major thrusts in medical computing translated into individual organizational goals? This question must be answered from three perspectives.

First, an organization should consider the appropriate strategic and tactical planning message of its leaders. As stated in Chapter 7, this is the most significant priority in systems planning and development. An individual hospital may elect to purchase an existing level I AHIS from a vendor to achieve a strategic goal. Second, organizations must determine whether to develop their own software or purchase existing software from vendors and other health organizations. Today there is an increasing recognition that selection of already developed software packages are a cost effective and operationally effective alternative to individually developed systems. In many organizations a combination of these approaches will be used.

Third, organizations must consider the practicality of investing in hospitalwide systems over selecting department-specific systems for their organizations. This consideration has direct impact on the MRD.

The fundamental need is provision for patient information whether through linking department-specific systems to a larger one, providing distributed processing such as PROMIS suggests, or developing a hospitalwide system that supports departmental functions. Recall the discussion of the R-ADT foundation in Chapter 7 to see how departmental applications can be developed from a central R-ADT system.

What exactly is available? How are vendors supporting organization-wide systems? How are they supporting department-specific systems? In a limited study that examined the current state of commercially available department specific systems sold by various vendors, four specific questions were addressed:

1. Are department-specific systems currently being acquired at a greater rate than hospitalwide systems?
2. Is there a trend for hospitals to acquire hospitalwide systems before acquiring department-specific systems?
3. Are currently active department-specific systems vendors providing greater interfacing capability than was reported in the previous study?
4. Do the problems facing potential users of the 1970s exist in the 1980s?

We quote from the conclusion.

> Although the sales of department-specific systems still are significant (slightly increasing), more vendors are currently marketing hospitalwide systems and the rate of acquisition of the latter is increasing more rapidly than that of department-specific systems. The opposite appeared to be the situation when the authors made the survey two years ago. The same problems that existed two years ago are still facing the hospital community when purchasing or using department-specific systems, although these problems tend to be less severe. The need for standards does not appear to be as urgent as previously thought in view of the trend for improved interfacing between systems.[16]

Tables 9–1 and 9–2 display the findings of this study in greater detail.

Table 9–1 Department-Specific Functional Applications

System Name	Operations Performed	Accomplishments	Major Vendor
Clinical Laboratory	Collecting, organizing, formatting final test results for medical staff, administration, and business office. Includes: recording of test requisitions, scheduling of specimen collections, calculating test results, periodically preparing summary reports of all test runs for a given patient, preparing statistical reports, keeping records for quality control, and administrative control of laboratory operations, and building a data base for patient care and research.[1,4,5]	Reduces clerical errors; results more rapidly available; increases productivity	Becton Dickinson; Technicon; Medlab; CHC Co.
Radiology	Scheduling patients; organizing and storing the dictated radiologist's report of findings; making the report available to the referring physician; tracking x-ray films; producing statistical, management, and financial reports.[1,4,5]	Better management; Increased productivity	Siemens; General Electric
Pharmacy	Controls dangerous drugs; provides for drug ordering and inventory control; controls drug distribution to patients; stores and retrieves drug information; constructs patient profiles; maintains hospital formulary; generates charges for patient billing.[1,4,5]	Reduces medication errors; provides inventory control.	Becton Dickinson IBM; SMS
Medical Records	Maintains and keeps track of medical charts; tracks the completion status of records; generates notices for the attending physician when final treatment information and diagnosis is needed as a reminder to complete documentation; indexes the record by patient identification number, by disease classification and other demographic factors for medical audit and research.[1,4,5]	Reduces lost medical records.	McDonell Douglas; Medicus

Source: Reprinted from Vincent Maturi and Richard Dubois, "Recent Trends in Computerized Medical Information Systems for Hospital Departments," in Proceedings, IEEE Fourth Annual Symposium on Computer Applications in Medical Care, Washington, D.C., 1980, © 1980 IEEE.

Table 9–2 Data on Commercial Information Systems
(Vendors Selling Department-specific Systems)

Name of System	Office	Typical Hospital Using System*	Does Vendor Provide Interface	Main Function	Estimated Units Sold** 1976–78	1979–80	Vendors Projection (5 yrs) on Rate of Acquisition
CHC[2]	Houston, Texas	Patnes, MO Montefiori, N.Y.	yes	Laboratory	14	16	Moderate
DATA STAT[3] Nat. Data Corp.	Atlanta, Georgia	Kaiser, California	yes	Pharmacy	180	220	Moderate
LCI[6]	Madison, Wisconsin	Ohio State Univ.	yes	Laboratory	35	37	Moderate
MEDLAB[7]	Salt Lake City, Utah	Hillcrest, Tulsa Oklahoma	yes	Laboratory	24	28	Moderate
RUBICON[8]	Dallas, Texas	Univ. of Chicago	yes	Laboratory	5	9	Moderate
SIEMENS[9] SIREP	Washington, D.C.	Johns Hopkins, MD.	no	Radiology	6	10	Moderate
BECTON DICKINSON[10] CLAS	Sharon, Mass.	Univ. of Washington Seattle	yes	Laboratory Pharmacy	12 5	2 9	Discontinuing Moderate

*Also the typical hospital contacted
**Estimates reported by manufacturer's representatives and from sales literature

Table 9-2 continued

(Vendors Selling both Department-specific and Hospital-wide Systems)

Name of System	Office	Typical Hospital Using System*	Does Vendor Provide Interface	Main Function	Estimated Units Sold** 1976-78	1979-80	Vendors Projection (5 yrs.) on Rate of Acquisition
HBO[11] MEDPRO	Atlanta, Georgia	St. Thomas, Nashville, Tenn.	yes	Pharmacy Hospital-wide	— 80	— 160	Slow Rapid
IBM[12]	Atlanta, Georgia	Ft. Myers Community, Florida Variety Children, Miami	yes	Laboratory-Pharmacy Hospital-wide	50 15	52 30	Slow Rapid
McDonnell Douglas	St. Louis, MO	D.C. General Washington, D.C.	yes	Pharmacy-Lab. Hospital-wide	232 18	260 36	Slow Rapid
MEDITECH[14] MR₁₁ HSCS	Boston, Mass.	Univ. of Rochester, N.Y. St. Joseph's, Redding, PA	yes	Lab.-Pharmacy Hospital-wide	30 2	31 4	Slow Rapid
SMS[15]	King of Prussia, PA	Woesthoff, Florida New York Hospital, N.Y.	yes	Pharmacy Hospital-wide	— 50	— 100	— Rapid
TECHNICON[16] LPM, TMIS	Tarrytown, N.Y. Madison, Wisconsin	Youngstown, Ohio El Camino, CA	yes	Laboratory Hospital-wide	10 6	15 9	Slow Moderate
Electronic[17] Data Systems	Dallas, Texas	North Arundel, MD. Holy Cross, MD	yes	Laboratory Hospital-wide	1 2	1 4	Slow Rapid

| Whittaker-MEDICUS SPECTRA | Evanston, Ill. | University of Texas | yes | Laboratory Pharmacy Hospital-wide | 10 6 | 11 12 | Moderate Rapid |

*Also the typical hospital contacted
**Estimates reported by manufacturer's representatives & sales literature

	Abbreviations		*Acronyms*
CHC	Community Health Computing, Inc.	CLAS	Clinical Laboratory System (Becton Dickinson, Inc.)
HBO	Huff, Barrington, Owen Corporation	HDCS	Hospital Data Collection System
IBM	International Business Machines	MR$_{II}$	Medical Record System (McDonnell Douglas)
LCI	Laboratory Computing, Inc.		
LDM	Laboratory Data Manager (Technicon)	MEDPRO	Medical Processing (HBO Corporation)
SMS	Shared Medical Systems	SIREP	Siemens Reporting Radiology (Siemens Inc.)
TMIS	Technicon Medical Information Systems		

Source: Reprinted from Vincent Maturi and Richard Dubois, "Recent Trends in Computerized Medical Information Systems in Hospital Departments," in *Proceedings, IEEE Fourth Annual Symposium on Computer Applications in Medical Care, Washington, D.C.,* 1980, © 1980 IEEE.

Vendor Commitments to Medical Information Systems

A schematic that provides data on the current offerings of various suppliers of computer hardware and software is provided in Tables 9–3–9–6,originally published in the proceedings of the IEEE Fourth Annual Symposium on Computer Applications in Medical Care, 1980.

These tables provide an excellent look at what is available and are a solid base of hard data that indicates how many hospitals in the country are using computers. Vendors are rapidly providing a number and variety of information processing applications that are impressive.

Readers can see that the current picture of computer technology as a vital force in modern medical information systems developments encompasses a myriad of directions, technologies, philosophies, and choices. It is important to understand the technological potential for organizations and individual professionals within the current trends. Not only are successful efforts now available for review, but the number and experiences of vendors is extensive. The information provided in the foregoing discussions should help practitioners identify a current picture of the state of the art in medical computing and associated patient data computerization.

Once a basic picture of the overall developments has been acquired, it is time to identify any significant information or technology issues that may affect this picture. Three key issues are featured in the next section of this chapter. They are issues of the future that have implications for practitioners in professional practice activities. Data quality, microcomputer technology potential, and programmerless systems design represent change and challenges for the future.

KEY ISSUES IN INFORMATION AND TECHNOLOGY

The Effect of Computers on Data Quality

A major information issue for health record practitioners is data quality.

The health care system in the United States has come to depend heavily on "data." There are many sources for this new emphasis. Driven by research, clinical medicine seeks quantitative measurements and concrete observables in an increasingly elaborate definition of diseases. Detailed requirements for documentation rest on medical, scientific, educational, and legal considerations, as well as those for reimbursement and the audit of quality care.

Table 9–3 Survey of Vendor Research and Demonstration Budgets for Hospital Information Systems by M. J. Ball, 1980*

Firm	R & D $ Invested to Date	R & D $ To Be Invested by 1983
Compucare (HIS)	3,500,000	2,000,000
Datacare (PCIS)	4,000,000	1,000,000 (vary by client needs)
HBO and Company	6,000,000 (approx.)	6,000,000 (approx.)
Hospital Data Center of Virginia (HDC)	Not available	250,000
Interpretive Data Systems (IDS)	1,000,000 (approx.)	300,000
McDonnell-Douglas Automation Co.	20,000,000	4,000,000 plus
Meditech	300,000/yr since 1969 (3,300,000)	600,000
Nadacom	15,000,000 on full line of patient data systems	Will continue to make a significant investment in new development
NCR Corporation	6,000,000	3,000,000
Shared Medical Systems	6,200,000 during 1979	No specific data, but in keeping with past efforts
Spectra (Medicus)	25,000,000	"Proprietary information"
Technicon Medical Information Systems	24,000,000 (incl. Lockheed)	4,000,000
Tymshare Medical Systems	"Well over the million dollar mark"	Not available
Burroughs	"Does not report R&D expenditure specifically for the health care industry"	"Does not report R&D expenditures specifically for the health care industry"
IBM (PCS)	"Not available for public disclosure"	"Not available for public disclosure"

*The data listed are intended to convey the scope of the aggregate funds currently spent on development activities. These data in no way should be used to compare individual vendors to each other since each of the vendors' budgeting and reporting conventions differs from the others'. The main point is that HIS development is not a minor expense.

Source: Marion J. Ball and Stanley Jacobs, "Hospital Information Systems as We Enter the Decade of the 80s," in *Proceedings, IEEE Fourth Annual Symposium on Computer Applications in Medical Care, Washington, D.C., 1980,* © 1980 IEEE.

Table 9–4 Present Vendors of Level 1 AHIS

Vendor	Number Installed or in Process—April 1980	Number of Beds Least–Most	Order Entry Terminal
HBO & Company (HBO)	237	100–629	CRT
Shared Medical Systems Corporation (SMS)	100	154–1,725	CRT
McDonnell-Douglas Automation Co. (McAuto)	53	103–650	CRT
IBM Corporation (IBM)	50 (est.)		CRT
NCR Corporation (NCR)	25	284–641	BCWR or OCR*
Burroughs Corporation	20	191–737	CRT
DATX Corporation (DATX)	10	176–534	Card Reader
Pentamation Enterprises	9	280–800	CRT and OCR
Tymshare Medical Systems	5	80–436	CRT
Compucare, Inc.	4	250–440	CRT
Space Age Computer Systems, Inc.	4	349–815	CRT
Hospital Data Center of Virginia (HDC)	4	256–800	BCWR†
Spectra Medical Systems, Inc.	2	140–600	CRT
Technicon Medical Information Systems Corporation	3	446–719	CRT
Interpretive Data Systems, Inc. (IDS)	1	250–250	CRT
National Data Communications Inc. (NADACOM)			CRT
Genitron, Inc.			BCWR
Informational Resource Electronics Corp. (IREC)			CRT (intelligent)

*BCWR = bar code wand reader; OCR = optical character reader.
†HDC also utilized CRTs for inquiry purposes.

Source: Marion J. Ball and Stanley Jacobs, "Hospital Information Systems as We Enter the Decade of the 80s," in *Proceedings, IEEE Fourth Annual Symposium on Computer Applications in Medical Care, Washington, D.C., 1980,* © 1980 IEEE.

The coordination of modern technological services has added further clerical tasks. It is no exaggeration to state that "clerical medicine" is the most rapidly growing aspect of medicine today. The volume of data now associated with medical practice has

Table 9–5 Vendors of Shared Business and Financial Systems Offering Level 1 HIS

Vendor	1979 Data Hospitals in Shared System	1980 Data Level 1 HIS Installations
McAuto	561	53
SMS	490	100
Tymshare	150	5
Technicon	45	3
Pentamation	31	9
Space Age	26	4
HDC	24	4
HBO (IFAS)*	14	237

*HBO introduced a new real-time terminal network-based financial system in 1979. Their new IFAS system directly interfaces in-house to HBO's hospital information system (MEDPRO).

Source: Marion J. Ball and Stanley Jacobs, "Hospital Information Systems as We Enter the Decade of the 80s," in *Proceedings, IEEE Fourth Annual Symposium on Computer Applications in Medical Care, Washington, D.C., 1980,* © 1980 IEEE.

Table 9–6 Vendors of Level 1 HIS Which also Offer Level 2 HIS

Vendor	Number of Hospitals with Level 2 HIS	Number of Hospitals with Level 1 HIS
Spectra	18	2
Technicon	17	3
NDC (NADACOM)	10	0
Burroughs	8	20
McAuto	6	53
IREC	4	0
IBM	Not available from IBM (50 est.)	50 est.

Source: Marion J. Ball and Stanley Jacobs, "Hospital Information Systems as We Enter the Decade of the 80s," in *Proceedings, IEEE Fourth Annual Symposium on Computer Applications in Medical Care, Washington, D.C., 1980,* © 1980 IEEE.

made manual methods of collection, manipulation, and display nearly untenable.[17]

Central to the issue of future computerized patient data systems is the fundamental need to ensure an acceptable level of data quality for clinical, analytical, and administrative use. Not only must appropriate data handling procedures be identified and followed, but accurate software protocols must be designed to provide the foundation for computer performance. The underlying question is, How can the computer positively affect data quality? This question represents the growing recognition that data management is a critical issue in modern medicine. The computer is a resource that can be used to resolve this issue.

Because computers have enabled us to look at large amounts of data in new ways, we have come to recognize the use of computer-processed data itself as a resource. For example, data analysis is performed via computer software. Computerized signal analysis brings continuous patient data more accurately to the attention of the clinicians to be integrated into treatment planning. Indeed, we are using computers today to evaluate past computer applications, to monitor present computer activities, and to determine future computer needs. Consider the following ways computers can affect data quality.

1. Computers are used *to streamline patient information handling.* Because effective decision making is dependent on information, which in turn is dependent on data, this technology has become a major vehicle in data collection and analysis. In addition, as evaluation of medical care directs health care providers into higher levels of expectation, demands for more accurate data handling will increase. Consider the power of computers to streamline data handling and reduce incidence of data transcription in the clinical laboratory.

> Data transcriptions introduce error. A 3 percent raw error rate in transcription has been found to be nearly irreducible minimum in an environment free of stress and distraction. Numerical indices are particularly difficult to proofread. . . . Error rates of such magnitude are not acceptable. Thus significant professional time must be allocated to procedures that attempt to catch such errors in an effort to reduce them to below 1 percent. What can a computer do to affect such error rates? A computer can

> - Structure and screen input
> - Handle the necessary bookkeeping
> - Organize, file, and retrieve data

- Call attention to unusual results
- Print specimen labels and completed reports
- Keep track of speciman status, laboratory decisions, workload, and quality control
- Post charges to patient accounts for work completed.[18]

2. Computers are also used *to communicate direct patient data*. Computerized readings of physical parameters provide more accurate data to the clinician to be used in the diagnostic process itself. "Data from high volume automated instruments can enter the system directly, without transcription, to be reviewed and verified on a formatted screen. Interactive video terminals at each work station tied to a common data base can provide consistent interactive communication."[19] Computers can improve data quality by applying protocols in investigative diagnostic activities such as laboratory test and EKG analysis. As such protocols are developed, increased consistency will be possible.

3. Computers can *direct and control data collection completeness*. Data input formats and program edit tests can be used to perform extensive validation and verification routines on data entry operations. Similarly, data use can be improved in computerized data bases by using well-designed protocols to search and match data elements against data retrieval attributes. This is accomplished through careful attention to data collection methods and data retrieval program design.

4. Computers can provide *consistent translation of medical terms into codes*. By producing a consistent coding assignment to medical diagnoses, a uniform data bank can be developed. In addition, the technical assignment of codes by computer program will improve data accuracy through consistent code assignment per disease and reduction of data transcription events.

5. Computers can also *improve data quality in primary care by extending the memory of the physician*. This can be done through real-time access to current medical practice results. The cardiovascular clinic at Duke University provides computerized information retrieval of hypertension therapies including medication results so that the participating cardiologists can employ the most recent knowledge possible in planning and evaluating treatment. For example, when an individual physician recognizes medication abuse effect based on information made available through computerized data retrieval and subsequently records it in the patient record, the quality of data has been improved in specificity. That same physician may then collect all such effects through the feedback and monitoring features of a computer.

Case Mix Analysis Is a Data Quality Issue for Health Record
  Practitioners

The current need to perform case mix analysis of hospital discharge data is a clear illustration that accurate, high-quality data can and will be necessary. Here the accuracy of information has direct impact on resource allocation in medical care. The use of the DRGs (diagnostic related groups) and similar systems requires that accurate data be provided in a uniform fashion. By performing edit checks and other data verification functions on discharge data, computers can help support effective standards of data quality. From these examples we can see that data accuracy and the methods for securing it will be an ongoing information issue in medical care. Computers are a needed resource.

Microcomputer Potential and the Health Record Professional

A major technological issue for health record professionals today is understanding and applying the appropriate potential of microcomputers. The rapid advent of microcomputers in the last five years represents another facet of the increased availability of computer technology to businesses and individuals. Reduced cost has been a major factor in the growing popularity of this technology.

Health record practitioners will see the impact of microcomputers as components in special purpose applications in medicine and health information, as components of office automation, and as personal tools for home and business applications. Medical research; medical data collection, storage, retrieval and manipulation; and medical decision making are examples in which microcomputers are employed by health care professionals. The most common example in an MRD is the microcomputer used in word processing systems.

In other examples, the powerful sensory, computational, memory, and display capabilities of microcomputer systems and their compact size have also offered new opportunities to relieve functional deficiencies associated with loss of limbs, paralysis, speech impediments, deafness, and blindness. Careful applications of these computers have improved the interpretation of diagnostic tests, such as the electrocardiogram, and monitoring of critically ill patients.[20]

Today microcomputers are used in physicians' offices to perform billing and accounting functions, clinical recordkeeping, word processing and inventory control.[21] The automated office system has the potential to change significantly the ways we handle our working lives. Advances in electronics and computer systems enable the restructuring of information handling to allow an immediacy of interaction not previously available.

This in turn sharply reduces the tedium of paperwork and facilitates collective work with others. The electronic desk can become a professional's link to a widely distributed array of information sources and services.[22]

What exactly is a microcomputer? The term *microcomputer* is a general term referring to a complete tiny computing system, consisting of hardware and software whose main processing blocks are made of semiconductor integrated circuits. The circuitry of this very small computer is usually contained on one or a few tiny semiconductor chips.[23]

In the 1970s integrated electronics developed so that several thousand electronic components could be integrated on a single chip of silicon. The level of this integration was sufficient to allow complete processors for calculators and small computers to be contained on a single chip measuring typically 3 to 5 millimeters on a side. These were termed *microprocessors*. As the technology improved so that significant amounts of memory could be included on the processor chip, the term *microcomputer* was applied to these tiny systems.[24]

Many functional operations and management tasks in medical record departments can be supported by this technology. These include but are not limited to the following examples:

- A department MPI capable of storing and retrieving patient names
- A chart tracking application that could be operated by an MRD file clerk
- A medical reports management information application that could maintain a record of all requests, responses, and activity of the medical reports unit
- A statistical display application in which data about risk management, quality assurance, general hospital statistics, and research findings could be entered through a terminal, manipulated by a statistical software package, and displayed on CRT and hard-copy output
- A discharge abstracting application in which the discharge abstract format is stored in the microcomputer system, used to enter discharge abstracts for future processing, and manipulated for appropriate statistics (could be developed from existing software packages such as a data base management systems for microcomputer use)
- An inventory control program that monitors departmental supplies according to budget categories
- A management information application that sets up JCAH standards and assists in maintaining ongoing status information for performance evaluation of the medical record services

- A tumor registry follow-up program
- An employee staffing program for use by departmental management

Some of these applications are available from commercial vendors. They may be marketed directly to health record managers as individual stand-alone applications that include both hardware and software. Initial outlay may be in the range of $5,000 to $14,000 with a yearly software use fee of approximately $2,000. Such products may enable an organization to acquire computer support for a given operation with less outlay than an in-house development approach.

In other cases these applications can be developed with existing commercial software packages that are marketed for use on the personal sized computers such as the TRS-80, PET, and APPLE. Generally speaking, the state of the art is not yet developed so that novice users can immediately use these machines to develop a particular application without a significant investment in time. Time is needed to become familiar with the machine. However, if a health information manager is interested in trying out the microcomputer, the outlay of approximately $3,000 may well justify some investment of time in the system.

Given a software data-base management package, students spent between 25 and 30 hours developing simple applications for use on an APPLE II system. Such systems and the associated user-developed applications can be further supported by participation in local computer user clubs that offer program exchanges. The largest computer exchange group in the country is the A.P.P.L.E., which is the Apple Puget Sound Program Library Exchange in the Seattle area. For about $25 per year, the 11,000 APPLE computer users who belong to the organization get access to software at a reduced rate, a subscription to the exchange's magazine, and use of its telephone hotline.[25]

The advent of microcomputers provides an opportunity for novices to invest a modest amount in a system and experiment with applications for departmental use. This experience helps introduce the potential of patient data computerization on a modest scale.

How can MRAs determine whether or not microcomputer applications such as these should be employed in their setting? They should proceed through the following steps:

1. Acquire a general understanding of microcomputers. Determine what they are, how much they cost, and how they are presently used in medical record or comparable department settings.
2. Determine the limitations of the microcomputers and question current users to verify this research. As sure as a limitation is identified

in one setting, a method to circumvent it is identified in another by some innovative user. Verification will help keep abreast of the potential.

3. Determine the overall automation plan in the organization to identify how the employment of departmental microcomputers is viewed and to identify those functions that will require information integration in the future.
4. Engage in strategic planning for the medical record or health information services department. As explained in Chapter 7, this is a critical process. Microcomputers offer a tactical approach to achieve a strategic goal.
5. Engage in systems analysis to identify and document clearly the specific requirements for the application in question.
6. Survey other area installations to determine which microcomputer system is in operation. This will be important if it later proves possible to agree to assist one another in software development. This is an important practical and political step that helps identify potential support areas when a microcomputer purchase is being considered.

Describing the Microcomputer

In addition to the specific applications described at the beginning of this section, microcomputers are also marketed as personal computers. Such microcomputers have flooded the consumer market to offer a variety of services and opportunities for the interested consumer. While the software developed for the personal sized computer offers products for the general consumer, much of it can be used in office settings as well. Table 9-7 provides a picture of the current microcomputer systems on the market today.

Typical Equipment Units

Equipment or hardware units common to most personal computers are the following:

- An input device used to transmit instructions, information, or programs to the computer. This is normally a keyboard with a typewriterlike design. It may be an integral part of the unit or connected to the unit by a 2- to 6-foot cable.
- A display device used to print out what the computer is communicating back to the user. This may be in the form of questions, answers, or status. It can be a CRT terminal or a standard black-and-white television.

Table 9–7 A Buyer's Guide to Personal Computers

Buyer's Guide to Personal Computers, February 22, 1982	
COMPANY/MODEL/PRICE*	**COMMENTS**
APPLE Apple II plus: $1,530–$5,000 Apple III: $3,500–$8,000	There is more software made for Apple II than for any other computer, and it has superb color graphics. The Apple III is primarily for business. New models are expected soon.
ATARI Atari 400: $399–$2,000 Atari 800: $899–$4,000	Both Ataris are extremely popular, known especially for their color graphics and sound—both of which enhance their copious game programs. They are designed almost wholly for homes and schools.
COMMODORE VIC 20: $299–$1,000 Commodore 64: $595–$4,000 PET: $955–$6,000	The Vic 20 has color and sound that rival the Ataris. Pet is the first Commodore effort. The 64, coming out in April, is intended as a competitor for Apple II. Software for the machines is limited.
HEWLETT-PACKARD HP-85: $2,750–$6,000	The HP-85 is directed primarily at scientists and engineers. While there are other HP models, this is the most capable, boasting a built-in thermal printer. HP just cut the price by $500.
IBM Personal Computer: $1,565–$6,000	The IBM is potentially unmatched in its class, with its powerful 16-bit microprocessor, color graphics and sound. It is highly versatile. The number of programs is limited, although more are expected soon.
OSBORNE Osborne 1: $1,795	First of a new breed of portable computers, the Osborne comes complete with disk drives, a word-processor and spreadsheet program and folds up like an attaché case to fit under a plane seat.
TANDY TRS-80 Color Computer: $399–$2,000 TRS-80 Model III: $699–$5,000 TRS-80 Model II: $3,499–$8,000 TRS-80 Model 16: $4,999–$10,000	Tandy's Model III is the largest-selling personal computer, particularly popular in small business and education. Only the Color Computer offers color graphics. Model 16 is a brand new, superpowerful business machine.

TEXAS INSTRUMENTS 99/4A: $525–$3,000	After a false start in 1978, TI has redesigned its keyboard and added hundreds of software programs. It is especially attractive to children because of its LOGO program.
XEROX 820: $2,995–$6,000	The 820 is a solid business machine, backed by the Xerox reputation. It has good word-processing capability. The basic price of $2,995 includes two 5¼-inch disk drives.

*Prices shown range from the suggested retail prices for basic units to estimates for complete systems. Many retailers offer discounts on certain machines.

Source: "To Each His Own Computer," *Newsweek* (February 22, 1982).

- The CPU consists of the logical and arithmetic units and a permanent main memory in which the operating system, diagnostics, and other core programs are stored. This part of the microcomputer is the actual microprocessor part of the machine. The CPU also contains a control component called a clock. The clock controls the speed at which instructions are executed by the configuration. The memory here is often read-only memory (ROM). ROM can be read from but not written to. This type of memory is preprogrammed. It contains a fixed set of instructions. It contains programs that include simple routines for interacting with a computer system or for examining computer random access memory (RAM) locations.

- The memory unit is separate from the CPU. It may consist of 5¼-inch floppy disk drives, which allow a maximum capacity of 256 kilobytes of RAM in 16 kilobyte steps. If external pluggable disk storage units are used, up to 2 million characters can be made available. Random access memory can be read from and written to. It is the kind of memory that can be used to hold application programs, compilers, transient data (data that are being transferred from one peripheral to another and being manipulated in the process) and other software. RAM is measured in bytes (8 bits) or in K bytes (1024 bytes).

- A printer may be used for system outputs. Printers can be either matrix or impact type.

The basic systems containing the units just described are available from just under $3,000 to slightly over $5,000.[26]

Uses of Microcomputers

In addition to stand-alone business or home uses and the examples listed at the beginning of this section, microcomputers are also being used in the following ways. These uses illustrate how this technology is being integrated into computer systems generally.

- They are being incorporated into terminal devices, such as CRTs, as intelligent terminals. These terminals have the capacity to perform some processing tasks at the terminal level before transmitting the information to another setting.
- Microprocessors are having an impact on the way computing is done. Distributed computing is a concept that is emerging because the microprocessors are being embedded in the circuitry of larger computers.
- Microprocessors are being used in the field of data communications. They are used in the circuitry of communications control units. This means that tasks that historically have been performed by the main computer can be off-loaded to a unit that contains a microprocessor.
- Microprocessors are also being used in programmable controllers. These are used in automative engineering, for example, and other manufacturing tasks.
- In the consumer's world, microprocessors are used in pocket calculators, electronic games, and electronic home surveillance systems.[27]

Limitations of Microcomputers

If all this is possible with microcomputers, then why not use them for all patient data computerization? Because of inherent limitations with these machines, this is not yet possible. Like all computers, they are tools that are selected to perform the task as efficiently as possible. To make judicious use of microcomputers, the limitations need to be understood. There are several.

While the initial hardware costs appear economical, it normally requires additional equipment to make the CPU effective for use in a hospital department. Enough memory to permit the use of high-level languages, interface boards, etc., are additional expenses. The peripheral input/output devices also drive up the cost. Big-name computer companies can often supply such peripherals at less cost than the local microcomputer dealer. Further, the personal computer hardware that is intended for hobby use might not hold up to the more rigorous demands of a hospital department. It is important to research hardware needs adequately and compare the microcomputer solution with alternatives provided by computer system manufacturers. Even though the microcomputer may perform the task, it may do so in an inefficient manner.

With systems acquired from developers of traditional computer systems, maintenance contracts can be obtained at approximately 10 percent of the total capital equipment cost. Few microcomputer dealers offer maintenance on a comprehensive basis.

In addition to general hardware limitations, there are serious software limitations as well. This is the area that costs the most in personnel time and frustration. While there are some software packages on the market that can be used for hospital applications, these may be limited in volume-handling capability and/or expansion capability. Here, a clear adherence to initial effective analysis activities will be necessary to be sure the requirements are accurately determined. In addition, programs that can be used for file management activities and other software utilities are tools that system programmers depend on for effective system performance. Software utilities are not generally available on the personalized computers.

Once an application is established and upgrading is required, hospital computer staff may need to become involved, which can result in the microcomputer system becoming part of the overall hospital computer system. Further, any software application will require modification as information needs change. This means additional time and effort on the part of the individual responsible.

Finally, as with any computer, the output is only as good as the input. There is a need to be sure that the data used in the microcomputer application is current, timely, and accurate. Here breakdown in data accuracy can occur if there is no regular access to organization statistics or no planned maintenance such as files updating. These things would usually be routine features of a larger, generalized computer system. This is another issue in data quality.

The following excerpts illustrate how microcomputer use is viewed as concerns by high-level organization management:

> If personal computers are the future, data processing departments are going to have to put the future on hold. Because data processing departments can't answer the question "What if all those tiny systems ask our mainframes 'what if'?" It's not only the difficulty of supporting hundreds of terminals. It's not only the lack of software to handle inquiries from nontechnical users. It's the problem that comes up right after somebody gets an answer and decides he could get a better one by adding more facts to the available data. For organizations just beginning to experience the

costs and the benefits of information that is recognized as a resource, the implications of the personal systems can be terrifying.[28]

When a manager with a microcomputer calls up and asks for access to hospital data, he or she is liable to be disappointed: "Hooking personal computers into a large system is just not as easy as it sounds."[29]

"To the professional, a micro is a computer that fits only when it can be brought under control of the central data processing organization. And it's not necessarily good for everyone to have a little system on a desk if it is going to be used incorrectly."[30]

"Real disservice can be done to the health-care environment by amateur computing systems that may contain naively configured hardware; poorly engineered, badly documented, difficult-to-modify, and inadequately tested software; and that may have inadequate vendor support. . . . If a computer application is worth developing at all, it is worth developing properly."[31]

This discussion is directed to the health record practitioners as an information source to be used as they move through the six steps outlined before. Reconsider the applications listed earlier in the light of the uses and limitations that have been featured. Notice that a final decision will ultimately rest with each organization or department and will be dependent on the careful progression of the practitioner through a basic planning and systems analysis process.

Microcomputers offer a unique opportunity to develop individual applications at low cost. This can enable MRAs to streamline operations and develop specific management information applications within the department.

Let us conclude this discussion of microcomputers with a reiteration of the steps an MRA should take in deciding whether to install microcomputers and also with a fundamental principle of patient data computerization. Consider again the following:

- Determine how a microcomputer will fit into the overall organization goals.
- Be sure that the microcomputer is a tactical choice for computerizing a particular application that fits into the strategic goals for automation in the medical record or health information department.

- See that the microcomputer decision is based on sound systems analysis activities that document the application needs thoroughly.
- Check out the environment to see what other microcomputer resources are available now or will be in the future.

To these steps we add the fundamental principle of patient data computerization: *Wherever patient information is automated, there must be provision for integration with a hospitalwide or clinicwide medical data base as one becomes available.* We are committed to the development of a fully integrated, longitudinal record for all patients.

As evidenced by the discussions of the R-ADT foundation in Chapter 7, we believe that an integrated patient record is the primary component of all medical information systems. This means a patient record that is available to all clinical and analytical users. Indeed computer technology enables the development of unique segments of the record for specialty areas, such as renal dialysis, capable of transfer into a medical data base designed to maintain a complete patient record structured to support all authorized user inquiries. As health record professionals, we have an opportunity to further that goal by supporting developments in automation that work toward that objective.

Prototyping: A Design Issue of the 1980s

The concept of prototyping or modeling is important. In data processing developments, it is an alternative design approach that will enable applications to be directed by the end user, the ultimate user of the application. Accountants, health record administrators, admitting clerks, hospital administrators, physicians, nurses, statistical technicians, and allied health professionals and personnel are all examples of end users.

In complex engineering, a prototype is created prior to development of the final product. This is done to test the principles, to ensure that the system works, and to obtain feedback on the design so that adjustments can be made before big money is spent. Prototype building is also carried out in airplane construction, shipbuilding, and industrial plant construction.[32]

Complex data processing systems such as medical and clinical information systems need the prototyping process even more than these examples. There is much to learn from a pilot operation. Many significant changes are likely to be made. Prototypes would help solve many of the design difficulties now encountered in working with the requirements definition process explained in this text. Data processing prototypes were not generally used until the 1980s because the cost of programming a prototype was about as high as the cost of programming the live system itself. Today, software has developed to such a point that prototypes can be created inexpensively. Some examples of the software developments that have led to this point are:[33]

- Simple-query facilities that enable stored records to be printed or displayed in a suitable format
- Complex-query languages that are data base user languages that permit the formulation of queries that may relate to multiple records. These queries sometimes invoke complex searching of the data base or the joining of multiple records. Many data base user languages now exist. Some are marketed by the vendors of their host data base management systems. Others are marketed by independent software houses.
- Report generators for extracting data from a file or data base and formatting it into reports. Good report generators allow substantial arithmetic or logic to be performed on the data before it is displayed or printed. Hospital statistics and management information reporting for a department could be enhanced by this feature.
- Graphics languages now exist that enable users to ask for data and specify how they want it charted. They can search files or data bases and chart information according to different criteria. Think how effective this capability would be in quality assurance data display.
- Application generators are designed to permit an entire application to be generated automatically. The input can be specified. Its validation can be specified. The arithmetic and logic functions performed on the input data and the output created can also be specified. Most of these systems operate with data bases. Readers may recall the description of TEDIUM presented in Chapter 5.
- Very high level programming languages are now designed for end users. These are designed so that a smaller number of instructions are required. BASIC, APL, and NOMAD are all examples of such languages.
- Parameterized application packages can be purchased for running some applications. These are preprogrammed packages. Although they often require a considerable amount of tailoring to fit the organization and system requirements, they are actually designed with parameters that can be chosen to modify their operation. Some application packages are marketed directly to end users so that they can avoid involvement with the data processing department. Some are

designed for end users' minicomputers. Some are marketed with turnkey installation so that the users need no installation skills.

While these developments are targeted for the end user, there are many cases in which the end user and the data processing professional would work together to apply them. In many cases, the users are not going to have the skill and training necessary to work with these features. Even with these innovations, such systems still require some knowledge and skill. These systems do mark the path of the future, however. Consider the following scenario.

An MRA or user-manager can work with a systems analyst to create and demonstrate dialogues for data base inquiries, report generation, and manipulation of screen information for a case mix analysis function. The needs can be discussed with the analyst and a specimen or sample dialogue can be shown on a terminal. This might take a few hours or a few days, depending on the complexity and the language that is used. The health record user is shown the preliminary dialogue and trained to use it. Usually, suggestions for changes or improvements can be made at this time. New calculations, formats, and data display charts might be selected. The user-manager might select from the variety of chart types available in the system. As the discussion continues, the focus of discussion is the prototype or model that is emerging. As changes are suggested, new screens are displayed so they can be reviewed by the user. The analyst may also continue to work on the screens adding new features based on his experience. This process continues until the user has the model desired.[34] In some cases the application can then be developed quickly because a data base already exists. In other cases more extensive design work has to be completed in an overall system development process. In the latter example, the prototype serves as the requirements document for the application programming.

The reader can see the potential of this design approach. As it becomes more available, more specific requirements determination can be made and applications can be developed more quickly. This design approach has the potential to bring the user into the design process more concretely than the traditional methods. Furthermore, many of the delays and misunderstandings inherent in the lengthier systems life cycle development approach because of time factors and the dynamic information changes in any information system can be eliminated. The prototyping process may well allow accomplishment of user-driven computing. In Exhibit 9–1 and Figure 9–1, a comparison of prespecified computing via the life cycle model and user-driven computing are made. Readers will note some commonality among the models represented. Perhaps the accomplishment of the user-directed design is in the foreseeable future.

Exhibit 9–1   Comparison of User-Driven Computing with Prespecified Computing

---

Prespecified Computing

- Formal requirements specifications are created.
- A development cycle is employed.
- Programs are formally documented.
- The application development time is many months or years.
- Maintenance is formal, slow, and expensive.

Examples: Compiler writing, airline reservations, air traffic control, missile guidance, software development.

User-Driven Computing

- Users do not know in detail what they want until they use a version of it, and then they modify it quickly and often frequently. Consequently, formal requirement specification linked to slow application programming is doomed to failure.
- Users may create their own applications, but more often with an analyst who does this in cooperation with them. A separate programming department is not used.
- Applications are created with a generator or other software more quickly than the time to write specifications.
- The application development time is days or at most weeks.
- Maintenance is continuous. Incremental changes are made constantly to the applications by the users or the analyst who assists them.
- The system is self-documenting, or interactive documentation is created when the application is created.
- A centrally administered data base facility is often employed. Data administration is generally needed to prevent chaos of incompatible data spreading.

Examples: Administrative procedures, shop floor control, information systems, decision support, paperwork avoidance.

Source: Reprinted from *Application Development without Computers* by James Martin by permission of Prentice-Hall Publishing Company, Englewood Cliffs, N.J., © 1982.

The demand for new applications is rising faster than the information services or data processing departments can supply them. The backlog of requests shows no signs of abating. If prototyping can be developed adequately, then the strategic objectives for the balance of the 1980s as identified by James Martin can be accomplished.[35] These are

Figure 9–1 Systems Life Cycle Development Model

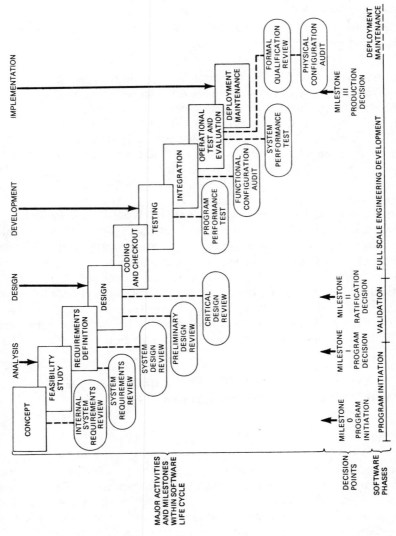

Source: Reprinted from *Application Development without Computers* by James Martin by permission of Prentice-Hall Publishing Company, Englewood Cliffs, N.J., © 1982.

- End users can access and manipulate information required to solve their problems 50 percent of the time without technical data processing help.
- Thirty percent of the applications can be generated or supplied directly from vendors without programming.
- Seventy percent of applications can be supplied with no more than 30 percent procedural code written.

Imagine the enhanced use of computer technology to the health record professional when this technology is available.

What about the process of systems analysis in all this? Prototyping does not eliminate the need for effective systems analysis. The same information will still be required from the user-managers. Detailed, well-documented user requirements must still be the foundation for all applications developments. Prototyping will streamline the process from that point. As it becomes more available, it will mean a revolution in data processing management.

How can these information and technology issues be incorporated into future directions? What specific role should be adopted by health information practitioners as participants in future developments? We have attempted to challenge the readers to a participatory leadership role. Strong management skills and techniques in introducing change are important. In particular, a fundamental philosophy of the right of the patient to a record of integrity is the underlying theme of this text. We will turn now to consider the philosophy of the patient record itself in planning for future medical computing and patient data computerization.

PLANNING FOR FUTURE MEDICAL COMPUTING AND PATIENT DATA COMPUTERIZATION

Philosophy of the Patient Record

Today, those who are developing or selecting a computerized patient data system recognize that much has been accomplished. They also may recognize that much is yet to be done. A review of the process of patient care is an essential first step in analyzing the flow of information that comprises one of the major communication aspects of patient care. The process of analysis must begin with the patient. The patient and the patient record are inseparable. They are the origin of all medical information systems. Until system designers build systems that recognize, account for, and plan for patient interface through the evolution of medical information, progress, like medical care, will remain fragmented.

> We can describe the patient in many ways, according to many needs, according to many characteristics, yet in so doing we will inevitably compile a description that presents a set of information inseparable from a particular individual. We can also compile a set of facts that, when applied to a particular patient, will describe the condition and care or treatment of that patient at a specific time. It is impossible, then, from our point of view, to separate the patient from the patient's medical record. There is, of course, the obvious physical separateness of the two, but this is of small importance when one considers the overwhelming significance of the unity of the patient and the recorded information that exclusively describes what has occurred in the health care of an identified, unique patient.[36]

This unity of patient and information relates to the first step in systems design because interaction with the patient, who is the source of all data entries, provides an opportunity for accuracy verification, spontaneous input and immediate feedback in the documentation of the interaction and its results. By following a patient through an episode of care, the systems design team will see the natural evolution of data entries and information flow. When the episode of care is completed, the processing of data continues as the information is retrieved and made available for all uses. This completes the cycle of patient-based data and prepares the way for computerization of patient information in its many extensions.

Just as patients and their records are inseparable, so the clinical and analytical uses of the patient information are inseparable. For each instance of use of medical information by clinicians to provide direct care to patients, there is a corresponding use of that same medical information by analysts, researchers, evaluators and so on. Whether addressing administrative analysis, cost accounting, medical audit, clinical research, or health services resource planning and allocation, the analytical use of the patient's medical information will continue to develop side by side with the clinical use.

Let us interject here a brief discussion of the information concepts inherent in the clinical data entries of patient records. Clinical data entries have traditionally ranged from narrative to alphanumerical, from graphic displays to machine-produced tracings, and from handwritten to typed. Currently, data entries take the form of computer printouts, microfiche, and, of course, we still have paper documents that contain handwritten or typewritten entries. For computerization, data have to be in such a form that they can be read, interpreted, collapsed into computer language, processed, stored, and retrieved. Weed's problem-oriented medical record (POMR) was the first format to address the resolution of the dilemma presented when computerization of traditional medical records was attempted. Through the POMR all clinical data entries are reordered to provide a logical and computer-compatible format not provided by any traditional format. It identifies problems, assigns them a number, dates the problems, and organizes the clinical data entries so that branching of entries is possible. This organization capability encourages the clinician to establish relationships among the data entries that would usually go unnoticed in other formats. The POMR is the most advanced method used to address the clinical uses of medical information employing maximum information organization of data entries and information technology.

Recommendations for the Future

This brings us to some important recommendations. These ideas express our view of the current state of the art of computerization of medical record applications within developing medical information systems. They are based on our experience with the problems of the early years of computer development and an awareness of the proliferation of systems implementation in the 1980s. We recommend that medical information systems designers do the following:

1. *Designate the patient record as the appropriate source of all clinical information.* Systems analysis that recognizes the role of the patient in the flow of information is the basis of this recommendation. In practice, this recommendation provides a directive to all clinicians, especially those whose major interest is research. All data derived from a patient's participation in diagnostic, therapeutic, rehabilitative, palliative, or research programs is most appropriately preserved on the patient's record in the facility where the patient participated in the program or encounter. Single-purpose records filed in specialty areas and not accessible to all patient care providers or records documented and retained in such a manner that the patient does not know of the existence of the record contradict the intent of this recommendation. Computer technology should enable patient information integration to satisfy all users.

2. *Require medical conditions and procedures to be categorized in a way that is understood by both clinical and analytical users.* Coding systems such as SNOMED, ICDA-CM, CPT, International Classification for Health Problems in Primary Care (ICHPPC), and established uniform data sets in hospitals, ambulatory care, and long-term care settings provide

an avenue to communicate standardized data. When describing, entering, or retrieving complete records or cases for care or research, it is imperative that all providers and users communicate through a language method that has been proven effective and achieved peer acceptance.

3. *Develop a precise method of information organization in computerized medical records that is compatible with differences in data acquisition, data transmittal, data storage, and retrieval.* The variances in individual medical record systems and the options offered by current computer technology provide many different methods to store, process, and retrieve patient data. This variety of options and methods is vital to the ultimate development of methods and systems that are universally recognized as high quality. It is also possible, because of the unique characteristics of health care facilities and the patient data systems utilized in the facilities, that there will always be a need for individual applications that are not of universal recognition or acceptance. Still, there must be a move toward the development of a computerized medical record that organizes the information components and data elements into a format that can be used by clinicians, administrators, and all other primary care data users. Included in this recommendation is the continued research and development of a computerized medical record that includes the study of various models of artificial intelligence. Research studies that investigate the interaction of human beings and computers in the acquisition, processing, and use of computerized medical records should also be carried out.

As pointed out earlier, Weed's POMR was the first format that addressed the issue of computerizing traditional medical records by providing a computer-compatible format. This design focuses on a dynamic, problem-oriented, precisely reasoned care approach that directs clinicians to work with a logical, mathematical rationale in reviewing and acting on past and present patient needs. It is not merely a redesigned chronological record but has far more depth of penetration into the medical arena than a simple format change. The POMR has serious implications regarding the way medicine is practiced. It restructures the method by which physicians acquire and use data when they treat a patient. As a byproduct of this restructured data capture and use, the format of the medical record is recognized from its chronological sequence and ordered into a more useful, easily identified source of patient data.

4. *Design a system that focuses on the patient/client as a primary user of medical information.* The Report of the Privacy Protection Study Commission, like others, recommends that "there are steps that can and should be taken: (a) to improve the accuracy, timeliness, and completeness of the information in a medical record; (b) to enhance the individual patient's awareness of the content and uses of a medical record about himself; and

(c) to control not only the amount and type of information that is disclosed to other types of users, but also the conditions under which such disclosures are made.''

CRTs are excellent for use in patient education and planning; personally carried patient health records are becoming increasingly available with improved technology; and, most important, patients should participate in data verification in an ongoing mode.

5. *Demand transportability by requiring that system information summaries be structured in such a way that they can be used in alternative settings*. This is directed to software development. If an individual application has unique design, provision of conversion to a transportable information structure on a technical plane can be accomplished. This can be carried out through adoption of formats that provide information summaries. *Transportability* of key data elements that provide linkage of care from one facility to another should be a major concern of all who are familiar with the need for concise and accurate data. Continuity of care depends on such transportability. Cost is one reason that this issue should be addressed. Cost in personnel time for communicating what has already been recorded, cost in repetition of tests that have already been completed, and cost to the patient for any complications or delays that are a result of unavailability of pertinent data for coordinated care.

6. *Promote and facilitate systems development with software and hardware, already developed and essentially ready to use*. This standard is a fundamental philosophy of the PROMIS program. It calls for using simple, recognizable programming techniques wherever possible and off-the-shelf hardware in equipment specifications. It is logical, simple, pragmatic, and must be adopted if extensive medical information system design is to occur in a timely and cost-responsible manner.[38]

7. *Design dynamic data security in programming and hardware design and delegate to a qualified individual responsibility for control of data access*. The federal Privacy Act[39] permits individuals to determine what records are maintained, for what purpose; recent amendments permit individuals to obtain copies of their own medical records.

Informed consent, in a study by Alan Westin has been found to be sorely lacking in all areas of health care delivery. Hardware design alone, not even in combination with software, is not able to resolve the privacy issue. This standard demonstrates quite readily that an educated, designated individual is needed to coordinate hardware and software, create and control policies, and implement user and consumer education regarding computerized information retention and retrieval.

8. *Require concurrent development for clinical and analytical users*. Adhering to precise systems analysis techniques in information documen-

tation should identify concurrent users of patient information. This is a measurable, performance-directed analysis.

The computer-stored ambulatory care record (CoStar) is designed to meet both medical/clinical and the financial/administrative analytical information needs.[40] The system objectives clearly specify that data retrieval and analysis capability required for ongoing operation, budgeting, and planning activities be incorporated in the system. Associated management reports and user-specified inquiry and report generation are also defined as integral components of the program. Clinical computing should draw from this model to initiate and support long-range planning that recognizes and promotes this capability as a functional standard of developing computer applications.

9. *Define and establish methods that ensure long-range retention and retrieval.* In traditional patient record systems, medical record practitioners plan, develop, and control a variety of long-range retention and retrieval systems. Microfilm, commercial storage, purging for volume reduction are some of the methods employed. As computerized systems develop, hard decisions will have to be made to weight and prioritize particular data entries for retention and retrieval needs. Systems must be designed that provide for initial weighting of information and predefined purging policies and schedules. The critical and historical information elements must remain readily accessible and retrievable within a reasonable turnaround time. The Ottery Saint Mary Clinic, in London, has a magnetic tape back-up capability.[41] Clinicians determine what information is transferred from on-line to back-up. They direct the system to transfer information to magnetic tape. The system is designed to provide for continuity of retention and retrieval by maintaining an avenue for the clinician to access the back-up files on request. Magnetic tape, floppy disk, and computer output microfilm techniques are all potential media. A directed, comprehensive systems analysis foundation will enable systems designers to plan appropriately for this standard.

10. *Build for linkage.* Networks are on the horizon. Transfer forms were introduced with Medicare legislation in the mid-1960s to provide for linkage of acute hospital care information to long-term care facilities. Transfer forms are now an integral part of patient transfer. The same capability must be established in computerized patient record systems. If we accept the reality that the only continuing source of patient care information is the patient, we must develop methods that will facilitate this source into an ongoing resource. Consider building information systems that will provide a bridge to and from medical data base networks. This could mean establishment of a microprocessor in individual clinical settings to satisfy the information needs of that setting and enter information to be transferred

based on predetermined data entries. These entries would be transmitted at intervals to a network or central data base. A transfer prograrm could be written to periodically scan the individual data base for these predetermined data entries and add any additional information that the clinician deems significant. This approach would allow users to work with their unique information needs and still provide for networking of appropriate information. As indicated in previous standards discussion, the role and future of microprocessing, floppy disks, and patient-centered information structures would be integral elements in meeting this standard.

Questions and Problems for Discussion

1. What is the general impact of computers in medical care?
2. What are some of the influences on the development of medical information systems (MISs)? What is the relationship of these influences to efforts in patient data computerization today?
3. Describe the major trends in medical computing and illustrate your answer with examples.
4. What is the significance of the CoStar program to developing patient data computerization? Are such programs likely to increase or decrease? Why?
5. How do data banks assist in the development of more sophisticated levels in MIS developments?
6. What kinds of strategies should be employed by health organizations in determining future directions in MIS developments and acquisitions?
7. Why is data quality a key issue in patient data and information technology? Why is this issue of particular importance to health record practitioners?
8. Prepare a report for a hospital computer planning committee on the potential of microcomputer technology for the MRD. Include illustrations of microcomputer use for specific functional operations.
9. What impact will prototyping have on future patient data computerization? Why?
10. Comment on the following statement: "The primary purpose of the health information profession is to develop and manage longitudinal, integrated patient records for all authorized users."
11. Discuss some ways the recommendations featured at the end of the chapter could be supported and developed in today's emerging medical and patient data computerization.

NOTES

1. Richard M. Dubois, "Clinical Information Systems: Current Trends Add Outlook for the 80s," *Computers in Hospitals* (January-February 1981).

2. E.R. Gabrieli, "Potential of Medical Computing," *Journal of Clinical Computing* (1975).

3. Donald A.B. Lindberg, *The Growth of Medical Information Systems in the United States* (Lexington, Mass.: D.C. Heath, Lexington Books, 1979).

4. Morris Collen, *Hospital Computer Systems* (New York: Wiley Biomedical Health Publishing, 1974), pp. 20, 21.

5. Ibid., p. 4.

6. Marion J. Ball and Stanley Jacobs, "Hospital Information Systems as We Enter the Decade of the 80s," in *Proceedings, IEEE Fourth Annual Symposium on Computer Applications in Medical Care, 1980* (New York: Institute of Electrical and Electronic Engineers, 1980).

7. Ibid.

8. Richard Fiddleman, "Preliminary Findings of an Evaluation of CoStar in a Community Health Setting," in *Proceedings, IEEE Fourth Annual Symposium on Computer Applications in Medical Care, 1980,* (New York: Institute of Electrical and Electronic Engineers, 1980, p. 1377.)

9. "Demonstration and Evaluation of a Total Hospital Information System," National Center for Health Services Research, DHEW Pub. No. 77-3188 (HRA).

10. Dubois, "Clinical Information Systems."

11. Ibid.

12. Ibid.

13. Ibid.

14. Ibid.

15. Ibid.

16. Vincent Maturi and Richard DuBois, "Recent Trends in Computerized Medical Information Systems for Hospital Departments," in *Proceedings, IEEE Fourth Annual Symposium on Computer Applications in Medical Care, Washington, D.C., 1980* (New York: Institute of Electrical and Electronic Engineers, 1980).

17. Counterpoint, *AMRA* (1980).

18. Ibid.

19. Ibid.

20. James D. Deindl, "Microelectronics and Computers in Medicine," *Science* 215, no. 4534 (February 1982).

21. John Ashton, David Brinkman, and Jeff Balsam, "Choosing a Medical Office Computer," *Interface Age* (September 1981).

22. R.J. Spinard, "Office Automation," *Science* 215, no. 4534 (February 1982).

23. Charles J. Sippl, *Computer Dictionary* (Indianapolis: Howard W. Sams, 1980).

24. Kensall D. Wise, Kan Chen, and Ronald E. Yokely, *Microcomputers—A Technology Forecast and Assessment to the Year 2000.*

26. Ruth Davis, "Computers and Electronics for Individual Services," *Science* 215, no. 4534 (February 1982), and Myles E. Walsh, *Understanding Computers—What Managers and Users Need To Know.*

27. Walsh, *Understanding Computers.*

28. Peter Krass and Hesh Wiener, "You Mean I Can't Just Plug It In?," *Datamation* (September 1981).

29. Ibid.

30. Ibid.

31. Ibid.

32. James Martin, *Application Development without Programmers* (Englewood Cliffs, N.J.: Prentice-Hall, 1982).

33. Ibid.

34. Ibid.

35. Ibid.

36. Kathleen Waters and Gretchen Murphy, *Medical Records in Health Information* (Rockville, Md.: Aspen Systems Corp., 1979).

37. D. Linouwer, chairman, *Personal Privacy in an Information Society,* Report of the Protection Study Commission, 1977.

38. PROMIS Lab PB-263-578, National Center for Health Services Research, Rockville, Md., 1976.

39. Ibid.

40. CoStar, Functional Specifications (Version 5.5), Laboratory of Computer Science, Massachusetts General Hospital, National Center for Health Services Research, Rockville, Md., 1976.

41. Waters and Murphy, *Medical Records in Health Information.*

Appendix A

Glossary

Access time—the time interval between the instant at which information is called for from storage and the instant at which delivery is completed, e.g., the time it takes to fetch a unit of data from memory and to move it to the CPU (central processing unit).

Activity list—an enumeration of major activities that are performed or that should be performed to fulfill the objectives of a department.

Activity ratio—when a file is processed, the ratio of the number of records that have activity to the total number of records in that file.

Address—the numeric designator for a location in memory where an element of data is or may be stored.

ALGOL—the first major language to be designed by an international committee of people from different organizations. Used for numerical and logical problems, particularly in Europe. Available on many computers.

Algorithm—a defined process or set of rules that leads and assures development of a desired output from a given input. Algorithms may contain a number of branch points where the next step taken depends on the outcome of the preceding step. Clinical algorithms are applications of this procedure to medicine. Also called a "decision support system." Examples can be found in radiology and other diagnostic areas.

APL—a general language with complex notation and powerful operators, difficult for those unfamiliar with it. Subsets available on IBM machines.

Application program—a special purpose program where the user must acquire and have loaded into the computer a program that will carry out specific tasks to be accomplished by using specific data, such as calculating the daily census.

ASCII (American Standard Code for Information Interchange)—a standard 8-bit information code used with most computers and data terminals.

Assembler—a computer program that operates on symbolic input data to produce from such data machine instructions by carrying out such functions as translation of symbolic-operation codes into computer-operating instructions, assigning locations in storage for successive instructions, or computation of absolute addresses from symbolic addresses. Generally translates input symbolic codes into machine instructions, item for item, and produces as an output the same number of instructions or constants that were defined in the input symbolic codes.

Audit trail—the trail or path left by a transaction when it is processed. The trail begins with the original documents, transactions entries, and posting of records and is complete with the report. Validity tests of records are achieved by this method.

Automated Hospital Information System (AHIS)—an on-line, real-time, computer-based system that receives, stores, processes, transmits, and displays information from and for all major patient care, administrative, and financial departments in a hospital.

Auxiliary storage—supplements primary or main storage—usually of higher capacity and lower speed or longer access time. Because central memory is limited and expensive relative to the size of programs and the data sets requiring processing, most computers have devices, usually disks, attached to them that will store large amounts of data (millions or billions of characters). Data are then transferred in blocks between auxiliary storage and central memory as the data in central memory are exhausted. Transfer of data between auxiliary storage and central memory is slower than the rate at which data may be moved between the CPU and memory. Can also include magnetic tape, magnetic drum, and card readers.

BASIC—a simple and particularly learnable language with some character-manipulation capabilities. Available on many computers, usually in a considerably extended form such as BASIC-PLUS on the PDP-11.

Batch processing—a systems approach to processing where a number of similar input items are fed into the computer in related groups for processing during the same machine run, e.g., discharge abstracting programs such as PAS and MEDART. An in-house computer system would use batch processing to prepare discharge analysis of services each month.

BCD (Binary Coded Decimals)—a numerical representation in which decimal digits are represented by binary numerals. The most common binary code is the 8-4-2-1. In binary coded decimals, the number 14 would be 0001 0100.

Bit—a binary digit; hence, a unit of data in binary notation. In the binary numbering system, only two bits (0 and 1) are used.

Block—a collection or group of words, records, or characters that are handled as a single unit.

Bubbles—along with CCD (charged coupled devices), the latest form of main memory technology. Bubble technology has slower data movement than MOS technology but up to ten times the storage density at a lower cost. In addition, bubbles are nonvolatile; if the power fails, the data in main memory are not lost.

Buffer—a temporary data storage device or location (usually fairly limited capacity) used to receive or hold data that are to be moved to some other part of the system. Required whenever the rate at which data are received by the receiving part of the system is faster than the transmitting equipment can handle it. Data can emerge from the CPU at the rate of several million characters a second. Telephone lines to terminals can transmit data at only a few hundred characters a second. The CPU, therefore, dumps a user's data into a buffer and goes on to other tasks while the data are trickled out a character at a time to the terminal.

Buffer storage techniques—methods using devices that are capable of compensating for differences in rate of flow of data or time of occurrence of events when transmitting data from one device to another. The buffer storage, or simply buffer when the meaning is clear from the context, serves as a synchronizing element between two different storage devices or computing elements, usually between internal and external storage or other devices of differing speeds. As an output buffer, data are transferred from internal storage to the buffer and held for transmittal to an output device.

Buses—are high-speed communication channels.

Byte—a term to indicate a measurable portion of consecutive binary digits, e.g., an 8-bit or 16-bit byte. The number of binary bits or elements required to encode one character of information. A byte is a binary element string usually acted upon as a unit by the computer.

Card reader—a device that reads a deck of punched cards by moving each card past a light and a set of photoelectric cells in order to detect the pattern of punched holes.

Central Processing Unit (CPU)—the part of the computing system that contains the circuits that control the interpretation and execution of instructions, including the necessary arithmetic, logic, and control circuits to execute the instructions.

Channel—a path along which signals can be sent, e.g., data channel, output channel.

Character—one symbol of a set of elementary symbols such as those corresponding to the keys on a typewriter. Symbols usually include decimal digits 0 through 9, letters A through Z, punctuation marks, operation symbols, and any other single symbols that a computer may read, store, or write.

Clinical Information System (CIS)—an on-line, real-time, computer-based system that contains in its files an appropriate medical knowledge data base and the rule base or structure for creating higher level medical information from its resident clinical data base or from clinical data entered at the time of inquiry.

COBOL—a popular language similar to natural English for solving business problems. Available on some mini and most large computers.

Codefinder—a system in which users are led through a menu-directed process. The encoding process is semiautomatic. After entering key words of English text, and subsequent interaction with computer requests for detail, the appropriate codes (SNOMED, ICD-9-CM, or CPT) are placed in the computer record. As each letter of the diagnostic key word is entered on the terminal, a comparison is made with the computerized medical dictionary. When enough letters of the key word have been entered to provide a manageable list of choices, the computer displays a list of choices to the medical coder, thus eliminating typing in the entire lines of text.

Communications equipment—devices and material used to transmit signals from one location to another and for sending and receiving the transmissions and changing them into human/understandable form. Includes telegraph lines, telephone lines, radio links, coaxial cable, microwaves, satellites, laser beams, and waveguides.

Compiler—a program that translates instructions, usually called subroutines, from high-level language to machine language for direct use by the computer.

Compiling—the translating process of converting programs written in higher level languages into specific operational steps and finally into required binary codes. The program that does this is a compiler.

Computer—a piece of electronic equipment that (1) performs large numbers of mathematical calculations at very high speeds, (2) operates under the command of a set of changeable instructions called programs, and (3) stores both programs and data in electronic and electromagnetic devices called memories.

Computer application—the identification, selection, installation, and evaluation of a software or hardware component.

Computer output—a purported statement or representation (in written, pictorial, graphic, or other form) of fact produced by a computer or accurately translated from a statement or representation so produced.

Computer output microfilm—a medium produced when data output from the central processor is read into a microfilm recorder connected to a film developer. The final product may be either microfiche or roll film.

Computerized Lexicon for Encoding and Retrieval (CLEAR)—a lexicon-driven computer-based system for transforming natural language (English) terms into codes. These codes are (1) ICD-9 CM (International Classification of Diseases-9th Revision-Clinical Modification—W.H.O.), (2) SNOMED (Systematized Nomenclature of Medicine—College of American Pathologists), (3) CPT (Current Procedural Terminology—A.M.A.), (4) Drug coding (generic and brand drugs organized hierarchically according to clinical usage), and (5) Unique code (the exact wording used by the physician).

Computerized word processing and transcription management—transforms ideas and information into a readable form of communication through the management of procedures, equipment, and personnel.

Constraints—the boundaries or limitations, rules, and regulations, legal and organizational, that affect a particular operation.

Conversion—a process that provides change from one form or function to another. The alteration of a method or system for more effective utilization.

CP/M(Control Program, Micro)—a disk-operating system designed for use on the 8080- and z-80-based microcomputers. Has been called the "software bus."

CRT (Cathode Ray Tube)—a terminal equipped with a television-like screen on which alphanumeric characters are displayed.

Data—a general term used to denote any or all facts, numbers, letters, and symbols that refer to or describe an object, idea, condition, situation, or other factors, e.g., letters in diagnosis, numbers in lab values, time delineation in length of stay.

Data base—a vast and continuously updated file of information, abstracts, or references on a particular subject or subjects. A highly structured data set organized for the efficient recovery and storage of data about cases on a random basis. This capability is obtained through a complex structure of reference tables and retrieval/storage programs in which data are stored so that they are independent of programs using the data.

Data base management—refers to the systematic approach to storing, updating, and retrieval of information stored as data items, usually in the form of a record in a file, so that many users, or even many remote installations, have access to common data banks.

Data coding—transplanting raw data into numeric or alphanumeric codes. The process usually results in considerable data compression; e.g., Wilbur Smith, M.D., is coded as number 306. Diagnoses are coded by numbers in ICDA coding; e.g., hypertension is represented as 437.1.

Data Communications Terminals—also called Video Display Terminals (VDTs) and Cathode Ray Terminals (CRTs). Typewriter or touch screen and televisionlike screen devices located away from the central computer.

Data element analysis—a tool used to identify all the data elements necessary for a given product. Points out data elements to be used for output and specifies format and size of information that must be stored.

Data security—the protection of computerized patient data/information from accidental or intentional disclosure to unauthorized persons and/or from unauthorized modification.

Debug—to detect, identify the source, and fix errors in a program.

Decision table—a tabulation or array of possible logical courses of action, selections, or alternatives that can be considered in the analysis of various problems, i.e., a graphic aid to problem description, flow, and potential results, having much the same purpose as a flow chart.

Decode—to translate data expressed in one set of symbols into the original set of symbols. In computer systems data expressed in numbers are translated back into English words, e.g., on a discharge abstract form used to collect the minimum data set for hospital discharges, the physician code number would be translated back into the correct name.

Density—the closeness of space distribution on a storage medium such as a magnetic drum, magnetic tape, or cathode-ray tube.

Descriptive profile—a written document that describes a functional operation through a detailed explanation of the operation's characteristics. Includes the objective of the operation, the specific sequences including entry/exit points, required outputs, input techniques, data element descriptions, volume characteristics, and user identification.

Direct access—the ability to read or write information at any location within a storage device in a constant amount of time.

Direct communications channels—the circuits or lines used to provide transmittal of electrical impulses (either voltage or current), e.g., telephone lines.

Disk pack—a set of magnetic oxide coated disks mounted on a central spindle and housed in a dust-proof cover when not mounted on a disk unit on the computer.

DOS (Disk Operating System)—a collection of programs that facilitate use of a disk drive.

Downtime—the period during which a computer is malfunctioning or not operating correctly due to machine failures.

DRGs (Diagnosis Related Groups)—a concept in health care reimbursement designed to group hospital costs according to patient consumption of hospital resources and services. Initially developed at Yale University, the system uses discharge abstract data, hospital patient bills, and hospital financial reports to determine which of 83 Major Diagnostic Categories (MDCs) best represent the individual patient. These major categories are further broken down into 383 distinct diagnostic-related groups.

Dumb terminal—a terminal that can only send and receive data.

Dump—to transfer all or part of the contents of one section of computer memory into another section, or to some output device.

Duplex—the ability to transmit data in both directions. Half duplex transmission allows data to move only in one direction at a time but allows the direction to be reversed as is needed. Full duplex allows data to be moving in both directions simultaneously.

EBCDIC (Extended Binary Coded Decimal Interchange Code)—a coding scheme involving 8 bits to encode 1 character of information. The user manager is concerned with BCD and EBCDIC as they apply to: (1) storage of information as in noting the size of memory and (2) transportability of data from one system to another, e.g., whether data stored on one scheme are "readable" by the system or a conversion process is required.

Edit—to prepare for publication. To rearrange data or information. May involve the deletion of unwanted data, the selection of pertinent data, or the application of format techniques.

Encode—to apply the rules of a code often by representing individual characters with other characters. To translate data expressed in one set of symbols into a different set of symbols. In computer systems, data expressed in English words are translated into numbers.

Execute—to carry out an instruction or perform a routine. To interpret a machine instruction and perform the indicated operation(s) on the operand(s) specified.

External outputs—any routine products made by a computer system plus any special inquiry products. In a computerized clinical lab system, patient lab report summaries generated by the system would be an example of external reports.

Feasibility analysis—the comparative assessment among alternative solutions to determine which one is acceptable. Can occur at any point where alternatives must be evaluated.

Federal Privacy Act—federal law giving a person the right to inspect, copy, or contest inaccuracies in his or her own medical records, provided those records are for care received in a federal medical-care facility.

Field—an area in a system record that is set aside for a particular bit of data, e.g., a name field, a sex field, a diagnosis field.

File—a collection of related records treated as a unit, e.g., the current year admissions file, the 1975 discharge file, the doctor's incomplete records file, the master patient index file. A disease index stored on magnetic tape at a computer center is a "disease index file."

File backup (Father, Grandfather)—a data file created by taking a previous copy of a destroyed file and running all intervening transactions against it. The immediate past copy is a father file; two cycles back is a grandfather file. Most data files are constantly being updated and are potentially vulnerable to loss through machine malfunction, processing errors, and human error. Backup file copies are usually stored in a different location from the main computer facility. An example: all charges for patients in a hospital are maintained on a grand master file. Each night the master file is run onto a new tape and the current day's charges are added to it. The new tape becomes the son file. The master is stored as a backup and is referred to as the father file.

File maintenance—modification of a file to incorporate changes that do not involve arithmetical operations, e.g., insertions, deletions, transfers, and corrections.

File purging—destroying the contents of a file.

File update—to incorporate the latest set of additions, changes, deletions, into a data system. Such revisions may be accumulated over a period of time and done in a group or made to each case as they occur. For example, each patient admission in a hospital computer system may update the census count as it occurs.

Firmware—software that is contained in ROM.

Fixed length record—a record in which data entries, such as blocks, words, characters, or digits, are limited to a given number. The master patient index is comprised of individual fixed length records for each patient.

Floppy disk—a flat 8-inch flexible disk of plastic, coated on both sides with a magnetic oxide coating and enclosed in a protective envelope. The floppy disk and its envelope are mounted in a floppy disk unit; a spindle inserts itself through the hole in the center and the spindle then starts to rotate the plastic disk at high speed. Centrifugal force stiffens the disk and a read/write head inserts itself so as to read both surfaces of the disk. Data are then written on the floppy disk in the same manner as for a regular disk. About a quarter million characters of data may be written on one floppy disk.

Format—a predetermined arrangement of characters, fields, lines, punctuation, page numbers, etc. Also printed guides on an operative report.

Forms analysis chart—a graphic display used to identify which fields of information are common to a group of forms.

Forms distribution chart—a graphic display that identifies the number of copies of a particular report and the departments that input or use the copies.

FORTRAN—a language for scientific and technical work, particularly involving numerical computation. Available on most computers.

Free text—natural language terms entered into the computer in contrast to coded information.

Freedom of Information Act—federal law assuring the public that there will be no secrets regarding information that is stored, collected, and used in the conduct of public business.

Front end processors—minicomputers or programmable machines used to accept the data from communications lines in one or more locations, further organize them, and store them in the computer or route them over additional communications equipment to the host computer.

Functional operation—a special purpose or characteristic process that executes defined actions. In medical records, functional operations are those common processes used to achieve specific departmental objectives related to patient record development processing, retention, and retrieval.

Functional requirements document—presents the design of the system together with costs, benefits, and implementation plans such that both users and designers may understand, evaluate, and formally agree on the proposed system. Identifies how the proposed system is to operate, detailing purpose, information flow, input, points of communication with other operations, narrative procedures when available, and proposed outputs.

Gantt chart—a document that geographically displays the major activities of a project and its schedules over a time period of several weeks or months.

GEMISCH—a language developed for medical record applications. Implemented on only one computer, the PDP-11.

Hard copy—typewritten or printed characters on paper, produced at the same time information is copied or converted into machine language that is not easily read by a human.

Hardware—the electric, electronic, and mechanical equipment used for processing data, consisting of cabinets, racks, tubes, transistors, wires, motors, and the like. Terminals used for on-line patient registration are examples of hardware.

Hard-wired—directly connected to the computer via physical wires going from terminals in the unit to terminals in the computer. A device can be up to 2,000 yards from the computer. Hard wiring eliminates the need for expensive modems at both ends of the pathway and avoids tying up a telephone line.

Head—a device that reads, records, or erases information in a storage medium, usually a small electromagnet used to read, write, or erase information on a magnetic drum or tape, or the set of perforating or reading fingers and block assembly for punching or reading holes in paper tape.

Impact printer—a device for printing on paper by striking an inked ribbon and the paper with a raised metal or plastic imprint of the desired character. Alternatively, the paper may be struck from the rear by a hammer pushing the paper and the ribbon against the character outline.

By using interleaved carbon paper or NCR (No Carbon Required) paper, multiple copies of the data may be produced by an impact printer.

Information—an increased level of knowledge derived from processing data. Information can be derived only to the extent that the data are accurate, timely, unexpected, and relevant to the subject under consideration. For example, letters in diagnosis about individual patients become information.

Information structure—a framework that incorporates and supports data. The data represent events and elements that take place in a given operation for communication to a recipient.

Input devices—devices primarily responsible for making information (data and instructions) available to the system, transferring data from an internal to an external storage device. Included are optical scanners, card readers (for punched cards), paper tape readers, and tape drives (which read magnetic tape or write magnetic tape as output). This equipment operates by direction of the programmed instructions executed by the central processing unit.

Input/output chart—a graphic display used to analyze source documents and to identify their relationship to output forms or reports that contain data elements derived from the source document.

Intelligent terminal—a terminal possessing data processing capabilities. With a terminal having computational abilities, data may be checked for accuracy and logicalness, computations can be carried out, and data may be prepared for transmission to other parts of the system. For example, entering information on cancer patients treated during a certain period may be verified (checking to see if all fields are completed at the terminal before transmitting to a host computer).

Interactive processing—also called conversational mode. An operation of data processing systems such that the user, at an input-output terminal, carries on a "conversation" with the system. As each unit of input data is entered, a prompt response is obtained from the system. Thus, a sequence of runs can take place between the user and the system that is typical of a conversation.

Interface—electronic components that allow two different devices to communicate. Connecting one computer to another.

Interim record—a computer printout of a particular, traditional, paper record form. The original paper form, such as a patient history or laboratory report, when processed by the computer, is reformatted to allow automated processing; the printout is also a reformatted version of the original. A document that represents the meshing of the old, manual record and the new, computer record.

Internal memory—the storage facilities forming an integral physical part of the computer and directly controlled by the computer. Includes all memory of storage that is automatically accessible to the computer without human intervention.

Internal outputs—reports on the activity of the system itself, primarily used for the benefit of the data processing department. However, some internal reports will provide feedback on the use of a particular application program and, as such, are valuable to managers in various application areas.

Job—a specified group of tasks prescribed as a unit of work for a computer. It usually includes all necessary programs, linkage, files, and instructions to the operating system, e.g., the monthly disease index.

Key to disk—keypunching data directly onto disk. Key to disk with interactive edit checks is a good step up from keypunching cards.

Key to punch card—keypunching information on punch cards, often the initial introduction to computer processing for hospitals. Some medical record departments store disease indexes on punch cards and "sort" them on a mechanical card sorter for listing out for individual requests.

Key to tape—keypunching data directly onto magnetic tape. Offers no significant advantage over cards; speed increases, but data cannot be verified easily.

Keyboard-to-storage—a system in which several CRTs are connected to a minicomputer or programmed controller that collects the data input, verifies it, provides various other functions, and writes it on tape or disk for processing.

Keypunching—a process whereby information is punched into patterns on standard punch cards so that the information on the cards can be entered into the computer system through a card reader.

Language error—errors in a program that are violations of the rules of that language. Compilers and assemblers attempt to detect and write out explanatory error messages about all such errors. In most cases language errors are easily found and eliminated.

Large-Scale Integration (LSI)—also (Very Large Scale Integration). A procedure that, through a series of processes including photolithography, chemical etching, and diffusion, creates a microminiature pattern of hundreds of switching circuits with transistors and conductors along the surface of a silicon chip (10,000 storage and logic elements per square inch on a single chip).

Layout flow chart—a graphic display of the physical environment (specific desks and pieces of equipment) in which activities or operations are carried out within a department or unit.

Loading—the placing by a compiler of a binary program into a storage file rather than directly into the central memory of the computer ready to be executed. Since the compiler is an extensive program itself and is in central memory working on the program statements, there may not be enough room for both to occupy central memory at the same time. Once the entire program is compiled, the compiler can be erased from central memory and the object program moved in from its storage file. Note that binaries from a compilation done earlier may need to be loaded without the compiler being used at all. In moving programs from storage into central memory, the loader (another program) makes a number of last-minute adjustments to the binary instructions based on the now-known precise location in central memory in which the program is going to reside.

Logic error—when a program directs the computer to carry out instructions that produce an erroneous or undesired result, even though the program satisfies all the formal rules of a language, i.e., it compiles successfully. Such program logic errors may occur repetitively in every case or they may occur only when some particular rare combination of data is present. Logic errors are by far the hardest to debug, and there is no way of assuming that a program of any complexity at all has ever been completely debugged. One must be skeptical of computer-produced data, especially in new programs and in rarely occurring configurations of data, and examine output thoroughly! Health record managers should regularly monitor and audit computer applications as a basic control process of management. Evaluation is also necessary prior to continued commitment of resources.

Loop—the repeated execution of a series of instructions for a fixed number of times.

Machine language—a set of symbols, characters, or signs, and the rules for combining them, that conveys instructions or information to a computer.

Magnetic core—a doughnut-shaped ferromagnetic-coated material, vertically aligned and placed on a wire in a series similar to a string of beads. An electric current is sent through, magnetizing the individual cores. The direction of the current determines the polarity of the magnetic state of the core. By reversing the current, the magnetic state of the core can be reversed. In a binary system of information, only one of two conditions needs to be tested. Core storage lends itself perfectly to binary in that each position will have a reading of 1 or 0 depending upon the state of the magnetism.

Magnetic disk—a storage device on which information is recorded on the magnetized surface of a rotating disk. A magnetic-disk storage system is an array of such devices, with associated reading and writing heads that are mounted on movable arms. Magnetic disks are commonly referred to as DASD, or Direct Access Storage Device.

Magnetic drum—a storage device that makes use of a rotating cylinder coated with a magnetic material for storing data in the form of magnetized spots arranged in circular adjacent rings by a set of heads positioned near the surface.

Magnetic tape—a tape or ribbon of any material impregnated or coated with magnetic material on which information may be placed in the form of magnetically polarized spots. A magnetic tape may be erased and reused repetitively, which makes it an economical and versatile form of storage. One example: patient ambulatory care reason-for-visit codes can be stored on magnetic tape.

Magnetic tape unit—the mechanism, normally used with a computer, that handles magnetic tape. It usually consists of a tape transport, reading or sensing and writing or recording heads, and associated electrical and electronic equipment. Most units may provide for tape to be wound and stored on reels; however, some units provide for the tape to be stored loosely in closed bins.

Mass storage devices—relatively large-volume storage, on-line, and directly accessible to the central processing, arithmetic, logic, or control unit of a computer. Usually utilizing large quantities of on-line disks, drums, tapes, or magnetic cards.

Medical computing—computerization of all aspects of medicine including clinical medicine, clinical algorithms, administrative computer systems in health care organizations; and medical record departmental systems.

Medical information structure—a framework that incorporates data representing treatment modalities, diagnostic and therapeutic action, and the results of services provided to an individual patient during a medical or health care encounter.

MICR (Magnetic Ink Character Recognition)—a check-encoding system employed by banks for the purpose of automating check handling. Checks are imprinted (using magnetic ink) with characters of a type face and dimensions specified by the American Banking Association. There are fourteen characters—ten numbers (0–9) and four special symbols.

Microcomputer—a small, limited capability, relatively slower than a mini, low-cost computer made up of a microprocessor and memory and input/output devices. The microprocessor is the equivalent of a CPU on a semiconductor silicon chip about the size of a nail head. The microcomputer has extensive potential in physician office practice systems, medical record department operations, and other related medical tasks.

Minicomputer—a small computer, often weighing less than fifty pounds, with a relatively low price range of from $20,000 to $100,000, which has the same components as a full-sized system and can be programmed to perform many of the tasks of larger computers. Minicomputers generally have main storage capacities ranging between 16 thousand and 1 million characters and can be considered general purpose machines readily suitable for medical applications.

Modem (Modulator/Demodulator)—a device that accepts the electrical 0 and 1 bit pulses from a terminal and converts them to different frequency sound tones for transmission over telephone lines (modulating). It will also receive tones from the telephone and convert the tones back into electrical pulses (demodulating). In transmitting data between a terminal and a computer, the following setup is required: Terminal-Modem-Telephone Line-Modem-Computer. (One modem must be present to translate signals from the sending location and one present to translate signals at the receiving location.)

MOS (Metal Oxide Semiconductor)—refers to the three layers used in forming the gate structure of a field-effect transistor. Circuitry (as used in computers) based on this technology offers very low power dissipation and makes possible circuits that jam transistors close together before a critical heat problem arises.

Multiplexor—often a specialized computer, with stored program capability, for handling input/output functions of a real-time system. With an interactive system, a number of users may be using the main computer via terminals at the same time. The multiplexor stands between the users and the main computer to maintain orderly data flow by sequentially polling each user's incoming line buser and the location where the computer stores data for that user. If data need to be moved in either direction, the multiplexor moves it and then turns to the next user. The entire cycle over all users is completed several times a second. May be fed into a modem.

MUMPS—a simple but powerful language, strong in data-management and text-manipulation features, designed specifically for handling medical records. Available on about 20 types of computers, especially minis.

Object program—the binary form of a source program produced by an assembler or a compiler. The object program (the final translation) is composed of machine-word or machine-coded instructions that the specific computer can execute.

OCR (Optical Character Recognition)—the machine identification of printed characters through the use of light-sensitive materials or devices. It is an expensive process, has a relatively high error rate, and needs document control. Some health information applications have been designed in patient history and physical exam reports. One example shows several types of pelvic anatomical bone structures that can be marked with a No. 2 pencil.

Off-line—descriptive of a system and of the peripheral equipment or devices in a system in which the operation of peripheral equipment is not under the control of the central processing unit (CPU).

On-line—descriptive of a system and peripheral equipment or devices such as terminals in a system in which the operation of such equipment is under control of the central processing unit (CPU). Information reflecting current activity is introduced into the data processing system as soon as it occurs. It is directly in line with the main flow of the transaction processing.

Operating system—an organized collection of techniques and procedures for operating a computer.

Operational feasibility—the determination of the capability of a given computer-based system to perform the intended functional operation.

Operation flow chart—a graphic depiction of the sequence of activities and identification of decision points encountered in carrying out a function. Specifies how a function is performed, not what is performed. Whenever possible, a written explanation should be placed within the symbols used in this chart.

Operations personnel—nonmanagement employees whose job responsibility is on the day-to-day, task-oriented work level that achieves the objectives of individual job descriptions.

Organization chart—a graphic depiction of functions performed in a department or an organization. Through vertical or horizontal lines it demonstrates the authority/responsibility relationship of workers and administrators. Depicts line and staff relationships and identifies the titles of jobs in accordance with job descriptions.

Organization function list—a document prepared for each unit shown on an organization chart describing the specific major activities performed by that unit and keyed to the organization chart by use of the titles and classification numbers added to the organization chart.

Output devices—a device or set of devices used to take data out of another device. Equipment used to remove data from the computer.

Outputs—the end products of a computer-based system. May be written reports, screen displays, and improved operations for a given task.

PASCAL—a language of particular interest to computer scientists that is used increasingly for applications. Available on about 60 types of computers.

Performance definition—the precise description of what functional or systems requirements are necessary to accomplish solution of a problem. Provides a foundation for the design phase of the computer based solution as well as criteria for post-implementation audit activities.

Peripherals—auxiliary equipment placed under the control of the central computer, e.g., card readers, card punches, magnetic-tape feeds, CRT terminals, and high-speed printers. May be used on-line or off-line depending upon computer design, job requirements, and economics. The console and control unit, main internal storage, arithmetic and logic unit, and associated power supplies are not considered peripheral equipment. Anything else that handles the same data before or after processing is considered peripheral, whether it is or is not under the direct control of the central processing or control unit. Usually, peripheral equipment provides a means of communication between a computer and its environment. In a medical record department, a terminal that is used to operate on-line record control/location is a peripheral device.

Planning—the process of making decisions in the present to bring about an outcome in the future. The determining of proper goals and the means to achieve them, involving the statement of assumptions, the development of premises, and the review of alternate courses of action.

PL/I—an advanced, powerful programming language for scientific, commercial, and other applications. Available largely but not exclusively on large IBM computers.

Primary records—the source documents prepared in the care of the patient.

Primary or main storage—also referred to as main memory. Usually the fastest storage device of a computer and the one from which instructions are executed. (Contrasted with auxiliary storage.) Storage that is integral to the computing system. The device in a computer in which the binary bit representations of program instructions and data are stored so that individual program instructions and data elements may be obtained from memory very rapidly and moved to the arithmetic-logic portion of the CPU for execution.

Printer—a device that takes output data codes from the computer and converts to printed lines of data on continuous form paper. High-speed printers (printing several hundred to several thousand lines per minute) essentially print an entire line at a time. A hard copy terminal is not a printer.

Privacy—the right of an individual to be left alone; to withdraw from the influence of his or her environment; to be secluded, not annoyed, and not intruded upon by extension of the right to be protected against physical or psychological invasion or against the misuse or abuse of something legally owned by an individual or normally considered by society to be his property. A condition or status. The capacity to control information about oneself or one's experiences.

Process flow chart—a graphic display of the step-by-step details or series of activities that take place from start to finish in a particular process, or the work of one subject only, i.e., person, material, or paper form. Uses symbols, notes operation, transportation, inspection, delay and storage, distance covered, and time involved.

Program—a plan for the automatic solution of a problem. A complete program includes plans for the transcription of data, coding for the computer, and plans for the absorption of the result into the system. The list of coded instructions is called routine.

Program Evaluation and Review Technique (PERT)—a graphic display of the relationship between tasks and schedules. Indicates what activities can be done independently of other activities and those are interdependent, and when properly executed, will allow a comparison of scheduling of tasks in order to show the items that accomplish completion of the project in the least amount of time and at the lowest cost.

Programmable communication processors—usually mini/micro computers that perform high-speed, high-volume message switching and front-end processing over major communications trunk lines or for data communications with a computerized information system. Coordinators of data entry operations from many locations.

Programming language—a specific language used to prepare computer programs. Originally, computers were programmed by directly entering the binary codes for each operation into memory. This cumbersome system has evolved into development of higher-level languages where a single statement would represent a long series of single-step operations. Common languages at this higher level are FORTRAN, BASIC, COBOL, PL/I, ALGOL, AND MUMPS. Each language has a specific set of operational symbols and rules for combining these into program statements.

Punched card—a heavy, stiff paper of constant size and shape, suitable for punching in a pattern that has meaning and for being handled mechanically. The punched holes are sensed electrically by wire brushes, mechanically by metal fingers, or photoelectrically by photo-cells. There are 80-column cards and 90-column cards.

RAM (Random Access Memory)—stored information and programs that may be modified by a program. Information in RAM is lost when the power is turned off.

Random file—also referred to as Files Stored on DASD (Direct Access Storage Device). A file stored in such a way that data for a particular case may be accessed in a relatively direct fashion. Generally, this is done by developing a case data location table that is referenced when data regarding a particular case are desired. Random files are ordinarily stored on disks since all portions of a disk file are equally accessible. ''Random'' refers to the order of access, not to the order of the data in the file; the data order may be highly organized and not at all random. A master patient index that is on-line is accessed through a terminal for patient identification, and verification of previous admission is stored in a random file.

Real-time processing—a form of interactive processing in which the computer system stays abreast of all changes as they come into the system, e.g., airline reservation systems. Each change is immediately recorded, and all the necessary files, etc., are updated. The "real" world is approximated as closely as possible. In health information a computerized, on-line appointment system could be real-time. Intensive care computers that monitor patient heart functions, breathing, etc., are real-time systems.

Record—a set of one or more consecutive fields on a related subject. Records may be either fixed length or variable length. Patient medical records are one form of records; billing records are another.

Register—a device in the computer that is used to temporarily store a specific amount of information. Registers are placed where they have immediate access to the internal memory and any input/output bus structures.

Remote access teleprocessing—communicating with a computer via devices, usually terminals, that are located away from the immediate vicinity of the computer. Usually, the communication link is the public telephone system. A professional review organization (PRO) may enter hospital discharge data on terminals located in Seattle that transmit the patient information over phone lines to a host computer in another state.

Remote job entry—the capability of computer systems to accept data from remote (geographically distant from the CPU) locations and store it for later batch processing. Uses such devices as on-line CRTs and key-to-disk units to collect and organize the data and telephone or other communications equipment to transmit the data to the central processing unit.

Request for Information (RFI)—a formal document used to ascertain the marketplace for vendor supply of a computerized process. A screening mechanism that requires one to obtain a minimal set of information from vendors.

Request for Proposal (RFP)—a formal document that details the functional requirements of an intended computerized application so as to provide a vendor with a detailed profile.

Response time—the amount of time that elapses between generation of an inquiry at a terminal and receipt of a response at the terminal. Includes transmission time to the computer, processing time at the computer, access time to obtain any file records needed to answer the inquiry, and transmission time back to the terminal. Response times of under 1 second are usually unnoticeable. Between 1 and 3 or 4 seconds, response begins to seem sluggish, and over 5 seconds the delay begins to be annoying to completely unacceptable, depending on the type of task. A drug ordering system in a hospital would be concerned with this element in drug verification and/or potential contraindication before processing the order.

ROM (Read Only Memory)—a special kind of memory in which information is stored once and cannot be changed, e.g., diagnostic programs, language interpreters, and other utility functions.

RPG—a language developed to facilitate the specification of report formats to be printed from stored files. Available on several types of computers.

Secondary records—documents prepared for use in analyzing or summarizing primary data.

Secondary storage—storage that may be used to store data and programs permanently or by the CPU to temporarily transfer data to and from when carrying out operations.

Sequential file—a file stored in such a way that, in order to obtain data for a particular case, all preceding records in the file must be shuffled through in some manner in order to reach the required data. Data stored and retrieved from magnetic tapes are accessed in a sequential file manner.

Simplex—transmitting data in one direction only, i.e., from a device to a computer or from a computer to a device but not both. Special sensor devices such as ICU (Intensive Care Unit) monitors operate in simplex mode.

Software—programs that are sets of instructions that direct actual computer operating functions (called the operating system) or that do a particular operation for a unique application (called "application software"). Software packages are being increasingly developed by vendor companies to be purchased and run on an organization's existing computer.

Solid state storage—a storage device in which the memory elements consist of semi-conductor circuits, usually integrated circuits. Over 50 million work capacities and 300 nanosecond access times are practical using this type of storage construction.

Source document—a document originally used by a data processing system that supplies the basic data to be input to the data processing system, e.g., a medical/health record.

Source program—a program coded in other than machine language that must be translated into machine language before use. The set of original program statements.

Stored programs—coded sets of instructions that are entered and stored in the computer to wait until the planned data are entered for manipulation.

Strategic planning—the continuous process of making present entrepreneurial (risk-taking) decisions systematically and with the events and knowledge of their futurity; organizing systematically the efforts needed to carry out these decisions; and measuring the results of these decisions against the expectations through organized, systematic feedback.

System—related elements that are coordinated to form a unified result; or, specifically, people, activities, equipment, materials, plans, and controls, working together to achieve a unified objective or whole. Also defined as an array of components that interact to achieve some objective through a network of procedures that are integrated and designed to carry out a major activity. The components of the system may themselves be systems depending on the complexity of the parent system.

Systems analysis—a scientific method of problem solving. A process that can be used to examine an activity, procedure, method, technique, or business. Uses a structured order to determine what must be accomplished to achieve change, select the best method for achieving a goal, and provide direction and evaluation for implementing change or installing new methods or equipment.

Systems flow chart—a graphic depiction of the flow of data (information) through a system, the files being used, and the programs required to operate the system.

Tactical planning—the identification of the major tasks required to achieve strategic objectives. Supportive to strategic planning, it must in fact honor the directions given by strategic planners.

Task list—an enumeration of duties individuals actually perform, not what they should be doing, and the hours per week spent on the tasks.

Telecommunication—the merging of telephone and other communications with computer technology. The foundation of data communications devices operations.

Terminal: (CRT typewriter)—an input/output device designed to receive data in an environment associated with the job to be performed and capable of transmitting entries to and obtaining output from the system of which it is a part. Personal input and computer output usually appear sequentially on a CRT screen or are printed by some print device on paper.

Thermo printer—a device for printing characters by briefly exposing the chemically treated surface of the paper to the heat of small high-speed heating elements embedded in a 7×9 element array in the print/head. Letters are formed by turning on only selected elements in the array. A line of data is printed by moving the head across the page. With thermo printers, only one copy can be generated at a time.

Timeshare—a method of operation by which several users appear to be sharing the computer at the same time. In reality, the computer serves each user in sequence. Depending on the capacity of the computer and the number of users, timesharing can give the user apparent "real-time" response. Most large computers are designed for timeshare use.

Track—a series of parallel pathways of 1 and 0 bits created down the length of the tape as the tape unit writes a consecutive series of characters. Since BCD characters require 6 bits plus the parity bit across the face of the tape, there will be 7 tracks created. EBCDIC characters require 8 bits plus the parity bit; therefore, 9 tracks will be created.

Transaction file—a record of any action taken on the data of a file that results in a change in the file, e.g., additions, deletions, or changes of data. It would contain information about what was changed, who made the change, and what time it was made.

Turnaround time—the particular amount of time required for a computation task to get from the programmer to the computer, onto the machine for a test or production run, and back to the programmer in the form of the desired results. Can refer to batch programs such as those used in commercial discharge abstracting companies or to on-line operations as available in time-share option programs.

Turnkey system—a finished product that is delivered, often including hardware, software, documentation, and training components. Some are designed to be transported and immediately "turned-on" in the user setting.

User—an individual whose job responsibility requires him or her to use or employ the products of the computer system.

User manager—an individual who is a manager or administrator of a department in which at least one prime objective is the use of data or information.

User output—an output product for clerical or auditing purposes dealing with information specifically for the user of the system. Can be limited to those at the operations level such as data entry clerks.

User surveys—organized methods for communicating with people to find out how things operate now and how users might like to see things operate in the future. Carried out through interviews and questionnaires.

Variable length record—record in which the number of data units, i.e., blocks, words, etc., is not fixed. Placing an entire medical record on computer would require a careful consideration of the use of variable length records.

Vendor—an individual or company that, in this context, sells computer-related products. The term encompasses those who sell software and hardware and those who sell partial or complete systems.

Virtual storage techniques—also called virtual memory. Methods used to provide notional space on storage devices, appearing to a computer user as main storage. The size of virtual storage is limited only by the addressing scheme of the computing system, rather than by the actual number of main storage locations. Provides faster, more powerful use of primary storage.

Word—a set of characters that occupies one storage location and is treated by the computer circuits as a unit and transported as such. Ordinarily treated by the control unit as an instruction and by the arithmetic unit as a quantity. Word lengths are fixed or variable, depending on the particular computer.

Word length—the number of bits or characters that are handled as a unit by the equipment, as a size of the field.

Word processing—uses a computer-typewriter interface to prepare documents in such a way that they can be extensively edited, rewritten, combined with other documents, and rearranged internally, through a CRT.

Word processing microprocessor-based unit—an arrangement of elements consisting of a microprocessor and a video screen incorporated in a cabinet, a keyboard included in the cabinet or attached to it, a small printer separate from but attached to the cabinet, and one or more diskettes or floppy disk read-write units, either incorporated into the cabinet or separate but attached, like the printer.

Work distribution chart—a graphic display of work activities performed, the time it takes to perform them, the individuals who are working on these activities, and the amount of time spent by each person on each activity.

Index

I